T0180835

Use R!

Use R!

This series of inexpensive and focused books on R is aimed at practitioners. Books can discuss the use of R in a particular subject area (e.g., epidemiology, econometrics, psychometrics) or as it relates to statistical topics (e.g., missing data, longitudinal data). In most cases, books combine LaTeX and R so that the code for figures and tables can be put on a website. Authors should assume a background as supplied by Dalgaard's Introductory Statistics with R or other introductory books so that each book does not repeat basic material.

More information about this series at http://www.springer.com/series/6991

Vasilis Pagonis

Luminescence

Data Analysis and Modeling Using R

 Springer

Vasilis Pagonis
Physics Department
McDaniel College
Westminster, MD, USA

ISSN 2197-5736 ISSN 2197-5744 (electronic)
Use R!
ISBN 978-3-030-67310-9 ISBN 978-3-030-67311-6 (eBook)
https://doi.org/10.1007/978-3-030-67311-6

© The Editor(s) (if applicable) and The Author(s), under exclusive license to Springer Nature Switzerland AG 2021

This work is subject to copyright. All rights are solely and exclusively licensed by the Publisher, whether the whole or part of the material is concerned, specifically the rights of translation, reprinting, reuse of illustrations, recitation, broadcasting, reproduction on microfilms or in any other physical way, and transmission or information storage and retrieval, electronic adaptation, computer software, or by similar or dissimilar methodology now known or hereafter developed.

The use of general descriptive names, registered names, trademarks, service marks, etc. in this publication does not imply, even in the absence of a specific statement, that such names are exempt from the relevant protective laws and regulations and therefore free for general use.

The publisher, the authors, and the editors are safe to assume that the advice and information in this book are believed to be true and accurate at the date of publication. Neither the publisher nor the authors or the editors give a warranty, expressed or implied, with respect to the material contained herein or for any errors or omissions that may have been made. The publisher remains neutral with regard to jurisdictional claims in published maps and institutional affiliations.

This Springer imprint is published by the registered company Springer Nature Switzerland AG
The registered company address is: Gewerbestrasse 11, 6330 Cham, Switzerland

This book is dedicated to my scientific collaborators and good friends Dr. George Kitis and Dr. Reuven Chen. Their enthusiasm and friendship over the years have been a constant source of inspiration for my research.

Preface

About This Book

The past decade has seen rapid growth in the development and application of the programming language R, in the fields of radiation dosimetry, luminescence dosimetry, and luminescence dating. R is now widely used in these scientific areas with new packages becoming available and used regularly by students and researchers. The present book covers applications of R to the general discipline of radiation dosimetry, and the specific areas of luminescence dosimetry, luminescence dating, and radiation protection dosimetry. The book will be a useful tool for the broad scientific audience involved in luminescence dosimetry: physicists, geologists, archaeologists, solid state physicists, and scientists using radiation in their research.

The book features 99 detailed worked examples of R code fully integrated into the text, ensuring their usefulness to researchers, practitioners, and students. Users can run immediately the R codes and modify them for their experimental data and to explore the various luminescence models. In each chapter, the theory behind the subject is summarized, and appropriate references are given from the literature, so that researchers can look up the details of the theory and the relevant experiments. Each R code includes extensive comments explaining the structure and the various parts of the code. The chapters discuss how researchers can use the available R packages to analyze their own experimental data, and how to extract the various parameters describing mathematically the luminescence signals.

Using the R Codes in This Book

This book assumes some basic knowledge of R; however, I believe that it will be useful for both newcomers to R and experienced programmers who wish to learn more about the various luminescence phenomena. The R codes for all programs in the book are available for downloading at the GitHub website https://github.com/vpagonis/Springer-R-book. The codes are also available, at the author's personal website at McDaniel College: https://blog.mcdaniel.edu/vasilispagonis.

The various R codes are self-contained and ready to run and are based either on previously published R packages or represent new codes written by the author and collaborators. In the latter case, appropriate credit is assigned to the authors of the code. Several chapters are dedicated to Monte Carlo (MC) methods, which are used to simulate the luminescence processes during the irradiation, heating, and optical stimulation of solids, for a wide variety of materials. Throughout the book, both localized and delocalized transition models are simulated using the new R package *RLumCarlo*.

Experienced programmers of R will certainly find out that they can improve the R codes given here, and it is of course possible to make the codes more compact and elegant. However, I chose to provide R codes that are simple and clear and that can be easily modified for the purposes of the reader, rather than attempting to create compact codes that may be difficult to follow. I have kept the number of required external R packages intentionally at a minimum, so that newcomers to R can follow the R codes easily. All figures in this book were produced using the R codes in the book, so that users know immediately what to expect when they run the codes. Additional drawings for the various luminescence models in the book were drawn using *Inkscape*, and the book was produced overall using *Ly*X.

How This Book Is Organized

This book concerns analysis of experimental luminescence data, as well as phenomenological models used for explaining the experiments.

Overall the book is organized in four Parts I–IV and 12 chapters, as shown in the diagram below.

The presented models fall within two broad categories, based on delocalized and localized transitions. In delocalized transition models, the conduction and/or valence band participate in the luminescence process. By contrast, in the localized type of models the luminescence process does not involve the energy bands but rather takes place between the ground and/or excited state of the trapped electron–hole, and an energy level of the recombination center.

Chapter 1 introduces general examples of experimental luminescence data and discusses the various experimental techniques for measuring the luminescence signals. These signals include thermoluminescence (TL), optically stimulated luminescence (OSL), infrared stimulated luminescence (IRSL), as well as the commonly used experimental modes of continuous wave (CW-OSL or CW-IRSL), linearly modulated (LM-OSL or LM-IRSL), constant heating rate TL, isothermal TL (ITL), and time resolved (TR). In addition, Chap. 1 provides a general overview of localized and delocalized models, which will be studied in detail later in the book.

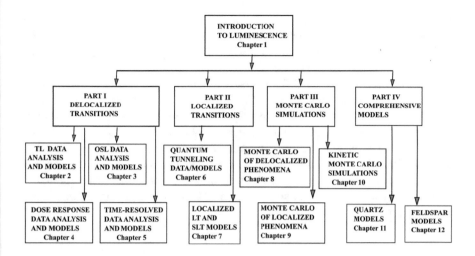

Part I of the book is titled *Luminescence Signals from Delocalized Transitions* and consists of Chaps. 2–5, which are a practical guide for analyzing luminescence signals having their origin in delocalized transitions involving the conduction and valence bands.

These four chapters provide a detailed presentation of various methods of analyzing and modeling experimental data for TL signals, OSL signals, TR-OSL signals, and the dose response of dosimetric signals. Analysis of the associated experimental data is based on a variety of available R packages (such as *tgcd*, *NumOSL, Luminescence, etc.*), as well as on my own custom-written programs.

Part II of the book is titled *Luminescence Signals from Localized Transitions* and consists of Chaps. 6–7, which are a practical guide for analyzing luminescence signals having their origin in localized transitions between energy states located between the conduction and valence bands. Chapter 6 contains a general introduction to quantum tunneling processes, as pertaining to dosimetric materials. Special emphasis in these chapters is on the analysis of luminescence signals from feldspars and apatites, which exhibit quantum tunneling luminescence phenomena.

Three types of models are considered in this part of the book: the localized transition (LT) model, the tunneling localized transition (TLT) model, and the semilocalized transition (SLT) model.

Part III of the book is titled *Monte Carlo Simulations of Luminescence Signals* and covers Chaps. 8–10. Chapter 8 provides a general description of luminescence phenomena as a stochastic process, by using Monte Carlo techniques for the description of TL and OSL phenomena. Special emphasis is given in the differences between the stochastic approach that uses Monte Carlo techniques and the deterministic approach that is based on differential equations. This chapter is of special interest to the research area of nanodosimetric materials, in which traps and centers may be spatially correlated, and the luminescence signals may be produced by a small number of these defects. A connection is also made between luminescence phenomena and the well-known *stochastic life and death phenomena*, which have been studied extensively in other areas of science. Chapters 8 and 9 are based on MC methods with fixed time intervals, while Chap. 10 presents some examples of kinetic Monte Carlo (KMC) methods.

Part IV of the book is titled *Comprehensive Luminescence Models*. Chapters 11 and 12 present two general classes of phenomenological models commonly used in luminescence dosimetry and dating. Specifically, these chapters present several commonly used comprehensive phenomenological models for quartz and feldspars, which are two of the best studied natural luminescence materials. These two chapters also simulate several commonly used experimental protocols for luminescence dating, including the very successful single aliquot regenerative (SAR) protocol. The simulations are based on the R program *KMS*, on the R package *RLumModel*, and on several new codes written by the author.

Acknowledgments

I am grateful to Dr. Reuven Chen and Dr. Sebastian Kreutzer for reading drafts of the book and providing me with extensive valuable comments about improving the presentation and the R codes. I also thank Dr. Adrie Bos, Dr. Arkadius Mandowski and Dr. George Kitis, for their useful comments and suggestions, during the initial and final stages of writing the book. Special thanks go to Dr. Christoph Schmidt and the students from Bayreuth University who contributed greatly to the development of the package *RLumCarlo*, and to Dr. Johannes Friedrich for re-writing several of my original *Mathematica* programs into R code. Finally, I thank Laura Briskman and the technical staff at Springer Nature, for their help and encouragement during the publication process.

Westminster, MD, USA   Vasilis Pagonis
December 2020

Contents

List of Codes

Acronyms

AF	Anomalous fading
CW-IRSL	Continuous wave infrared stimulated luminescence
CW-OSL	Continuous wave optically stimulated luminescence
EST	Excited state tunneling model
FOK	First order kinetics
GOK	General order kinetics
GST	Ground state tunneling model
GOT	General one trap model
IGST	Irradiation ground state tunneling model
IMTS	Interactive multi-trap system model
IRSL	Infrared stimulated luminescence
ITL	Isothermal luminescence
KP	Kitis–Pagonis general equation
KP-CW	Kitis–Pagonis CW-IRSL equation
KP-TL	Kitis–Pagonis TL equation
LM-IRSL	Linearly modulated infrared stimulated luminescence
LM-OSL	Linearly modulated optically stimulated luminescence
LT	Localized transition model
KMC	Kinetic Monte Carlo method
MC	Monte Carlo method
MOK	Mixed order kinetics
NMTS	Non-interactive multi-trap system model
OSL	Optically stimulated luminescence
OTOR	One trap one recombination model
PKC	Pagonis–Kitis–Chen superlinearity equation
POSL	Pulsed optically stimulated luminescence
PIRSL	Pulsed infrared stimulated luminescence
SLT	Semilocalized transition model
TA-EST	Thermally assisted excited state tunneling model
TL	Thermoluminescence

TLT	Tunneling localized transition model
TR-IRSL	Time resolved infrared stimulated luminescence
TR-OSL	Time resolved optically stimulated luminescence

Chapter 1
Introduction to Luminescence Signals and Models

1.1 Thermally and Optically Stimulated Luminescence Phenomena

In this introductory chapter we describe several types of thermally and optically luminescence signals, which form the basis of luminescence dosimetry and luminescence dating.

The interaction of ionizing radiation with both natural and artificial inorganic materials makes these materials suitable to act as radiation detectors. An important family of radiation detectors, called passive detectors, is based on the existence of localized energy levels within the forbidden energy band. These energy levels are created by the presence of imperfections and impurities in the crystal lattice and by the irradiation of these materials in nature or in the laboratory.

In the framework of the phenomenological energy band model of solids, the irradiation process creates electrons in the conduction band and holes in the valence band. The electrons from the conduction band are trapped in electron traps, and the holes are trapped in hole traps (which can also act as luminescence centers) within the forbidden band.

After irradiation, the system is in an excited state, and the lifetime of these excited states in nature varies widely from microseconds to billions of years. When the material is thermally or optically stimulated in the laboratory, the trapped electrons are released and may eventually recombine with holes at luminescence centers, thus causing the emission of light from the sample. The property making the material a dosimeter is the functional relationship between the number of electrons trapped in the material and the irradiation dose.

Various stimulation methods are commonly used in the laboratory to liberate the trapped electrons, and these methods generate different types of stimulated luminescence (SL) signals. In terms of the time scales involved in luminescence processes, one can distinguish two broad types of experiments. In the first category one studies phenomena such as thermoluminescence (TL) or optically stimulated luminescence (OSL) that take place in time scales of seconds. A general description

V. Pagonis, *Luminescence, Use R!*, https://doi.org/10.1007/978-3-030-67311-6_1

of TL signals and associated delocalized and localized transition models is given in Sect. 1.2, followed by a discussion of isothermal TL (ITL) signals in Sect. 1.3. An overview of OSL signals measured with visible and infrared light is presented in Sect. 1.4.

In the second broad category of experiments one studies time-resolved (TR) phenomena, where researchers use short light pulses to separate the stimulation and emission of luminescence in time. TR experiments usually involve much shorter time scales than TL/OSL, typically of the order of milliseconds or microseconds. An overview of TR signals is given in Sect. 1.5 of this chapter. Section 1.6 presents a brief description of electron spin resonance (ESR) signals and optical absorption (OA) signals, and their connection and importance in luminescence dosimetry. Finally, this chapter concludes in Sect. 1.7 with a brief discussion of what types of information researchers typically extract from the experimental data described in this chapter.

For a comprehensive review of the phenomena described in this book, the readers are directed to several available excellent textbooks and review articles on luminescence dosimetry and its applications (Chen and Pagonis [38], Yukihara and McKeever [195], Pagonis et al. [137], Böetter-Jensen et al. [19], Chen and McKeever [37]).

1.2   Overview of Thermoluminescence (TL) Signals and Models

When the stimulation of the sample in the laboratory is thermal, the phenomenon is called *thermoluminescence* (TL) or *thermally stimulated luminescence* (TSL). Commonly used experimental heating functions during a TL experiment are linear, hyperbolic, and exponential heating functions.

During a TL experiment, one records the light intensity as a function of temperature, and the total signal is termed a *TL glow curve*. An example of a TL glow curve measured with a constant heating rate is shown in Fig. 1.1a, for a synthetic quartz sample (Kitis et al. [76]). In Fig. 1.1b, the TL signal from the same sample is plotted as a function of the irradiation dose; this type of graph is usually referred to as the *TL dose response curve*.

The analysis of complex TL glow curves like the one shown in Fig. 1.1a is known as the deconvolution of the TL glow curve and is discussed in detail in Chap. 2. The analysis of dose response data similar to Fig. 1.1b is covered in detail in Chap. 4.

An example of TL glow curves measured after irradiation with different doses is given in Fig. 1.2, for the dosimetric material MgB_4O_7:Dy,Na or MBO (Pagonis et al. [124]). This data shows that as the irradiation dose is increased, the height of the TL peak increases proportionally to the irradiation doses of (1,3,4,6) Gy. At the same time, the shape of the TL peak remains unchanged. This is shown in Fig. 1.2b, where the TL peaks were normalized by multiplying each one with an appropriate scaling

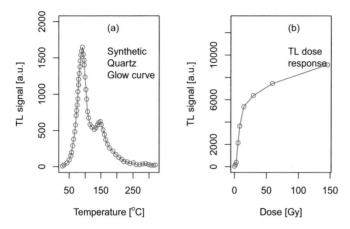

Fig. 1.1 (**a**) Example of a TL glow curve for synthetic quartz, after irradiation with a beta dose of 2 Gy. (**b**) The TL dose response of this quartz sample, obtained by plotting the maximum intensity of the TL signal as a function of the irradiation dose. For more details, see Kitis et al. [76]

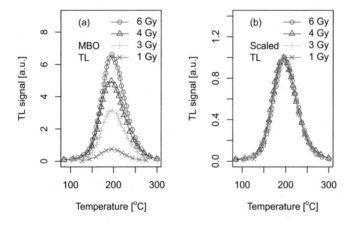

Fig. 1.2 (**a**) Example of a series of TL glow curves for sample MBO, at different beta doses. (**b**) The data in (**a**) is normalized to the maximum TL height. For more details, see Pagonis et al. [124]

factor, so that their maximum heights are 1. This type of behavior as a function of irradiation dose, in which the shape of the TL peak and the temperature of maximum height remain unchanged, is described as *first order kinetics*. The general theory of first order kinetics (FOK) and of other types of kinetics is discussed in detail in Chaps. 2 and 3.

The shape of the TL glow curves, the number of peaks, and their temperature are of particular interest to researchers developing new dosimetric materials. Figure 1.3 shows a comparison of the TL glow curves for the dosimetric materials MgB_4O_7:Dy,Na (MBO) and LiB_4O_7:Cu,In (LBO) (Kitis et al. [81]). The TL glow curves were obtained by irradiating the samples with the same dose and

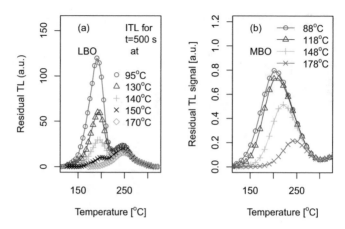

Fig. 1.3 (**a**) Example of a series of TL glow curves for sample LBO, after irradiation and heating up to the temperatures indicated in the legends. (**b**) A very similar series of TL glow curves for sample MBO. The different behaviors of these two samples indicate that different luminescence mechanisms are involved in each case. For more details, see Kitis et al. [81]

subsequently heating the sample up to the temperatures indicated in the legends of Fig. 1.3a, b. Note that the shapes and widths of the main dosimetric peaks around 200 °C in these two materials are very different. The overall shape of the TL peak in LBO is asymmetric, while the TL peak in MBO is very nearly symmetric.

In addition to the differences in their widths, the TL peaks behave very differently in Fig. 1.3a, b. The temperature of the maximum intensity (T_{max}) in the series of TL glow curves in Fig. 1.3a does not shift significantly, while the corresponding T_{max} changes continuously in the series of TL glow curves in Fig. 1.3b.

In the above example for materials LBO and MBO, the different behaviors of the TL glow curves point to the possibility that a different luminescence mechanism is involved in these materials.

In fact, the luminescence mechanism for LBO can be explained using a *delocalized transitions model* that involves the conduction and valence bands, while the mechanism for MBO is described by a *localized transitions model*.

In the next two subsections we will provide an overview of these two types of models that have been proposed in the literature, in order to describe the properties of luminescence signals.

1.2.1 Delocalized Transition Models of TL

Figures 1.4 and 1.5 show several types of *delocalized models* that have been studied extensively in the literature. The arrows in these figures indicate electronic transitions, some of which involve transport of electrons and holes via the conduction and/or valence band. The simplest possible model that has been studied extensively

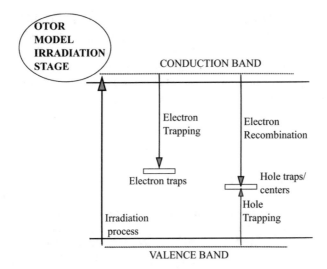

Fig. 1.4 The simplest OTOR model, showing the various electronic transitions during the irradiation stage. Irradiation creates electrons and holes in the conduction and valence bands, respectively. Electrons in the conduction band can subsequently either be trapped in electron traps, or they can be recombined with holes at the recombination centers/hole traps. Holes in the valence band can be trapped in the same recombination centers/hole traps

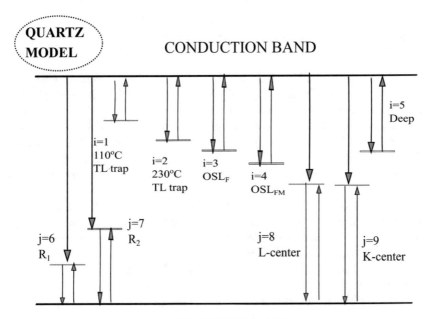

Fig. 1.5 Example of a complex *delocalized model* for quartz, involving multiple traps and centers (after Bailey [10]). Each arrow represents a transition between energy levels and the conduction and valence bands. These arrows also represent a mathematical term in a system of differential equations, as discussed in Chap. 11

in the literature consists of one trap and one recombination (OTOR) center and is shown in Fig. 1.4 for the irradiation process in a sample. The application of the OTOR model for TL and OSL data is presented in detail in Chaps. 2 and 3, in connection with analytical solutions based on the Lambert W function.

All differential equations in the models used to describe TL phenomena contain the Arrhenius thermal stimulation term $p(t)$ given by

$$p(t) = s \, \exp\left[\frac{-E}{k_B \, T(t)}\right] \tag{1.1}$$

where E (eV) is the thermal activation energy of the trap, $k_B = 8.617 \times 10^{-5}$ eV/K is the Boltzmann factor, $T(t)$ (K) is the time dependent temperature of the sample, and s (s^{-1}) is the associated frequency factor. Typical values of the frequency factor are $s = 10^8$–10^{12} s^{-1}, and for the activation energy $E = 0.7$–2 eV.

A more complex delocalized model that describes luminescence phenomena in quartz is shown in Fig. 1.5 and is presented in detail in Sect. 11.2 (Bailey [10]). This model is based on 5 electron traps indicated in the figure by the index $i = 1 - 5$ and on 4 hole traps/recombination centers indicated by the index $j = 6 - 9$ in this figure. Detailed applications of this model and its implementation in several R packages will be given in Chap. 11.

1.2.2   Localized Transition Models of TL

Figure 1.6 shows a comparison of the TL glow curves for the dosimetric materials MgB$_4$O$_7$:Dy,Na (MBO) and Al$_2$O$_3$:C (Pagonis et al. [119], Kitis et al. [81]). The shapes and widths of the main dosimetric peaks around 200 °C in these two materials are very different. The overall shape of the TL peak in Al$_2$O$_3$:C is asymmetric, while the TL peak in MBO is very nearly symmetric. The asymmetric shape for the peak in Al$_2$O$_3$:C is due to the luminescence mechanism involving first order kinetics (FOK) and *delocalized transitions*, as will be discussed in detail in Chap. 2.

The very wide TL peak observed for MBO is the result of a more complex underlying mechanism, which may involve an underlying distribution of activation energies and/or quantum tunneling mechanisms. These types of processes are discussed in Chap. 6. The luminescence mechanism in this material can be described by a *localized transition model*, similar to the one shown in Fig. 1.7.

In this book, we will look at several types of localized transition models that are organized in different chapters and sections, as shown in the chart of Fig. 1.8.

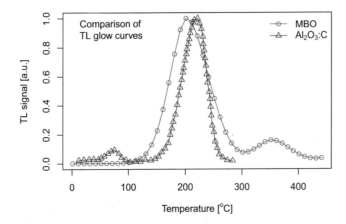

Fig. 1.6 Comparison of the TL glow curves MBO and Al_2O_3:C. The main dosimetric peak around 200 °C for MBO is very nearly symmetric, while for Al_2O_3:C the main peak is asymmetric. Notice also the difference between the widths of the main dosimetric peaks in the two materials. For more details, see Pagonis et al. [119] and Kitis et al. [81]

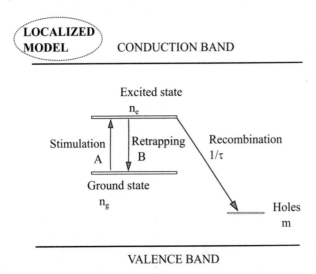

Fig. 1.7 A typical *localized transition model*, consisting of the ground state and the excited state of a trapped electron, and an additional energy level associated with a recombination center. Electrons are excited from the ground state (n_g) into the excited state (n_e) of the trap at a rate A and can also relax back into the ground state of the trap at the rate B. The transition indicated by the rate $1/\tau$ occurs from the excited state of the trap to an energy level of the recombination center (after Pagonis et al. [131])

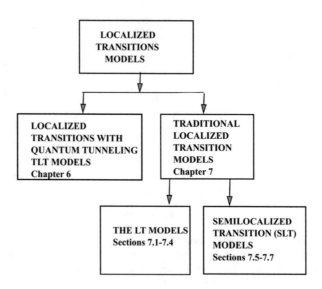

Fig. 1.8 The organization of localized transition models in Chaps. 6 and 7 of this book

1.3 Overview of Isothermal TL (ITL) Experiments and Models

When the thermal stimulation takes place at a constant temperature T_D, the stimulated luminescence signal is called *prompt isothermal decay* (PID), and the thermal stimulation function $p(t)$ is given by

$$p(t) = s \, \exp\left[\frac{-E}{k_B \, T_D}\right] \tag{1.2}$$

where E, s are the thermal kinetic parameters of the trap and k_B is the Boltzmann constant. An example of a PID curve is shown in Fig. 1.9a for the dosimetric material Magnesium tetraborate MgB_4O_7:Dy,Na (MBO) (Kitis et al. [81]). Figure 1.9b shows a series of remnant TL (RTL) glow curves for sample MBO, measured after an isothermal TL experiment for 500 s at the indicated temperatures. As the temperature of the isothermal TL experiment is increased, the maximum of the peaks decreases, and it also shifts toward higher temperatures. We will study this type of TL glow curve in connection with quantum tunneling and by using Monte Carlo methods in Chap. 6, and in connection with the R package *RLumCarlo* (Kreutzer et al. [87]).

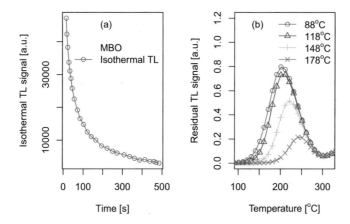

Fig. 1.9 (**a**) Isothermal TL signal for sample MBO and (**b**) a series of RTL glow curves, measured after an isothermal TL experiment for 500 s at the indicated temperatures. For more details, see Kitis et al. [81]

1.4   Overview of OSL, IRSL, and Their Models

When the stimulation of a sample is optical using visible light, one is dealing with *optically stimulated luminescence* (OSL). Typically, blue LEDs with a wavelength of 470 nm are used during these OSL experiments. When the stimulation is with visible light and also occurs with a source of constant light intensity, the stimulated luminescence is termed as *continuous wave optically stimulated luminescence* (CW-OSL), and the optical stimulation function $p(t)$ is given by

$$p(t) = \sigma\, I \qquad\qquad (1.3)$$

where $\sigma\,(\text{cm}^2)$ represents the optical cross section for the CW-OSL process, and $I\,\left(\text{photons cm}^{-2}\,\text{s}^{-1}\right)$ represents the photon flux. The units of the stimulation rate $p(t)$ are s^{-1}.

In a similar manner, when the stimulation takes place with infrared photons, this process is called *infrared stimulated luminescence (IRSL)*. Typically, infrared LEDs with a wavelength of 850 nm are used during these IRSL experiments. In this case $\sigma\,\left(\text{cm}^2\right)$ in Eq. (1.3) represents the corresponding optical cross section for the IRSL process. During CW-OSL or CW-IRSL experiments, the intensity of the light is kept constant, resulting in most cases in a monotonically decaying curve. An example of a CW-OSL curve measured with a constant optical excitation rate is shown in Fig. 1.10a, while Fig. 1.10b shows the CW-IRSL signal from the same sample (Kitis et al. [80]).

Figure 1.10 shows that the CW-OSL and CW-IRSL signals from the sample KT4 are very similar in shape. However, these signals are obtained with very different wavelengths of light (470 nm for blue light LEDs and 850 nm for infrared LEDs).

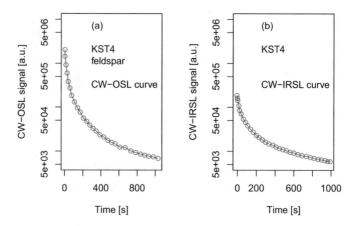

Fig. 1.10 Examples of (**a**) a CW-OSL curve and (**b**) a CW-IRSL curve, from the same geological feldspar sample KST4. For more details, see Kitis et al. [80]

Extensive research has shown that the mechanisms involved in the production of these signals are very different. In the case of the CW-OSL signal in Fig. 1.10a, the mechanism is believed to involve the conduction band and can be described by a *delocalized* model similar to the OTOR model shown in Fig. 1.4.

In the case of the CW-IRSL signal in Fig. 1.10b, the production mechanism is believed to involve localized energy levels located between the conduction and valence bands, as shown schematically previously in Fig. 1.7. Two versions of this type of *localized transition model* have been used in the literature, the simple localized transition (LT) model and the tunneling localized transition (TLT) model that is based on quantum mechanical tunneling processes. Both of these types of models are implemented in R in this book and are studied within the context of quantum tunneling, Monte Carlo methods, and the R package *RLumCarlo* in subsequent chapters.

Figure 1.11a shows the effect of different illumination powers on the CW-OSL signals for a quartz sample (Polymeris [154]). The stimulating power of the blue LEDs was varied in the range 50–90% of maximum power, and the data are normalized to the maximum intensity at time $t = 0$. The shape of the CW-OSL signal in Fig. 1.11a does not change significantly with the stimulating power of the blue LEDs. Figure 1.11b shows the CW-IRSL data obtained at different infrared light powers in the range 10–90% of maximum power, for a feldspar sample (Pagonis et al. [138]). This set of data has also been normalized to the maximum intensity, for comparison purposes. The shape of the CW-IRSL in Fig. 1.11b changes more significantly with the illumination power. This indicates either non-first order kinetics or the presence of multiple exponential components.

Note also that the OSL signal in Fig. 1.11a becomes negligible after 400 s of exposure to blue LEDs, while a large signal remains in Fig. 1.11b, even after exposure to 1000 s of infrared LED stimulation.

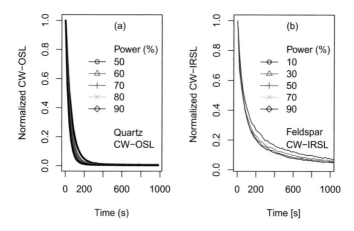

Fig. 1.11 (**a**) CW-OSL for quartz at different illumination powers in the range 50–90% of maximum power. (**b**) CW-IRSL signal at different illumination powers in the range 10–90% of maximum power, for a feldspar sample (laboratory code KST4). Both sets of data have been normalized to the maximum intensity. The shape of both signals changes with the illumination power. For more details, see Pagonis et al. [138] and Polymeris [154]

As we will discuss in later chapters, CW-OSL signals from quartz are described by *delocalized* transition models, while CW-IRSL signals from feldspars are described by *localized* transition models.

When the optical stimulation takes place using a source with an intensity that increases linearly with time, the stimulated luminescence is called *linearly modulated optically stimulated luminescence* (LM-OSL or LM-IRSL), and the stimulation rate $p(t)$ is given by

$$p(t) = \frac{\sigma I}{P} t \tag{1.4}$$

where σ, I have the same meaning as in Eq. (1.3), t is the elapsed time, and P (s) is the total illumination time.

During an LM-OSL or LM-IRSL experiment, the light intensity is recorded as a function of time, while increasing the stimulation linearly with time, thus obtaining a peak-shaped LM-OSL curve. The kinetics of such processes and the deconvolution of complex OSL signals into separate components by using available R packages are discussed in detail in Chap. 3.

Two examples of LM-OSL and LM-IRSL signals are shown in Fig. 1.12 for the dosimetric material CaF$_2$:N (Kitis et al. [78]) and for a K-feldspar (Bulur and Göksu [28]). The LM-OSL signal in Fig. 1.12a shows two peaks, and therefore, the luminescence mechanism for CaF$_2$:N involves at least two underlying components. By contrast, the LM-IRSL signal for the feldspar in Fig. 1.12b shows no apparent structure and may or may not involve a single luminescence component.

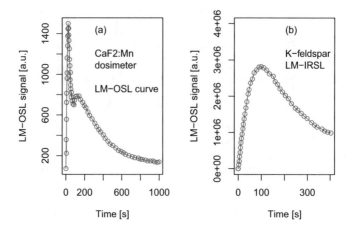

Fig. 1.12 Examples of (**a**) LM-OSL curve for the dosimetric material CaF_2:N (Kitis et al. [78]) and (**b**) LM-IRSL for a K-feldspar. For more details, see Bulur and Göksu [28]

1.5 Time-Resolved Luminescence Signals

In a time-resolved (TR) experiment, optical stimulation is used to separate in time the stimulation and emission of luminescence. The luminescence is stimulated using a brief light pulse (usually from bright LEDs), and the emission is monitored during the stimulation (when the LED is ON) and also during the relaxation stages of the experiment (when the LEDs are OFF). The details of TR experimental setups can be found in the review article by Chithambo et al. [41].

TR techniques have been very useful in the study of luminescence mechanisms for a variety of dosimetric materials. The optical stimulation is carried out in a pulsed mode, giving rise to *pulsed OSL* and *pulsed IRSL* signals (POSL and PIRSL).

Figure 1.13 shows two examples of TR-OSL curves, with very different time scales on the horizontal (time) axis. Figure 1.13a shows the TR-OSL signal from an Al_2O_3:C sample, while Fig. 1.13b shows the same experiment for a high purity synthetic quartz sample (Pagonis et al. [143]). The modeling of TR-OSL signals, and appropriate analysis of such experimental curves, is presented in Chap. 5.

Figure 1.14 shows two examples of TR-IRSL curves, from a microcline sample (laboratory code FL1, Pagonis et al. [120]). The modeling of these TR-IRSL signals, and appropriate analysis of such experimental curves, is presented in Chap. 5.

In many dosimetric materials, the pulsed OSL/IRSL signals depend strongly on the stimulation temperature. Specifically, one studies the temperature dependence of the luminescence intensity, as well as of the luminescence lifetimes determined from time-resolved luminescence spectra.

Figure 1.15 shows the TR-OSL signals from an Al_2O_3:C sample, measured at two different temperatures of 20 and 170 °C (Pagonis et al. [119]). One observes that as the stimulation temperature increases, the intensity of the TR-OSL signal

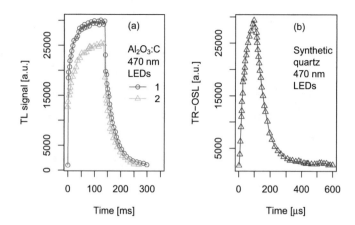

Fig. 1.13 Examples of TR-OSL curves for (**a**) Al_2O_3:C and (**b**) high purity synthetic quartz. Notice the very different time scales on the horizontal axis. The curves labeled 1 and 2 in (**a**) represent repeated measurements on the same sample, after the trapped electrons have been partially depleted. For more details, see Pagonis et al. [143]

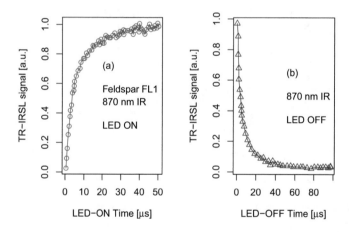

Fig. 1.14 Examples of TR-IRSL data for a feldspar sample FL1. (**a**) The TR-IRSL intensity as a function of the stimulation time during a $50\,\mu s$ LED ON pulse. (**b**) The TR-IRSL intensity as a function of the stimulation time during the $100\,\mu s$ LED OFF period. For more details, see Pagonis et al. [120]

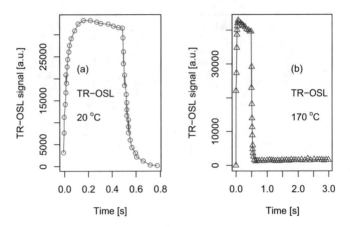

Fig. 1.15 Examples of TR-OSL data for Al$_2$O$_3$:C (**a**) at room temperature 20 °C and (**b**) at a stimulation temperature of 170 °C. For more details, see Pagonis et al. [119]

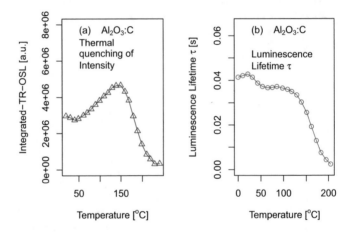

Fig. 1.16 Examples of TR-OSL data for Al$_2$O$_3$:C, showing the phenomenon of thermal quenching. (**a**) The integrated TR-OSL intensity as a function of the stimulation temperature. (**b**) The luminescence lifetime parameter τ as a function of the stimulation temperature. For more details, see Pagonis et al. [119]

varies, while at the same time the TR-OSL signal reaches its maximum value much faster. This phenomenon is referred to as *thermal quenching* of luminescence signals and will be discussed in some detail in Chap. 5.

Figure 1.16 shows the effect of thermal quenching on the TR-OSL signals from an Al$_2$O$_3$:C sample, as a function of the stimulation temperature (Pagonis et al. [119]). In Fig. 1.16a the intensity of the TR-OSL signal decreases as the stimulation temperature increases above 150 °C, while in Fig. 1.16b the lifetime τ of the TR-OSL signals overall decreases from about 40 ms at room temperature, to a few ms at stimulation temperatures higher than 200 °C.

1.6 Overview of ESR and OA Experiments: Correlations with Luminescence Signals

In this section we discuss briefly optical absorption (OA) and electron spin resonance (ESR) experiments and point out their relationship to TL and OSL data. For a more complete review of these types of experiments in connection with TL/OSL data, the reader is referred to the ESR review paper by Trompier et al. [181] and also to the extensive list of references in the book by Chen and Pagonis [38].

ESR and OA experiments are often measured simultaneously with TL and/or OSL with the same sample. These combined types of experiments produce important information about the underlying luminescence mechanisms, as well as about the nature of the trapping and luminescence centers (see, e.g. the quartz study by Yang and McKeever [193]).

The absorption technique of electron spin (paramagnetic) resonance (ESR, EPR) is used for the study of materials that exhibit paramagnetism because of the magnetic moment of unpaired electrons. ESR spectra are usually presented as plots of the absorption or dispersion of the energy of an oscillating magnetic field of fixed radio frequency, versus the intensity of an applied static magnetic field. Among the wide variety of paramagnetic substances to which ESR spectroscopy has been applied, the one that is of interest in the present context is that of impurity centers in solids, mainly single crystals. The same impurity centers may be associated with TL or OSL, either acting as the charge carrier traps or as recombination centers. They may also be responsible for optical absorption.

Since ESR is normally capable of identifying the impurities in the crystal, in cases where the simultaneous TL-ESR measurements show a direct relation between the two phenomena, the identification by ESR may serve as a direct proof for the identity of the impurity involved in the luminescence process. In many cases the OA and/or the ESR signal show a decrease at a certain temperature range, where the trapping state becomes unstable. The instability may be associated with either the thermal release of charge carriers from the paramagnetic impurity, or, alternatively, the filling of paramagnetic impurities that serve as TL recombination centers. In general, it is expected that the TL peak will resemble the negative derivative of the ESR signal, with a close resemblance to OA (Chen and Pagonis [38]).

Figure 1.17 shows an example of the dose dependence of an ESR signal in quartz from Duval [46]. This author measured the dose response curves of the Al center for 15 sedimentary quartz samples from the Iberian Peninsula. The samples were irradiated up to a maximum dose of 23–40 kGy. It was found that the ESR signal grows almost linearly with the absorbed dose for doses above about 4 kGy.

The dose responses of TL, OSL, OA, and ESR signals exhibit nonlinear regions, associated with the phenomena of superlinearity and sublinearity. These are discussed in detail in Chap. 4, in connection with recent theoretical research involving the Lambert W function (Pagonis et al. [135]).

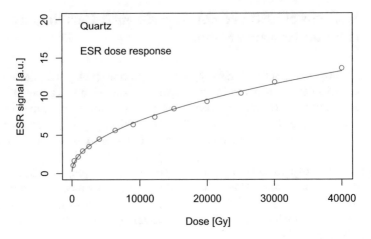

Fig. 1.17 Example of the dose response of ESR data from quartz. Redrawn from Pagonis et al. [135], original data from Duval [46]

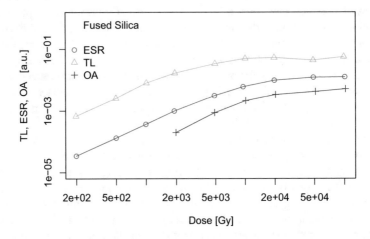

Fig. 1.18 Measurements of TL, ESR, and OA signals as a function of the irradiation dose, from a single sample of fused silica. Note the log scale in both axes. For more details, see Pagonis et al. [136], original data from Wieser et al. [189]

Figure 1.18 shows simultaneous measurements and possible correlations between the TL, ESR, and OA signals from a sample of fused silica (Wieser et al. [189]). This type of multifaceted experiment can be very useful in identifying the source of the various luminescence signals, i.e. the nature of the traps and centers in the dosimetric material.

1.7   What Information Can We Extract from TL, OSL, and TR Luminescence Signals?

The luminescence signals discussed in this chapter and in the rest of this book are obviously complex. The fundamental question for researchers is what kind of information they can extract from the experimental curves, and how.

In general, researchers would like to extract the following specific information:

- The intensity of the light emitted during each stimulated luminescence effect, as a function of time or as a function of temperature.
- The *dose response*, i.e. the intensity of the light emitted as a function of irradiation dose.
- The basic physical parameters of the energy level responsible for the emission of light. For thermally stimulated processes, these parameters are the activation energy E and the frequency factor s. For optically stimulated processes, they are the cross section σ for the corresponding process.
- Information on the physical mechanism behind the observed luminescence signals, and on the various processes involved.
- Information on whether the delocalized conduction and valence bands are involved in the luminescence processes, and whether localized energy transitions play a role in the production of the luminescence signals.

Luminescence signals from dosimetric materials are characterized by the presence of several components, originating from different traps or centers. These signals vary according to the preconditioning of the samples (irradiation dose, prior optical and thermal stimulation, radiation quality) and are described mathematically by phenomenological models based usually on systems of coupled differential equations.

As discussed in this chapter, two types of models that have been used extensively in the literature are *delocalized models* based on transitions involving the conduction and valence bands, and *localized models* usually involving different energy levels of the traps/centers. For a comprehensive historical summary of various types of luminescence models, the reader is referred for example to the book by Chen and Pagonis [38], and the review paper by McKeever and Chen [112]. The details of these models and their implementation with R are given in several later chapters.

The main goals of *modeling* studies are:

- to provide a quantitative description of these dosimetric signals;
- to develop methods for the accurate evaluation of parameters characterizing traps and centers;
- to search for general models that explain the behavior of preconditioned samples;
- to help researchers understand the underlying luminescence production mechanisms.

Throughout this book, you will find examples of using the R package *RLumCarlo*, to simulate various luminescence phenomena. This is a general purpose R package that can describe quantitatively a variety of luminescence signals (TL, CW-OSL, LM-OSL, IRSL, ITL, etc). The package contains a variety of luminescence models: delocalized transitions models (DELOC functions in *RLumCarlo*), as well as localized transitions models (LOC functions), and tunneling transitions models (TUN functions).

Part I
Luminescence Signals from Delocalized Transitions

Chapter 2
Analysis and Modeling of TL Data

2.1 Introduction

In this chapter we will summarize the simplest models used to describe TL signals, and also provide several examples of using R to analyze and model these signals. We start by presenting the simple OTOR model in Sects. 2.2–2.6, and the relevant differential equations for *first* and *second order kinetics*. We show how these equations can be integrated numerically using R.

This is followed in Sect. 2.7 by a discussion of the *initial rise method*, which is the simplest method of evaluating the activation energy E from an experimental TL glow curve. Next, in Sect. 2.8 we discuss and give examples of how to apply the *method of various heating rates*, which allows evaluation of both the activation energy E and the frequency factor s.

Next is a discussion of the popular empirical equation for *general order kinetics* (GOK) in Sect. 2.9, which has been used often to analyze complex TL glow curves.

This is followed in Sects. 2.10–2.11 by a discussion of the *general one trap* (GOT) differential equation and its analytical solution, which is based on the Lambert W function. Several examples are presented in Sects. 2.12–2.14 of using computerized glow curve analysis (CGCA) for single-peak and multiple-peak TL glow curves, based on the R-packages *tgcd* and the Lambert W function.

We also discuss the more general *interactive multi-trap system (IMTS)* model in Sects. 2.15–2.16, and how it leads to *mixed order kinetics* (MOK) for luminescence phenomena. Specific examples are given of how to analyze complex TL glow curves using the package *tgcd* and the Lambert W function, the empirical general order kinetics and the physically meaningful mixed order kinetics.

Section 2.17 contains a discussion and several examples of the new R package *RLumCarlo*, which is used to simulate TL glow curves with different kinetic parameters. This chapter concludes in Sect. 2.18 with a list of recommended experimental protocols, which experimentalists can apply when studying TL signals.

© The Author(s), under exclusive license to Springer Nature Switzerland AG 2021   21
V. Pagonis, *Luminescence*, Use R!, https://doi.org/10.1007/978-3-030-67311-6_2

2.2 The Simplest TL Model (OTOR)

The process by which materials emit light when heated can be understood by considering the simplest possible model consisting of two localized levels, an isolated electron trap (T) and a recombination center (RC), as shown in Fig. 2.1. This is commonly referred to as the *one-trap-one-recombination center model* (OTOR). We now discuss the differential equation approach to the OTOR model.

In the OTOR model, trapped electrons can be released by thermal or optical excitation into the conduction band. Once in the conduction band, electrons can be either captured into the recombination center, or can be retrapped into the electron trap. These transitions for the OTOR model are shown schematically in Fig. 2.1. Transition 1 represents thermal or optical excitation of trapped electrons, transition 2 is the capturing of an electron from the conduction band into the recombination center (RC), and transition 3 represents retrapping of an electron from the conduction band into the electron trap.

Let us denote by N the total concentration of the traps in the crystal (in cm^{-3}), by $n(t)$ the concentration of filled traps in the crystal (in cm^{-3}) at time t, and by $m(t)$ the concentration of trapped holes in the recombination center (in cm^{-3}). The initial concentration of filled traps at time $t = 0$ is denoted by n_0.

In a typical thermoluminescence experiment the sample is heated with a constant heating rate $\beta = dT/dt$ from room temperature up to a high temperature, usually around 500 °C. As the temperature of the sample is increased, the trapped electrons in T are thermally released into the conduction band, as shown by the arrow for transition 1 in Fig. 2.1. These conduction band electrons can either recombine with holes in the recombination center RC (transition 2) with a recombination coefficient A_m ($cm^3 \, s^{-1}$), or they can be retrapped into the electron trap T (transition 3) with a retrapping coefficient A_n ($cm^3 \, s^{-1}$), as shown in Fig. 2.1.

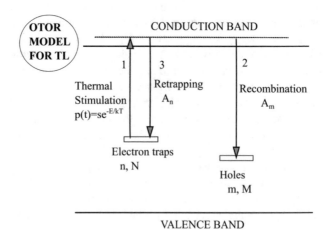

Fig. 2.1 Schematic diagram of the OTOR model for a TL process

The equations governing the process during thermal or optical stimulation are

$$\frac{dn}{dt} = A_n(N - n)n_c - n\,p(t), \tag{2.1}$$

$$\frac{dn_c}{dt} = n\,p(t) - A_n(N - n)n_c - A_m m n_c, \tag{2.2}$$

$$I(t) = -\frac{dm}{dt} = A_m m n_c, \tag{2.3}$$

$$\frac{dm}{dt} = \frac{dn}{dt} + \frac{dn_c}{dt} \tag{2.4}$$

The charge neutrality condition is

$$m = n + n_c \tag{2.5}$$

The emitted light is associated with the recombination of free electrons with holes in recombination centers, and the luminescence intensity $I(t)$ is given by Eq. (2.3). The term $p(t)$ represents the rate of thermal or optical excitation of trapped electrons, which in the case of TL is given by $p(t) = s\,\exp[-E/k_B T(t)]$.

The following R code solves the system of differential equations for the OTOR model, with activation energy $E = 1\,\text{eV}$, frequency factor $s = 10^{12}\,\text{s}^{-1}$, $A_m = 10^{-7}\,\text{cm}^3\,\text{s}^{-1}$, $A_n = 10^{-7}\,\text{cm}^3\,\text{s}^{-1}$, $n_0 = 10^9\,\text{cm}^{-3}$, $m_0 = 10^9\,\text{cm}^{-3}$, $N = 10^{10}\,\text{cm}^{-3}$, $n_c(0) = 0$, and the constant heating rate $\beta = 1\,\text{K/s}$. The R package *deSolve* is used to numerically solve the system of differential equations (Soetaert et al. [170, 171]). Specifically the function *ode* in the package *deSolve*, is called with appropriate numerical values of the parameters N, s, E, A_m, A_n, β contained in the parameter *parameters*. The additional variable *state* contains the initial number of filled traps n_0.

Figure 2.2 shows how the concentration of trapped electrons (n), the concentration of conduction band electrons (n_c), and the luminescence intensity (TL) vary during a typical TL experiment. As the temperature of the sample is increased, the concentration of trapped electrons n decreases, and the TL intensity initially increases as well. However, as the trapped electrons are thermally released into the conduction band, the traps are gradually emptied and the TL intensity will reach a maximum and then decrease to zero. In Fig. 2.2b the concentration $n_c(T)$ has been multiplied by a factor of 10 to make it visible, i.e. $n_c(T)$ is several orders of magnitude smaller that $n(T)$.

The parameter *out* in this code contains the numerical solution of the OTOR system, for example *out[, "n1"]* is the vector containing the values of $n(t)$, while *out[, "nc"]* contains $n_c(t)$, etc.

In the next two sections we discuss first order kinetics in luminescence processes.

Code 2.1: System of differential equations for OTOR

```
# Solution of the system of ODE's for the OTOR model
rm(list=ls())
library(deSolve)
TLOTOR <- function(t, x, parms) {
  with(as.list(c(parms, x)), {
    dn1 <- - n1*s*exp(-E1/(kb*(273+hr*t)))+ nc*An*(N1-n1)
    # n1=concentration of trapped electrons
    dnc <-   n1*s*exp(-E1/(kb*(273+hr*t)))-nc*An*(N1-n1)-m*Am*nc
    # nc=concentration of conduction band electrons
    dm <- -m*Am*nc
    # m=concentration of recombination centers
    res <- c(dn1, dnc, dm)
    list(res)
  })}
## Parameters
hr<-1   # heating rate in K/s
parms  <- c(E1 =1, s=10^12, kb=8.617*10^-5, hr=hr,
            An = 10^-7,  N1 = 10^10,Am = 10^-7)
## vector of timesteps
times  <- seq(0, 250)
temps<-times*hr
## Initial conditions for the system
y <- xstart <- c(n1 = 10^9, nc = 0, m = 10^9)
## Solve system of differential equations
out <-  lsoda(xstart, times, TLOTOR, parms)
## Plotting
par(mfrow=c(1,2))
plot(temps,out[,"n1"],xlab=expression("Temperature ["^"o"*"C]"),
ylab =expression("Filled traps n(T),cm"^-3*" "),ylim=c(0,1.2e9))
legend("left",bty="n",expression("OTOR","n(T)"))
legend("topleft",bty="n", expression("(a)"))
plot(temps,out[,"m"]*parms["Am"]*out[,"nc"],pch=1,col="red",
xlab=expression("Temperature ["^"o"*"C]"),
ylim=c(0,2.8e7),ylab="TL [a.u.] and nc (t)",xlim=c(0,250))
lines(temps,10*out[,"nc"],xlab=expression("Temperature [
"^"o"*"C]"),ylab="nc(t)",typ="o",pch=2,col="blue")
legend("topright",bty="n", expression("(b)"))
legend("topleft",bty="n", pch=c(1,2),expression("TL","ncx10"),
col=c("red","blue"),lwd=1)
```

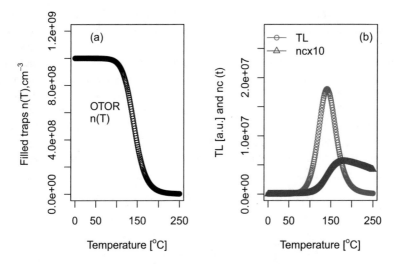

Fig. 2.2 Numerical solution of the system of differential equations for the OTOR model. (**a**) The concentrations of trapped electrons $n(T)$ and (**b**) of the conduction band electrons $n_c(T)$ (triangles), and luminescence intensity $I(T)$ (circles), during a typical TL experiment. Note that $n_c(T)$ has been multiplied by a factor of 10, for display purposes

2.3   The Prevalence of First Order Kinetics in Luminescence Experiments

During optical or thermal excitation in a luminescence experiment, we are interested in the rate of change of the concentration of trapped electrons in the dosimetric trap (dn/dt). In most luminescence dosimetric materials, the intensity of the luminescence signal $I(t)$ follows the differential equation:

$$I(t) = -\frac{dn}{dt} = n\,p(t) \tag{2.6}$$

where $p(t)$ (s^{-1}) is the rate of thermal or optical excitation. This equation expresses the well-known *first order kinetics* of the luminescence process. For example, in the case of TL this equation becomes

---

Differential equation for first order TL kinetics

$$I_{TL} = -\frac{dn}{dt} = n\,s\,e^{-\frac{E}{kT}} \tag{2.7}$$

It is important to keep in mind that the vast majority of TL dosimetric materials exhibit first order kinetics processes, and that there have been very few reports of non-first order kinetics behavior. This overwhelming prevalence of first order kinetics in nature has been explained by Pagonis and Kitis [133], by simulating TL glow curves in the presence of many competitors. These authors showed that the physical origin of first order kinetics is the presence of competitors in a dosimetric material. As the number of competitors increases, the simulations by these authors showed that the kinetic order b tends towards the first order kinetics value of $b = 1$.

Assuming a constant heating rate $\beta = dT/dt$, the solution of Eq. (2.7) is the well-known TL equation by Randall and Wilkins [161] for first order kinetics:

Analytical solution for first order TL kinetics

$$I_{TL}(T) = \frac{n_0 \, s}{\beta} \, e^{-\frac{E}{kT}} \, \exp\left[-\frac{s}{\beta} \int_{T_0}^{T} e^{-\frac{E}{kT'}} \, dT' \right] \qquad (2.8)$$

---

where n_0 is the initial value of the concentration of trapped electrons at $t = 0$, T_0 is the initial temperature and T' is a dummy integration variable representing temperature. The units of the *temperature* dependent TL intensity $I(T)$ in Eq. (2.8) are $\mathrm{cm}^{-3}\,\mathrm{K}^{-1}$. However, note that the units of the *time* dependent luminescence intensity $I(t)$ are $\mathrm{cm}^{-3}\,\mathrm{s}^{-1}$. These two quantities are related by $I(T) = I(t)/\beta$ where β is the constant heating rate.

This expression for $I_{TL}(T)$ is the product of two expressions: the exponential function $\exp(-E/kT)$ which increases as the temperature T of the sample is increased, and of the function $\exp\left[-\frac{s}{\beta} \int_{T_0}^{T} e^{-\frac{E}{kT'}} \, dT' \right]$ which decreases to zero as T increases. The result of multiplying these two expressions is the peak-shaped TL glow curve.

As in the previous section, the following R code solves the differential equation for first order kinetics, with activation energy $E = 1\,\mathrm{eV}$, frequency factor $s = 10^{12}\,\mathrm{s}^{-1}$, $N = 10^{11}\,\mathrm{cm}^{-3}$ and the constant heating rate $\beta = 1\,\mathrm{K/s}$. The R package *deSolve* is used again to numerically solve the differential equation, and the R-code *abs(diff(num_ODE[,2]))* is used to evaluate the TL intensity $I(t) = -dn/dt$, by using the approximation $\Delta n/\Delta t$. In this example $\Delta t = 1\,\mathrm{s}$ and $\Delta T = 1\,\mathrm{K}$, since the heating rate is $\beta = 1\,\mathrm{K/s}$.

As can be seen from the graph produced by the R code in Fig. 2.3, a first order TL peak is not symmetric.

The physical origin of first order kinetics in luminescence processes can also be explained by considering the OTOR model discussed in the previous section. The OTOR model leads to first order kinetics under certain physical conditions, and these are discussed in Sect. 2.10.

Code 2.2:   ODE for TL: First order kinetics

```
# Numerically solve the ODE for TL: first order kinetics
rm(list=ls())
library(package ="deSolve")
# Define Parameters
k_B <- 8.617e-5  # Boltzmann constant
E <- 1 # electron trap depth [eV]
s <- 1e12 # frequency factor [1/s]
delta.t <- 1
t <- seq(0, 200, delta.t)
n.0 <- 1e10
N.traps <- 1e11
ODE <- function(t, state, parameters){
        with(as.list(c(state, parameters)),{
            dn <-   -s*exp(-E/(k_B*(273+t))) * n
    list(c(dn)) })}
parameters <- c(N.traps = N.traps, s = s, E = E, k_B = k_B)
state <- c(n = n.0)
num_ODE <- ode(y = state, times = t, func = ODE,
parms = parameters)
# Plot remaining electrons and TL as a function of temperature
par(mfrow=c(1,2))
plot(x =num_ODE[,1],
y = num_ODE[,2], xlab=expression("Temperature ["^"o"*"C]"),
 ylab = expression("Filled traps n(T), cm"^-3*" "),
col = "red")
legend("topright",bty="n",c("(a)","n(T)"))
plot(num_ODE[-1,1], y = abs(diff(num_ODE[,2])),
xlab=expression("Temperature ["^"o"*"C]"),
ylab = "TL signal [a.u.]",xlim=c(0,170))
legend("topleft",bty="n",c("(b)","TL"))
```

2.4   Simulation of the Dose Response of TL Peaks Following First Order Kinetics

Under certain physical assumptions, the concentration of filled traps (n_0) at the end of the irradiation process is proportional to the radiation dose D received by a sample. When the sample is heated after the end of the irradiation, the observed TL intensity may also be proportional to n_0 (again under certain physical assumptions).

We can then simulate the dose response of a dosimetric trap by plotting TL glow peaks with different initial concentrations n_0, when all other parameters in the model are kept fixed.

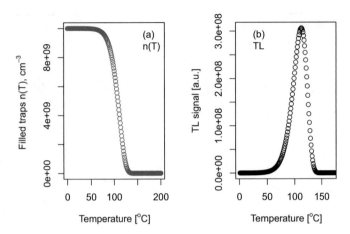

Fig. 2.3 Numerical solution of the Randall and Wilkins Eq. (2.7), for first order kinetics TL. (**a**) Remaining trapped electrons $n(T)$, and (**b**) The corresponding TL signal

We will use the function *simPeak* which is part of the package *tgcd* developed by Peng et al. [150]. The function *simPeak* can simulate first order, second order, or general order glow peaks. The arguments of the *simPeak* functions are *temps* (a vector containing the temperature values in K), *n0* (a numeric containing the initial concentration of trapped electrons e.g. in cm^{-3}), *Nn* (a numeric containing the total concentration N of traps in the crystal e.g. in cm^{-3}), *bv* (a numeric between 1 and 2, containing the kinetic order parameter b for the general order glow peak), *ae* (containing the thermal activation energy E), and *ff* (a numeric containing the frequency factor s in s^{-1}).

The following R-code shows the effect of the initial concentration of filled traps (n_0) on a TL peak which follows first order kinetics. The function *simPeak* is called three times to plot the TL glow curves. The height of the first order TL peak is proportional to n_0, and (under certain physical conditions) this height is also proportional to the irradiation dose D received by the sample before the heating process.

Another important characteristic of the first order peaks seen in Fig. 2.4 is that the location of the maximum of the TL peak does not change at different values of n_0.

Code 2.3: First order TL by varying the initial trap concentrations (tgcd)

```
# Simulate first-order glow peaks with various
# initial electron trap concentrations (n0).
rm(list=ls())
# library(tgcd)
library(package ="tgcd")
temps <- seq(300, 440, by=1)
peak1 <- simPeak(temps,n0=0.2e10,Nn=1e10,ff=1e12, ae=1.0, hr=1,
                 typ="f",plot=FALSE)
peak2 <- simPeak(temps,n0=0.4e10,Nn=1e10,ff=1e12, ae=1.0, hr=1,
                 typ="f",plot=FALSE)
peak3 <- simPeak(temps,n0=0.6e10,Nn=1e10,ff=1e12, ae=1.0, hr=1,
                 typ="f",plot=FALSE)
peaks<-cbind(peak1$tl, peak2$tl, peak3$tl)
matplot(temps, peaks, type="o",  pch=c(0,1,2),
        col=c("black","red","blue"),lwd=2,
        xlab="Temperature (K)", ylab="TL intensity [a.u.]")
legend("topleft",bty="n", pch=c(NA,NA,NA, 0,1,2),
       c(expression('First order peaks',' ','n'[o]*'/N'),
"0.2","0.4","0.6"),col=c(NA,NA,NA,"black","red","blue"))
```

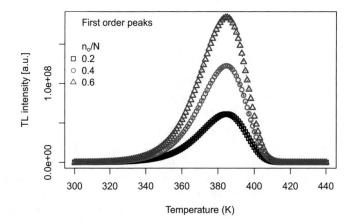

Fig. 2.4 Simulation of first order TL glow peaks with three different initial electron trap concentrations $n_0 = 0.2, 0.4, 0.6 \times 10^{10}$ cm^{-3}. The location of the maximum TL intensity does not shift significantly as n_0 changes, and the peak height is proportional to n_0

2.5   Second Order Kinetics in Luminescence Processes

As discussed previously, most luminescent dosimetric materials exhibit first order
kinetics. However, there are some reports of materials which follow *second order
kinetics*. As we will see in Sect. 2.10, the OTOR model leads to second order kinetics
under certain physical assumptions. The following differential equation describes
second order kinetics in luminescence processes:

$$I(t) = -\frac{dn}{dt} = \frac{n^2}{N} p(t) \tag{2.9}$$

In the case of TL this equation yields

Differential equation for second order kinetics TL

$$I_{TL}(t) = -\frac{dn}{dt} = \frac{n^2}{N} s\, e^{-\frac{E}{kT}} \tag{2.10}$$

Assuming again a constant heating rate, the solution of Eq. (2.10) is the well-
known second order kinetics equation by Garlick and Gibson [51]:

Analytical solution for second order TL kinetics

$$I_{TL}(T) = \frac{n_0^2 s}{\beta N} e^{-\frac{E}{kT}} \left[1 + \frac{n_0 s}{\beta N} \int_{T_0}^{T} e^{-\frac{E}{kT'}} dT' \right]^{-2} \tag{2.11}$$

where all variables have the same meaning as in the equation for first order kinetics.
As mentioned previously, the units of the *temperature* dependent TL intensity $I(T)$
in Eq. (2.11) are $cm^{-3}\, K^{-1}$, while the units of the *time* dependent $I(t)$ are $cm^{-3}\, s^{-1}$
since $I(T) = I(t)/\beta$, where β is the constant heating rate.

The following R code shows the effect of the initial concentration of filled traps
(n_0) on a TL peak which follows second order kinetics.

As seen in Fig. 2.5, an important characteristic of second order TL peaks is that
the location of the maximum of the TL peak changes for different values of n_0. This
type of behavior is seen very rarely in dosimetric materials; in most natural and
synthetic luminescence materials, the temperature of the TL maximum stays the
same, and the shape of the TL glow curve also remains unchanged (see the detailed
discussion in Pagonis and Kitis [133], and also the papers by Sunta et al. [173, 174]).

Code 2.4:   Second order TL by varying the initial trap concentrations (tgcd)

```
# Simulate second order glow peaks with various
# initial electron trap concentrations (n0).
rm(list=ls())
library(tgcd)
temps <- seq(300, 500, by=1)
peak1 <- simPeak(temps,n0=0.2e10,Nn=1e10,ff=1e12, ae=1.0, hr=1,
typ="s",plot=FALSE)
peak2 <- simPeak(temps,n0=0.4e10,Nn=1e10,ff=1e12, ae=1.0, hr=1,
typ="s",plot=FALSE)
peak3 <- simPeak(temps,n0=0.6e10,Nn=1e10,ff=1e12, ae=1.0, hr=1,
typ="s",plot=FALSE)
peaks<-cbind(peak1$tl, peak2$tl, peak3$tl)
matplot(temps, peaks, type="o",  pch=c(0,1,2),
        col=c("black","red","blue"),lwd=2,
        xlab="Temperature (K)", ylab="TL intensity [a.u.]")
legend("topright",bty="n", pch=c(NA,NA,NA,0,1,2),
       c(expression('Second order peaks',' ','n'[0]*'/N'),
"0.2","0.4","0.6"),col=c(NA,NA,NA,"black","red","blue"))
```

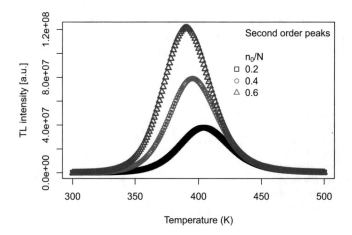

Fig. 2.5 Simulation of second order TL glow peaks with three different initial electron trap concentrations $n_0 = 0.2, 0.4, 0.6 \times 10^{10} \, \text{cm}^{-3}$. The location of the maximum TL intensity shifts with n_0, and the *area* under the peak is proportional to n_0

2.6 The Geometrical Shape Factor

A useful method of characterizing the shape of TL peaks is with the *geometrical shape factor* μ_g. If we denote the maximum temperature by T_m and the low and high half-intensity temperatures by T_1 and T_2 respectively, the low temperature half width is defined as $\tau = T_m - T_1$, the high temperature half width is defined as $\delta = T_2 - T_m$, and the full-width half intensity by $\omega = T_2 - T_1$, then the geometrical shape factor is defined by

---

The geometrical shape factor for TL glow curves

$$\mu_g = \delta/\omega \qquad (2.12)$$

These widths are shown in Fig. 2.6. The first order peak is rather asymmetric with $\mu_g \approx 0.42$, while a second order peak is very nearly symmetric with $\mu_g \approx 0.52$. Intermediate μ_g shape factors have been observed in experimental TL peaks between these two values, and occasionally also out of the range of $\mu_g = 0.42$–0.52 [37].

The following R code simulates and plots together a first and a second order glow peak, by using the same kinetic parameters. The solid line represents the results for first order kinetics, and the dashed line represents second order kinetics. The R package *tgcd* is used, and the function *simPeak* is called to calculate the first and second order TL peaks, with the parameters $typ = "f"$ and $typ = "s"$, and the logical parameter *plot* determines whether to produce a plot or not.

The graph in Fig. 2.7 shows that a second order TL glow peak (dashed line) is clearly wider than the first order TL peak with the same parameters (solid line). The second order TL glow peak is also almost exactly symmetric with $\mu_g \approx 0.52$. The

Fig. 2.6 Schematic diagram of TL peak. The half-widths τ, δ, ω define the geometrical shape factor $\mu_g = \delta/\omega$. First order peaks have an asymmetric form, while second order peaks are very nearly symmetric

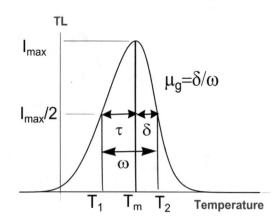

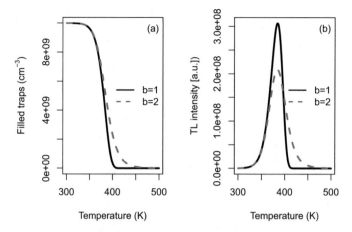

Fig. 2.7 Comparison of simulated first and second order TL peaks (solid and dashed lines respectively), evaluated with the same kinetic parameters. The second order process is slower, and the peak shape for second order is almost symmetric. (**a**) Filled traps $n(T)$. (**b**) The corresponding TL signal

R-code prints out the values of the parameters T_1, T_2, T_m, $d_1 = T_m - T_1 (= \tau)$, $d_2 = T_2 - T_m (= \delta)$, $thw = T_2 - T_1 (= \omega)$, and $sf (= \mu_g = \delta/\omega)$.

From a physical point of view, second order kinetics is a slower process than first order kinetics.

Code 2.5:   First and second order TL with the same parameters

```
# Simulate a first and a second  order TL glow peak
# with the same kinetic parameters
rm(list=ls())
library(package ="tgcd")
temps <- seq(300, 500, by=.2)
peak1 <- simPeak(temps, n0=1e10, Nn=1e10, ff=1e12, ae=1, hr=1,
                 typ="f",plot=FALSE)
peak2 <- simPeak(temps, n0=1e10, Nn=1e10,  ff=1e12, ae=1, hr=1,
                 typ="s",plot=FALSE)
n<-cbind(peak1$n, peak2$n)
par(mfrow=c(1,2))
matplot(temps, n, type="l", lwd=3,lty=c(1,2),
xlab="Temperature (K)",
```

```
ylab=expression("Filled traps (cm"^"-3"*")"))
legend("topright",bty="n", expression("(a)"))
legend("right",bty="n", lty=c(1,2),expression("b=1","b=2"),
       col=c("black","red"),lwd=2)
peaks<-cbind(peak1$tl, peak2$tl)
matplot(temps, peaks, type="l", lwd=3, lty=c(1,2),
xlab="Temperature (K)",ylab="TL intensity [a.u.])")
legend("right",bty="n", lty=c(1,2),expression("b=1","b=2"),
       col=c("black","red"),lwd=2)
legend("topright",bty="n", expression("(b)"))

print.noquote("Parameters for first order TL peak")
print(c(peak1$sp[1],peak1$sp[2],peak1$sp[3]))
print(c(peak1$sp[4],peak1$sp[5],peak1$sp[6]))
cat("\nShape factor=",round(peak1$sp[7],3))

print.noquote("Parameters for second order TL peak")
print(c(peak2$sp[1],peak2$sp[2],peak2$sp[3]))
print(c(peak2$sp[4],peak2$sp[5],peak2$sp[6]))
cat("\nShape factor=",round(peak2$sp[7],3))

   ## [1] Parameters for first order TL peak
   ##       T1        T2        Tm
   ## 367.2004 397.0405 384.6000
   ##       d1        d2       thw
   ## 17.39956 12.44053 29.84009
   ##
   ## Shape factor= 0.417[1] Parameters for second order TL peak
   ##       T1        T2        Tm
   ## 363.5701 405.6734 383.8000
   ##       d1        d2       thw
   ## 20.22992 21.87338 42.10330
   ##
   ## Shape factor= 0.52
```

2.7 The Initial Rise Method for Estimating E

The *initial rise method* of analyzing TL data is the simplest method for estimating the activation energy E. In this method one plots the natural logarithm of the TL intensity $\ln(TL)$ on the y-axis, and the quantity $1/(k_B T)$ on the x-axis, where

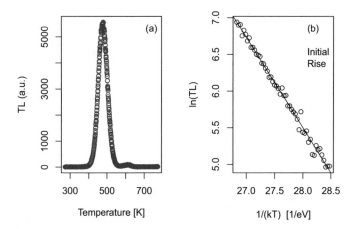

Fig. 2.8 Example of applying the initial rise method to find the activation energy E. (**a**) The red line indicates the initial rise area being analyzed; (**b**) The slope of the best fit line gives the activation energy $(-E)$

$k_B = 8.617 \times 10^{-5}$ (eV K^{-1}) is the Boltzmann constant. The slope of the resulting linear graph is the activation energy E. The following R code shows an example of estimating E using the initial rise method.

The TL data being analyzed is part of the package *tgcd*, and is loaded with the line *data("kitis")*. The first set of data is Reference glow curve #1 from the intercomparison project GLOCANIN (Bos et al. [21]). This is a synthetic TL glow curve with an activation energy $E = 1.182$ eV and a frequency factor $s = 8.06 \times 10^{10}$ s^{-1}. The parameters *initialPos, finalPos* in the code select the range of data points in the file, for which the initial rise method will be applied. As a rule of thumb, it is best to select TL points located lower than 5–10% of the maximum intensity of the TL peak.

The *lm* function evaluates the best fitting line in Fig. 2.8, and *plot*, *abline* are used to plot together the data and the best fit line. The code line *summary* produces a summary of several statistical quantities, including the adjusted coefficient $R^2 = 0.984$.

In this example the activation energy from the initial rise method is $E = 1.23$ eV, which is about 4% larger than the expected value of $E = 1.182$ eV for this synthetic TL glow curve.

Code 2.6: The initial rise method: find energy *E* from TL data

```
# Apply the initial rise method to find the activation energy E
# Load the data from txt file, which  contains pairs of

# data  in the form: (Temperature_in_K,TL_Intensity (any units)
rm(list=ls())
library(tgcd)
data("Kitis")
x<-Kitis$x001[,1]
y<-Kitis$x001[,2]
mydata<-data.frame(x,y)
kB<-8.617*1e-5 # Boltzmann constant in eV/K
initialPos<-270   #analyze data points from #270 to #320
finalPos<-320
x<- mydata[,1][initialPos:finalPos]
y<- mydata[,2][initialPos:finalPos]
rangeData<-cbind(x,y)
y<-log(y)
x<-1/(kB*x)
bestfit<-lm(y~x)
summary(bestfit)
coefficients(bestfit)
par(mfrow=c(1,2))
plot(mydata,col="blue",
xlab=expression("Temperature [K]"),ylab = "TL (a.u.)")
lines(rangeData,col="red",lwd = 3)
legend("topright",bty="n"," (a)")
plot(x, y, xlab = "1/(kT)  [1/eV]",ylab = "ln(TL)")
abline(lm(y~x))
legend("topright",bty="n",c(" (b)"," ","Initial","Rise"))

    ##
    ## Call:
    ## lm(formula = y ~ x)
    ##
    ## Residuals:
    ##       Min       1Q    Median        3Q       Max
    ## -0.158165 -0.030093  0.001636  0.027663  0.172387
    ##
    ## Coefficients:
    ##             Estimate Std. Error t value Pr(>|t|)
    ## (Intercept) 39.97106    0.44678   89.46   <2e-16 ***
    ## x           -1.23111    0.01616  -76.17   <2e-16 ***
    ## ---
    ## Signif. codes:  0 '***' 0.001 '**' 0.01 '*' 0.05 '.'
    ##                     0.1 ' ' 1
    ##
    ## Residual standard error: 0.05592 on 49 degrees of freedom
    ## Multiple R-squared:  0.9916,Adjusted R-squared:  0.9915
    ## F-statistic:  5802 on 1 and 49 DF,  p-value: < 2.2e-16
    ##
    ## (Intercept)            x
    ##    39.971063    -1.231108
```

2.8   The Heating Rate Method of Finding the Kinetic Parameters E, s

When the sample is heated with different heating rates, the TL glow curve shifts towards higher temperatures, while the area under the TL glow curve remains the same.

This is shown in the following R code, where the *simPeak* function in the package *tgcd* is used to plot four first order TL glow curves measured with heating rates $\beta = 1, 2, 3, 4 \, \text{K/s}$. The command *sapply* is used to apply the function *findTL* to the vector *hRates* which contains the values of the heating rates.

An important point to remember when comparing signals measured with different heating rates is that one should plot TL/β as a function of temperature, i.e. the y-axis should be in counts/K and *not* in counts/s. When the TL data is plotted in this way, one expects that for first order kinetics the area under the curves would stay the same, and that the height of the TL peak would decrease, as shown in Fig. 2.9.

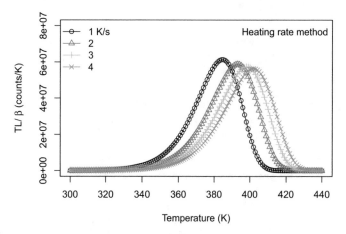

Fig. 2.9 Simulation of the same TL glow curve measured with four different heating rates $\beta = 1, 2, 3, 4 \, \text{K/s}$. As the heating rate increases, the TL peak shifts to the right towards higher temperatures, and the intensity decreases. The area under the curves stays the same. Notice that the y-scale is TL/β, i.e. in counts/K, and not in counts/s

Code 2.7: TL glow curve for four different heating rates

```
# Plot the same TL glow curve for four different heating rates
rm(list=ls())
library(tgcd)
temps <- seq(300, 440, by=1)   # temperature in K
hRates<-c(1,2,3,4)
## function to calculate TL for different hehating rates####
findTL<-function(hRate)
{peak<-simPeak(temps,n0=0.2e10,Nn=1e10,ff=1e12,ae=1.0, hr=hRate,
typ="f",plot=FALSE)
peak$tl}
### Calculate TL with different heating rates
TLs<-sapply(hRates,findTL)
matplot(temps,TLs,type="o",  lty=c(1,1,1,1),lwd=1,  pch=1:4,
col=1:4,xlab="Temperature (K)",  ylim=c(0,8e7),
ylab=expression(paste("TL/ ",beta," (counts/K)")))
legend("topright",bty="n","Heating rate method")
legend("topleft",bty="n",  pch=c(1:4,NA),expression(
  "1 K/s","2","3","4"," "),col=c(1:4,NA),lwd=1)
```

By taking the derivative of Eq. (2.8), we find the following well-known relationship between the temperature of maximum TL intensity (T_m), and the heating rate β used during the TL experiment:

The heating rate equation for first order kinetics

$$\ln\left(\frac{T_m^2}{\beta}\right) = \frac{E}{k_B T_m} + \ln\left(\frac{E}{k_B s}\right) \qquad (2.13)$$

This equation allows us to evaluate the frequency factor s for a first order TL glow curve, when the values of E and T_m are known.

This equation is also the basis of the *variable heating rate method*, which is used to calculate both the activation energy E, and the frequency factor s.

When we plot $\ln\left(T_m^2/\beta\right)$ as a function of $1/(k_B T)$, the slope of the linear graph represents the activation energy E, and the intercept is equal to $\ln[E/(k_B s)]$. Once the value of E is estimated from the slope, the frequency factor s is found from Eq. (2.13) to be

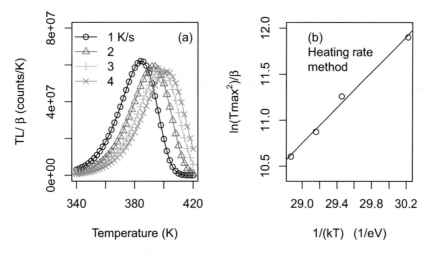

Fig. 2.10 Applying the heating rate method to obtain both the activation energy E and the frequency factor s. (**a**) The simulated TL glow curves. (**b**) The slope and intercept of the best fit line yield both parameters (E, s)

Equation to find s in the heating rate method

$$s = \frac{E}{k_B} e^{-intercept} \qquad (2.14)$$

The following R code demonstrates how to use the heating rate method to find E, s. The code uses the function *simPeak* in the R package *tgcd*, to simulate four first order TL peaks measured with different heating rates β, as in the previous code. The function *findTmax* evaluates the temperature T_m for the maximum TL intensity in each peak.

Figure 2.10 shows that the activation energy $E = 0.96\,eV$, which is within 4% of the expected value of $E = 1.00\,eV$. Similarly, the frequency factor is found from Eq. (2.14) to be $s = 3.12 \times 10^{11}\,s^{-1}$, instead of the expected value of $s = 1.0 \times 10^{12}\,s^{-1}$. This discrepancy is due to the term $e^{-intercept}$ appearing in Eq. (2.14) for s; because of the exponential factor, small changes in the value of the *intercept* result in large changes in the calculated value of s.

Code 2.8:   Apply heating rate method to TL data, to find E, s

```
# Apply the heating rate method to find E,s
rm(list=ls())
library(tgcd)
library(scales)
kB<-8.617*1e-5 # Boltzmann constant in eV/K
temps <- seq(340, 420, by=2)
#### function to calculate TL for different heating rates####
findTL<-function(hRate)
{peak<-simPeak(temps,n0=0.2e10,Nn=1e10,ff=1e12,ae=1.0, hr=hRate,
typ="f",plot=FALSE)
peak$tl}
######## function to find Tmax ######
findTmax<-function(hRate)
{peak<-simPeak(temps,n0=0.2e10,Nn=1e10,ff=1e12,ae=1.0, hr=hRate,
typ="f",plot=FALSE)
temps[[match(max(peak$tl),peak$tl)]]}
########## calculate 1/kTmax and log(Tmax^2/beta)
hRates<-c(1,2,3,4)
maxTL<-sapply(hRates,findTmax)
TLs<-sapply(hRates,findTL)
####### plots
par(mfrow=c(1,2))
matplot(temps,TLs,type="o",lty="solid",lwd=1,pch=1:4,
  col=c(1:4),xlab="Temperature (K)",  ylim=c(0,8e7),
ylab=expression(paste("TL/ ",beta," (counts/K)")))
y<-log(maxTL^2/hRates)
x<-1/(kB*maxTL)
coefficients(lm(y~x))
energy<-coefficients(lm(y~x))[[2]]
intercept<-coefficients(lm(y~x))[[1]]
cat('\nFrequency factor s=',
    scientific(exp(-intercept)*energy/kB),' s^-1')
legend("topright",bty="n","(a)")
legend("topleft",bty="n", pch=c(1:4,NA),expression(
  "1 K/s","2","3","4"," "),col=c(1:4,NA),lwd=1)
plot(x,y,xlab="1/(kT)    (1/eV)",
ylab=expression(paste('ln(Tmax'^'2'*')/',beta)),ylim=c(10.4,12))
abline(lm(y ~ x))
legend("topleft",bty="n",legend=c("(b)","Heating rate",
"method"))

    ## (Intercept)              x
    ## -17.1459974    0.9619712
    ##
    ## Frequency factor s= 3.12e+11   s^-1
```

2.9   The Empirical Nature of the General Order Kinetics (GOK) Equations

The TL glow peaks in many dosimetric materials do not have the exact shape of first order peaks. Instead, their shapes fall in-between the value of the geometrical shape factor $\mu_g = 0.42$ for first order kinetics, and $\mu_g = 0.52$ for second order kinetics. These in-between shapes can be described empirically by the general order kinetics.

The TL glow peaks in many dosimetric materials can be described by the well-known *empirical general order differential equation* (GOK). The differential equation for this empirical GOK processes is

---

Differential equation for general order kinetics

$$I = -\frac{dn}{dt} = s'n^b \exp(-E/k_B T) \tag{2.15}$$

---

where n (cm^{-3}) is the concentration of electrons in traps, $T =$ temperature (in K), $k_B =$ Boltzmann constant (in eV $\cdot$ K^{-1}), and E(eV), s' are respectively the activation energy and pre-exponential factor associated with the thermal release of electrons from traps. The dimensionless constant b is the kinetic order of the TL process, and for many TL peaks it has a value between 1 and 2 (May and Partridge [110]).

The units of the "pre-exponential factor" s' in this equation are cm$^{3(b-1)}$ s^{-1}, therefore s' cannot have the direct physical meaning of a frequency factor. These units reduce to cm^3 s^{-1} for second order kinetics with $b = 2$, and to s^{-1} for first order processes with $b = 1$. For a linear heating function β (in K/s) such that the time variation of the temperature is $T = T_o + \beta t$, the solution of Eq. (2.15) is

---

Analytical solution for general order kinetics

$$I = s''n_0 \exp(-E/kT)\left[1 + (b-1)(s''/\beta)\int_{T_0}^{T} \exp(-E/kT')dT'\right]^{-b/(b-1)} \tag{2.16}$$

---

where $s'' = s'n_0^{b-1}$ and n_0 (cm^{-3}) is the initial concentration of electrons in traps, and T' is a dummy integration variable representing temperature.

Since the GOK equation is purely empirical, one finds different versions of Eq. (2.15) in the literature. For example, Rasheedy [162] suggested the following GOK differential equation:

$$I = -\frac{dn}{dt} = \frac{s}{N^{b-1}}n^b \exp(-E/kT). \qquad (2.17)$$

The advantage of this specific form of the differential GOK equation is that the frequency parameter s appearing on the right hand side of Eq. (2.17) has the appropriate units of s^{-1}, for all values of the GOK kinetic order b.

The GOK equation is widely used in the TL literature to describe the shape of TL glow curves in a variety of materials. However, GOK has been criticized because it is an empirical equation, and therefore the GOK parameter b has no direct physical meaning. As we will discuss later in this chapter, a much more useful parameter to describe the shape of a TL glow curve is the retrapping ratio R in the OTOR model. This ratio has a direct physical meaning, as discussed in Sect. 2.10.

Even though the GOK kinetic parameter b has been used widely in the TL/OSL literature, I recommend that researchers use the retrapping ratio R instead of b, in order to describe the shape of TL glow curves.

Another approach is to use *mixed order kinetics* (MOK), which is discussed later in this chapter. When using MOK, the appropriate parameter is the dimensionless mixed order parameter α, which can be used instead of the kinetic order b and instead of the retrapping ratio R, to describe the shape of a TL glow curve.

Perhaps the most useful aspect of the GOK equations is that they can be used to estimate the activation energy E of the dosimetric trap, by using experimental values of the widths τ, δ, ω and the geometrical shape factor μ_g.

Halperin and Braner [55] developed shape methods based on the measurement of the low temperature half width τ of a TL glow curve, or the high temperature half width δ, or the full width ω. These temperature widths are shown in Fig. 2.6.

These equations were amended by Chen [31, 32]. For example, the Chen equations for *first order peaks* are

---

Chen equations for E and for first order kinetics

$$E = 1.52kT_m^2/\tau - 1.58(2kT_m), \qquad (2.18)$$

$$E = 0.976kT_m^2/\delta, \qquad (2.19)$$

$$E = 2.52kT_m^2/\omega - (2kT_m) \qquad (2.20)$$

---

and similar equations with other coefficients for second order peaks.

In another work, Chen [31] used interpolation on the shape (or symmetry) factor $\mu_g = \delta/\omega$, for evaluating the activation energy of general order kinetics peaks. He first showed that for first order peaks $\mu_g \approx 0.42$, and for second order $\mu_g \approx 0.52$.

All the equations of this form were combined into one, by using α to denote δ, τ, or ω:

Chen equations for E and for GOK

$$E_\alpha = c_\alpha(kT_m^2/\alpha) - b_\alpha(2kT_m), \tag{2.21}$$

$$c_\tau = 1.52 + 3(\mu_g - 0.42); \ b_\tau = 1.58 + 4.2(\mu_g - 0.42), \tag{2.22}$$

$$c_\delta = 0.976 + 7.3(\mu_g - 0.42); \ b_\delta = 0, \tag{2.23}$$

$$c_\omega = 2.52 + 10.2(\mu_g - 0.42); \ b_\omega = 1 \tag{2.24}$$

2.10   The General One Trap Equation (GOT)

We now show how the system of differential equations of the OTOR model can be approximated by a single differential equation, which is known as the *general one trap (GOT) equation*. We will also see that under certain physical assumptions, the GOT equation can lead to first and second order kinetics during the luminescence process.

The development of the GOT equation requires some simplifying assumptions, with the most important one being the quasi-equilibrium (QE) assumption, expressed as

$$\left|\frac{dn_c}{dt}\right| << \left|\frac{dn}{dt}\right|, \ \left|\frac{dm}{dt}\right| \tag{2.25}$$

The QE assumption requires that the free electron concentration in the conduction band is quasi-stationary (Chen and Pagonis [38]). In practice, this means that the concentration of n_c in the conduction band is always much less than the concentration of n, or that $n \simeq m$. This allow us to set

$$\frac{dn_c}{dt} = 0 \tag{2.26}$$

and the luminescence intensity is given by

$$I(t) = -\frac{dm}{dt} \simeq -\frac{dn}{dt} \tag{2.27}$$

Based on Eqs. (2.26) and (2.2), the n_c value under the QE conditions is

$$n_c = \frac{n \, p(t)}{A_n(N - n) + m \, A_m} \tag{2.28}$$

By replacing n_c from Eq. (2.28) into Eq. (2.3) and since $n \simeq m$, the following is obtained:

The GOT equation in the OTOR model

$$I(t) = -\frac{dn}{dt} = p(t) \frac{n^2 \, A_m}{(N - n)A_n + n \, A_m} \tag{2.29}$$

Equation (2.29) is termed the *general one trap (GOT) equation*. When the GOT equation was first introduced in 1960 (Halperin and Braner [55]), it was not possible to solve it analytically. The analytical solution was obtained close to 50 years later by Kitis and Vlachos [82], and will be presented in Sect. 2.11.

The OTOR model leads to first order kinetics, in cases of very weak retrapping ($A_n/A_m << 10^{-2}$), or in cases of very strong retrapping. For example, by assuming that $(N - n)A_n \ll n A_m$, Eq. (2.29) leads to the following differential equation:

$$I = -\frac{dn}{dt} = n \, p(t) \tag{2.30}$$

which is of course the differential equation for first order kinetics. The OTOR model also leads to second order kinetics, at the boundary condition $A_n = A_m$. In this case Eq. (2.29) leads to the following differential equation for second order kinetics:

$$I = -\frac{dn}{dt} = \frac{n^2}{N} \, p(t) \tag{2.31}$$

which is of course the differential equation for second order kinetics.

Figure 2.11 shows an example of solving the GOT equation, for specific values of the parameters. The package *deSolve* and the function *ode* are used to numerically evaluate and plot the concentration $n(t)$ and the intensity of the glow curve $I(t) = -dn/dt$.

The general analytical solution of Eq. (2.29) is studied in detail in the next section, by using the Lambert function.

Code 2.9: The GOT equation for TL in OTOR (deSolve)

```r
# Numerical solution of the GOT equation for TL
rm(list=ls())
library("deSolve")
# Define Parameters
A_n <- 1e-10 # coefficient of retrapping
A_m <- 1e-8 # coefficient of recombination
k_B <- 8.617e-5  # Boltzmann constant
E <- 1 # electron trap depth [eV]
s <- 1e12 # frequenc factor [1/s]
delta.t <- 1
t <- seq(0, 300, delta.t)
# time = temperature, i.e. heating rate= 1 K/s
n.0 <- 1e10  # initial concentration of filled traps
N.traps <- 1e11   # total concentrations of available traps
# Calculate numerical ODE solution
ODE <- function(t, state, parameters){
with(as.list(c(state, parameters)),{
dn <- -s*exp(-E/(k_B*(273+t))) * n^2 * A_m / ((N.traps - n) *
A_n + n * A_m)
list(c(dn))   })}
parameters <- c(N.traps = N.traps, s = s, E = E, k_B = k_B,
A_m = A_m, A_n = A_n)
state <- c(n = n.0)
num_ODE <- ode(y = state, times = t, func = ODE,
parms = parameters)
# Plot filled traps n(T) and TL as a function of temperature
par(mfrow=c(1,2))
plot(x = num_ODE[,1],
    y = num_ODE[,2],xlab=expression("Temperature ["^"o"*"C]"),
    ylab =expression("Filled traps (cm"^"-3"*")"),col = "red")
legend("topright",bty="n",expression('(a)',' ','GOT','n(T)'))
plot(x = num_ODE[-1,1],  y = abs(diff(num_ODE[,2])),
xlab=expression("Temperature ["^"o"*"C]"),
    ylab = "TL signal [a.u.]")
legend("topright",bty="n",c("(b)"," ","TL"))
```

2.11   Analytical Solution of the GOT Equation Using the Lambert Function *W*

Kitis and Vlachos [82] and later Singh et al. [169] were able to solve analytically the GOT equation, which can be written in the form:

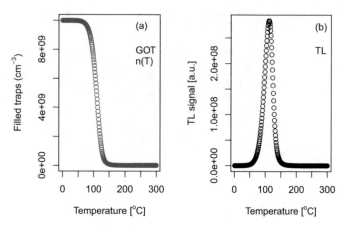

Fig. 2.11 Numerical solution of the GOT equation for TL. (**a**) Filled traps $n(T)$. (**b**) The TL glow curve

$$I = -\frac{dn}{dt} = p(t)\,\frac{n^2}{(N-n)\,R+n} \qquad (2.32)$$

where we defined the dimensionless *retrapping ratio* as

The retrapping ratio in the GOT/OTOR model

$$R = A_n/A_m \qquad (2.33)$$

When $A_n < A_m$ i.e. $R < 1$, Kitis and Vlachos obtained the following general analytical expression [82]:

Analytical solution of GOT with the Lambert W function

$$I(t) = \frac{N\,R}{(1-R)^2}\,\frac{p(t)}{W[e^z]+W[e^z]^2} \qquad (2.34)$$

---

where $W[e^z]$ is the Lambert W function (Corless et al. [44, 45]). This function is the solution $y = W[e^z]$ of the transcendental equation $y + \ln y = z$. The concentration of trapped electrons $n(t)$ is equal to

$$n(t) = \frac{N\,R}{(1-R)}\,\frac{1}{W[e^z]} \qquad (2.35)$$

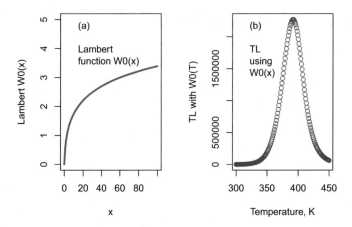

Fig. 2.12 (**a**) Plot of the Lambert function $W(x)$ between $x = 0$ and $x = 100$. (**b**) Plot of the analytical solution of the GOT Eq. (2.34), using the Lambert W function

The analytical Eq. (2.34) is termed a *master equation* in the review paper by Kitis et al. [78], because it can describe a wide variety of stimulated luminescence phenomena (TL, LM-OSL, CW-OSL, ITL signals, etc.). The only thing needed is to specialize the function $p(t)$ for the desired stimulated luminescence effect. As we will see in later chapters, the Lambert function also describes several other luminescence phenomena, including the irradiation of samples.

For TL processes, the function z has the following form (Kitis and Vlachos [82]):

$$z_{TL} = \frac{1}{c} - \ln(c) + \frac{s}{(1-R)\,\beta} \int_{T_0}^{T} \exp\left[-\frac{E}{k\,T'}\right] dT' \tag{2.36}$$

where we define the dimensionless positive constant $c > 0$ by

$$c = \frac{n_0}{N} \frac{1-R}{R} \tag{2.37}$$

and n_0 is the initial number of filled traps at time $t = 0$.

Figure 2.12a shows an example of the W Lambert function, and also implements Eqs. (2.34)–(2.37), by using the function *lambertW0(x)* within the R-package *lamW*. For more details on the Lambert W and its real branches, the reader is referred to the review paper by Kitis et al. [78]. In this code the function k is used to numerically evaluate the integral $\int_{T_0}^{T} \exp[-\frac{E}{k\,T'}]\,dT'$ appearing in Eq. (2.36). The command *lapply* applies this integration function k to the vector $x1$ containing the temperatures in the range 300–450 K.

A more extensive implementation of these functions for TL glow curves can be found within the R package *tgcd* developed by Peng et al. [150]. The *tgcd* package is used for computerized analysis of TL glow curves, and is discussed later in this chapter.

Code 2.10: Plot the W0-Lambert solution of GOT equation

```
rm(list=ls())
library(lamW)
## Example of plot for Lambert W-function from x=0 to x=100
xs <- seq(0, 100, by=1)
ys <- lambertW0(xs)
## Plot the analytical solution of GOT, using Lambert W-function
x1<-300:450    # temperatures in K
kB<-8.617E-5
no<-1E8
N<-1E10
R<-.01
c<-(no/N)*(1-R)/R
En<-1
s<-1E12
beta<-1
k<-function(u) {integrate(function(p){exp(-En/(kB*p))},
                       300,u)[[1]]}
y1<-lapply(x1,k)
x<-unlist(x1)
y<-unlist(y1)
zTL<-(1/c)-log(c)+(s*no/(c*N*R))*y
# plots
par(mfrow=c(1,2))
plot(xs, ys, type="l", col="red", lwd=3, ylim=c(0,5),
     xlab="x", ylab="Lambert W0(x)")
legend("topleft",bty="n",c("(a)"," ","Lambert",
"function W0(x)"))
plot(x,(N*R/((1-R)^2))*s*exp(-En/(kB*x))/(lambertW0(exp(zTL))
+lambertW0(exp(zTL))^2),xlab="Temperature, K",col="blue",
     ylab="TL with W0(T)")
legend("topleft",bty="n",c("(b)"," ","TL", "using", "W0(x)"))
```

2.12 Computerized Glow Curve Analysis of Single TL Peak

In this section, we use the R-package *tgcd* developed by Peng et al. [150], to perform a CGCD analysis of a single TL peak. This package consists of a function called *tgcd*, which is used for deconvolving thermoluminescence glow curves by using either the Lambert *W* function, or by using the empirical general order kinetics equation, as discussed later in this section.

The algorithm in package *tgcd* is based on analytical equations developed by Kitis et al. [69, 70]. These authors developed several new analytical expressions for use in computerized glow curve deconvolution (CGCD) analysis, by using an approximation of the TL integral $\int \exp\left(-E/kT'\right) dT'$. In the case of general order kinetics, these authors derived the following expression:

$$I(T) = I_M b^{\frac{b}{b-1}} \exp(u) \left[Z_M + (b-1)(1-\Delta)\left(\frac{T^2}{T_M^2} \exp(u)\right) \right]^{-\frac{b}{b-1}} \qquad (2.38)$$

In this expression b is the kinetic order of the TL process, I_M = maximum intensity of the glow peak (in counts per K), T_M = temperature at peak maximum (in K), $\Delta = 2kT/E$, $Z_M = 1 + (b-1)2kT_M/E$, and $u = (E/kT)(T - T_M)/T_M$. The accuracy of this expression has been tested widely by researchers for both synthetic and for experimental glow curves, and it was found that it can describe accurately glow peaks for first, second, and general order kinetics (Bos [20]).

The advantage of using this equation to approximate the TL intensity is that it involves two quantities which are measured experimentally, namely the maximum TL intensity (I_M) and the corresponding temperature (T_M). The activation energy E is treated in this expression and during the computerized fitting procedure as an adjustable parameter.

The package uses the Levenberg–Marquardt algorithm , but also allows for constraining and fixing the parameters in the model. Specifically the Levenberg–Marquardt algorithm was modified [150], so that the user can enter numerical constraints in the parameters, and also can use fixed values for some of the parameters. The least squares fitting procedure minimizes the objective:

$$f = \sum_{i=1}^{n} | y_i^{expt} - y_i^{fit} |, \qquad i = 1 \ldots n \qquad (2.39)$$

where y_i^{expt} and y_i^{fit} are the i-th experimental point and the fitted value respectively, and n is the number of data points. It is assumed that the background counts have been subtracted from the experimental data before using the *tgcd* function. The code consists of a call to the function *tgcd*. The parameters in this function are *Sigdata* (a 2-column matrix, containing the temperature values in K, and thermolumines-cence signal values in a.u., in the first and second column respectively), *npeak* (an integer representing the number of glow peaks, with the allowed maximum equal to 13), *model* (a character identifying the model used for glow curve deconvolution, with a value of "f1," "f2," and "f3" for first order models, "s1," "s2," and "s3" for second order models, "g1," "g2," and "g3" for general order models, "lw" for the Lambert W, "m1," "m2," and "m3" for the mixed order models), *edit.inis* (a logical parameter determining whether the user wants to further modify, constrain, or fix the initialized parameters through a Dialog Table), *inisPAR* (a prompt for the user to click with a mouse in order to locate each glow peak, and a matrix with 3 or 4 columns containing the initial kinetic parameters I_m, E, T_m and b, R, a for GOK, the analytical Lambert W and the MOK model respectively).

The goodness of fit of the equation to the data is expressed by the Figure of Merit (FOM) which is defined as follows (Balian and Eddy [13]):

$$FOM = \frac{\sum_{i=1}^{n} | y_i^{expt} - y_i^{fit} |}{\sum_{i=1}^{n} | y_i^{expt} |}, \qquad i = 1 \ldots n \qquad (2.40)$$

where y_i^{expt} and y_i^{fit} were defined above.

The first example in Fig. 2.13 shows the deconvoluted Reference glow curve #1, from the project GLOCANIN, which was an interlaboratory comparison of different TL deconvolution methods (Bos et al. [21]). In this example the deconvolution method uses the Lambert W function, which was discussed in the previous section. Later in this chapter, we will see how the *tgcd* package can also be used with the empirical general order kinetics equation.

The results of the following R code show that the activation energy $E = 1.182\,eV$, which is essentially the same as the expected value of $E = 1.18\,eV$. In this example which uses the analytical solution with the Lambert W, we obtain a best fitting parameter R which is practically zero, indicating that this is indeed a first order TL peak, as expected. The FOM = 0.0097, indicating an excellent fit to this simulated TL peak.

Code 2.11: Deconvolution of Glocanin glow curve (tgcd)

```
# Deconvolution of Reference glow curve #1 (project GLOCANIN)
rm(list=ls())
library("tgcd")
data(Refglow)    # Load the data
# Deconvolve data with 1 peak using the LAMBERT W function
startingPars <-
    cbind(c(15.0),c(1.0),c(520), c(0.1))    # Im, E, Tm, R
invisible(capture.output(TL1 <- tgcd(Refglow$x001, npeak=1,
model="lw",inisPAR=startingPars,nstart=10,edit.inis=FALSE)))
print.noquote("Best fit parameters")
TL1$pars
cat("\nGeometrical shape factor=",
round(TL1$sp[,7],2))
cat("\nFigure Of Merit FOM=",round(TL1$FOM,4))

## [1] Best fit parameters
##            INTENS(Im) ENERGY(E) TEMPER(Tm) rValue(r)
## 1th-Peak    10968.39  1.182269   490.4689    1e-16
##
## Geometrical shape factor= 0.43
## Figure Of Merit FOM= 0.0097
```

Fig. 2.13 Deconvolution of TL data containing a single peak, using the Lambert W function. The data is from Reference glow curve #1 in the intercomparison project GLOCANIN (Bos et al. [21])

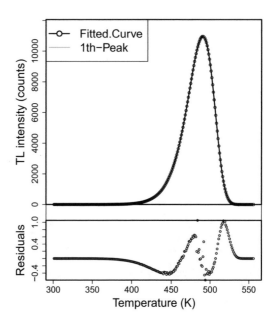

2.13 Deconvolution of Complex TL Glow Curves

In the following example, the package *tgcd* is used to fit a TL glow curve with two peaks, by using the Lambert W function, as discussed above.

The code reads the *.txt* file containing pairs (T,TL) of temperature (in K) and TL intensity (in counts or counts/s, etc.), and the user provides initial estimates of the maximum TL intensity I_m, of the activation energy E, of the temperature of maximum TL intensity T_m, and of the recombination ratio R. The experimental data in Fig. 2.14 are for sample LBO, from the experimental work by Kitis et al. [81].

When applying CGCD analysis, the estimated values of the retrapping ratios R and the activation energies E should be examined carefully and critically, and we should not accept blindly the results of the CGCD analysis. For example, if one finds that the *higher* temperature peak has $E_2 < E_1$, then the experimental data should be further analyzed using different methods. For a more complete analysis of this experimental data for sample LBO using several methods of analysis, see Kitis et al. [81].

Code 2.12: Deconvolution of TL user data (.txt file, tgcd)

```
# Deconvolve data with 2 peaks using the LAMBERT W function
rm(list=ls())
library("tgcd")
# Load the data
mydata = read.table("lbodata.txt")
startingPars <-
cbind(c(105.0,5.0),c(1.1,1.4),c(460,550),c(0.01,.01)) #Im,E,Tm,R
invisible(capture.output(TL1 <- tgcd(mydata, npeak=2,
model="lw",inisPAR=startingPars, nstart=10, edit.inis=FALSE)))
print.noquote("Best fit parameters")
TL1$pars
cat("\nGeometrical shape factors"," ")
round(TL1$sp[,7],2)

   ## [1] Best fit parameters
   ##            INTENS(Im)  ENERGY(E)  TEMPER(Tm)    rValue(r)
   ## 1th-Peak  116.41492   1.221954    464.1371  1.574355e-02
   ## 2th-Peak   22.67729   1.001351    519.3871  3.353241e-06
   ##
   ## Geometrical shape factors  1th-Peak 2th-Peak
   ##        0.45       0.45
```

Fig. 2.14 Deconvolution of experimental data for dosimetric material LBO with two peaks (Kitis et al. [81]), by using the Lambert *W* function

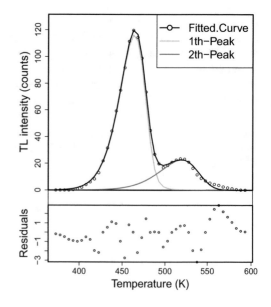

2.14 Computerized Glow Curve Analysis with General Order Kinetics

In this section we use the R-package *tgcd* with the option of using the general order empirical equation [70]. The four parameters for the GOK model are I_M, T_M, E, and *b*. As discussed previously, the package uses the Levenberg–Marquardt algorithm and also supports constraining and/or fixing the parameters.

The code consists of a call to the function *tgcd*, with the specification model = "*g*1" which refers to the GOK model represented by Eq. (2.38). The data analyzed in the example of Fig. 2.15 are obtained with the code line *data(RefGlow)*, and represent typical TL glow curve for the well-known material LiF (TLD700).

The R-code line *TL$pars* creates a Table of the best fit parameters (Im, E, Tm, and b), obtained by the Levenberg–Marquardt algorithm for the 9 peaks in the TL glow curve.

The line *TL$sp*, creates a Table of the parameters (T1, T2, Tm, d1, d2, thw, sf) obtained by the algorithm. Here T1, T2 are the half-maximum TL intensity temperatures on the low and high temperature sides of each glow peak, Tm is the temperature of the maximum TL intensity, d1 and d2 are the low and high temperature half-widths, thw are the full half-widths of each peak, and *sf* is the shape factor μ_g. As usual, a value of $sf = 0.42$ indicates a first order kinetics peak, and a value of $sf = 0.52$ corresponds to a second order kinetics peak.

The R-code line *TL$FOM* prints the FOM value for the best fit, and the value $FOM < 1$ obtained here indicates an excellent fit to the experimental data. The lines *plot* and *hist* create graphs of the residuals *TL$residuals*, which are the

Fig. 2.15 Deconvolution of a glow curve from the GLOCANIN project using 9 peaks with a GOK model, and with user-supplied initial kinetic parameters (Bos et al. [21])

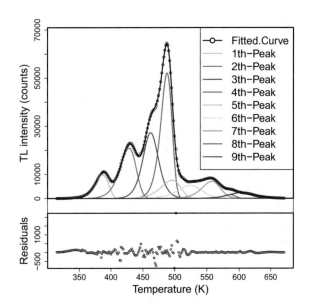

Fig. 2.16 Histogram of residuals $y_i^{expt} - y_i^{fit}$ from the best fit shown in Fig. 2.15, for the nine-peak TL glow curve of the GLOCANIN project, with a GOK model

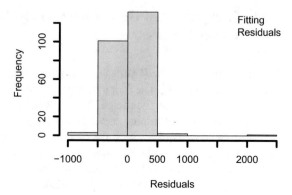

difference between the experimental and fitted values $(y_i^{expt} - y_i^{fit})$. A histogram plot of the distribution of the residuals of the fitting process is shown in Fig. 2.16.

Code 2.13: Deconvolution of 9-peak Glocanin TL data (tgcd)

```
# Deconvolve TL signal using 9 peaks (no background subtraction)
# a GOK model using user-supplied initial kinetic parameters.
rm(list=ls())
library("tgcd")
data(Refglow)
knPars <-
cbind(c(9824,21009,27792,50520,7153, 5496,6080,1641,2316), # Im
c(1.24, 1.36, 2.10, 2.65, 1.43, 1.16, 2.48, 2.98, 2.25), # E
c(387, 428, 462, 488, 493, 528, 559, 585, 602), # Tm
c(1.02, 1.15, 1.99, 1.20, 1.28, 1.19, 1.40, 1.01, 1.18)) # b
invisible(capture.output(TL1 <- tgcd(Refglow$x009, npeak=9,
model="g1",inisPAR=knPars, nstart=10, edit.inis=FALSE)))
print.noquote("Best fit parameters")
TL1$pars

## [1] Best fit parameters
##           INTENS(Im) ENERGY(E) TEMPER(Tm) bValue(b)
## 1th-Peak   9820.004  1.237294   387.3132  1.025077
## 2th-Peak  20959.877  1.361432   428.1756  1.149763
## 3th-Peak  27324.883  2.058686   462.1433  1.825707
## 4th-Peak  52194.441  2.498238   488.1280  1.119436
## 5th-Peak   7519.826  1.411680   495.8060  1.045013
## 6th-Peak   5351.507  1.474229   524.2461  1.388807
## 7th-Peak   7077.083  2.177145   557.5652  1.334783
## 8th-Peak   1547.388  3.343393   586.0483  1.057578
## 9th-Peak   2311.923  2.249152   602.6770  2.000000
```

```
# This is a continuation of R code above)
hist(TL1$residuals,lwd=3,xlab="Residuals",
ylab="Frequency",main=NULL)
legend("topright",bty="n",c("Fitting","Residuals"))
```

2.15 Mixed Order Kinetics and the IMTS Model

In previous sections we presented the OTOR model as a basic luminescence model, which can be approximated by the GOT equation. We also saw that under certain physical assumptions the OTOR leads to first order and second order kinetics. An analytical solution was described for the GOT/OTOR equation, which involves the Lambert W function.

In order to cover the gap between first and second order kinetics, researchers introduced the *mixed order kinetics (MOK)*, which is a linear combination of the equations for first and second order kinetics. Specifically Chen et al. [34] introduced the MOK model for TL in 1981, and later Kitis et al. [68] introduced the MOK model for OSL.

This section presents the general *interactive multi-trap system model (IMTS)*, and its more restricted version known as the non-interacting multi trap system (NMTS). Under special physical assumptions, the NMTS model is shown to lead to the MOK model. The IMTS is shown schematically in the energy band model of Figs. 2.17 and 2.18.

A complete presentation of any luminescence model must include three stages which correspond to events taking place in a typical luminescence experiment. These stages are *irradiation stage*, *relaxation stage*, and *read-out stage*. We now describe the differential equations for the irradiation and read-out stages in the IMTS model.

The system of differential equations describing the traffic of charged carriers during the irradiation stage in the IMTS model are

$$\frac{dn_i}{dt} = -n_i\,p(t) + A_i\,(N_i - n_i)\,n_c \qquad i = 1,\ldots,n \tag{2.41}$$

$$\frac{dn_d}{dt} = A_d(N_d - n_d)\,n_c \tag{2.42}$$

$$\frac{dm}{dt} = A_h\,(M - m)\,n_v - A_m\,m\,n_c, \tag{2.43}$$

$$\frac{dn_v}{dt} = X - A_h\,(M - m)\,n_v, \tag{2.44}$$

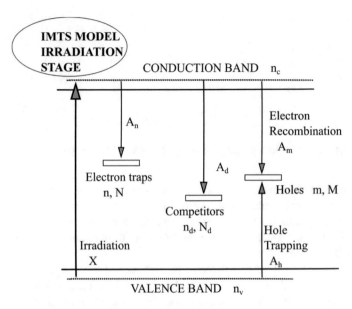

Fig. 2.17 The general phenomenological interactive multi-trap system (IMTS) model, describing stimulated luminescence effects, during the irradiation stage

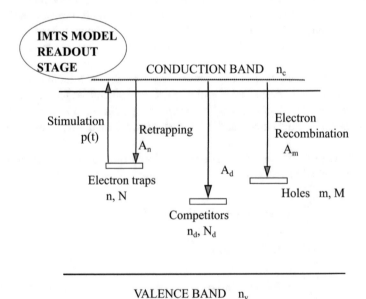

Fig. 2.18 Schematic diagram of the IMTS model, describing stimulated luminescence effects during the thermal or optical stimulation stage

$$\frac{dn_c}{dt} = X - \sum_i \frac{dn_i}{dt} - \frac{dn_d}{dt} - A_m \, m \, n_c \tag{2.45}$$

Here the index $i = 1, \ldots, n$ stands for the electron traps, N_i (cm^{-3}) is the concentration of available electron traps, n_i (cm^{-3}) the concentration of trapped electrons, M (cm^{-3}) is the concentration of available luminescence centers, m (cm^{-3}) concentration of trapped holes. N_d, n_d (cm^{-3}) are the concentrations of available and occupied traps in a thermally disconnected deep trap (TDDT), n_c (cm^{-3}) and n_v (cm^{-3}) are the concentration of electrons in the conduction and holes in the valence band, A_i (cm^3 s^{-1}) are the trapping coefficients in electron traps n_i, A_m (cm^3 s^{-1}) is the recombination coefficient, A_h (cm^3 s^{-1}) the trapping coefficient of holes in luminescence centers, A_d (cm^3 s^{-1}) trapping probability in the TDDT, β (K/s) the heating rate, and X is the rate of production of ion pairs (i.p) per second and per unit volume, which is proportional to the dose rate.

In the above model all traps having the index i can trap electrons from the conduction band, but can also release electrons by thermal or optical excitation, and they are referred to as active traps. On the other hand, the trap with index d can only trap electrons but thermal stimulation is not allowed, so these traps are called thermally disconnected deep traps (TDDT).

At the end of excitation, we end up with finite concentrations of the free electrons (n_c) in the conduction band, and free holes (n_v) in the valence band. If we wish to mimic the experimental procedure of TL or OSL, we have to consider a relaxation time between the end of excitation and the beginning of heating or exposure to stimulating light. This is done by setting X to zero and solving Eqs. (2.41)–(2.45) for a further period of time, so that at the end of this time period both n_c and n_v are negligibly small. The final values of n, m, n_c, and n_v at the end of excitation are used as initial values for the relaxation stage. Similarly, we take the final values of these four functions at the end of relaxation period as initial values for the heating stage. One utilizes a certain heating function, e.g. linear heating in the form $T = T_o + \beta t$, where β is the constant heating rate. The same set of equations is being solved numerically with $X = 0$ and with the $-np(t)$ term on the right hand side of Eq. (2.41) being more and more important as the temperature rises. On the other hand, it is usually assumed that during heating, $n_v \approx 0$.

During the heating stage and for a single active trap these equations become

$$\frac{dn_1}{dt} = -n_1 \, p(t) + A_n \, (N_1 - n_1) \, n_c \tag{2.46}$$

$$\frac{dn_d}{dt} = A_d(N_d - n_d) \, n_c \tag{2.47}$$

$$\frac{dm}{dt} = \frac{dn_1}{dt} + \frac{dn_d}{dt} + \frac{dn_c}{dt} \tag{2.48}$$

$$I(t) = -\frac{dm}{dt} = A_m \, m \, n_c, \tag{2.49}$$

where $p(t)$ depends again on the experimental stimulation mode. The concentration of holes m in the recombination centers (RC) is equal to

$$m = n_1 + n_c + n_d \tag{2.50}$$

so that the charge balance of the system is maintained.

When during the irradiation stage the TDDT (N_d) becomes saturated i.e. when $n_{d0} = N_d$, this trap cannot capture free carriers from the conduction band during the stimulation stage. This is known as the *non-interactive multi-trap system (NMTS)* limit of the IMTS model.

As in the case of the OTOR model, it is possible to arrive at an analytical expression for the NMTS model, by using the QE conditions. In the IMTS model the QE is expressed by the relation:

$$\left|\frac{dn_c}{dt}\right| \ll \left|\frac{dn_1}{dt}\right|, \left|\frac{dm}{dt}\right| \tag{2.51}$$

Since the TDDT in the NMTS model is saturated, we have $dn_d/dt = 0$. By proceeding as in the GOT equation, Eq. (2.48) can be solved with respect to n_c, and then can be replaced in Eq. (2.49) to obtain

---

The GOT/IMTS differential equation

$$I = p(t)\,n_1 \frac{(n_1 + N_d)\,A_m}{(N_1 - n_1)\,A_n + (n_1 + N_d)\,A_m} \tag{2.52}$$

In the case of slow retrapping $(N_1 - n_1)A_n \ll (n_1 + N_d)A_m$, and Eq. (2.52) leads directly to the first order kinetics expression.

Alternatively, in the case of fast retrapping, the condition $(N_1 - n_1)A_m \gg (n_1 + N_d)A_n$, along with $n_1 \ll N_1$ for low irradiation doses, leads to the following expression for the stimulated luminescence intensity:

$$I = \frac{A_m}{N_1 A_n} p(t)\,n_1\,(n_1 + N_d) \tag{2.53}$$

For the special case of $A_n = A_m$, Eq. (2.52) becomes

$$I = \frac{1}{N_1 + N_d} p(t)\,n_1\,(n_1 + N_d) \tag{2.54}$$

Both Eqs. (2.53) and (2.54) can be presented in the form:

Differential equation for the mixed order kinetics (MOK) model

$$I = g\, p(t)\, n_1\, (n_1 + N_d) \tag{2.55}$$

---

with the constant $g = A_m / (N_1\, A_m)$ or $g = 1/(N_1 + N_d)$ correspondingly. Equation (2.55) is the differential equation for the *mixed order kinetics (MOK)* model, which is the sum of the two terms $g\, p(t)\, n_1\, N_d$ and $g\, p(t)\, n_1^2$, corresponding to first and second order kinetics respectively [34]. In the next section we present the analytical solution of Eq. (2.55).

2.16   Analytical Solution of the MOK Model

In order to obtain an analytical expression for the MOK model, we use $I = -dm/dt \approx -dn_1/dt$ and write Eq. (2.55) in the form:

$$-\frac{dn_1}{dt} = g\, p(t)\, n_1\, (n_1 + N_d)\,. \tag{2.56}$$

At this point Chen et al. [34] defined the mixed order parameter α as

$$\alpha = \frac{n_{10}}{n_{10} + N_d} \tag{2.57}$$

From Eq. (2.56) it is found that

$$n_1\,(t) = \frac{\alpha\, N_d}{F(t) - \alpha} \tag{2.58}$$

with $F(t)$ given by

$$F(t) = \exp\left(g\, N_d \int_0^t p(t')\, dt'\right) \tag{2.59}$$

By inserting $n_1\,(t)$ from Eq. (2.58) in Eq. (2.56), the following expression describing the intensity of the stimulation phenomena in the MOK model is obtained:

The analytical solution of the MOK model

$$I(t) = g \, N_d^2 \, \frac{\alpha \, p(t) \, F(t)}{\left[F(t) - \alpha\right]^2} \tag{2.60}$$

This equation joins together several expressions previously available in the literature for TL in the MOK model (Chen et al. [34]), and also for OSL (Kitis et al. [68]).

Equation (2.60) is a *master equation for MOK* containing the function $p(t)$ which describes the various experimental luminescence modes. The function $F(t)$ in Eq. (2.60) has the following forms for TL, LM-OSL, ITL, and CW-OSL processes:

$$F_{TL}(t) = \exp\left(g \, N_d \, \frac{s}{\beta} \int_{T_0}^{T} e^{-\frac{E}{kT'}} \, dT' \right) \tag{2.61}$$

$$F_{LM-OSL}(t) = \exp\left(g \, N_d \, \frac{\lambda \, t^2}{2P} \right) \tag{2.62}$$

$$F_{ITL}(t) = \exp\left(g \, N_d \, \lambda \, t \right) \tag{2.63}$$

Equation (2.63) also applies for CW-OSL signals, with the appropriate value of the constant excitation constant λ.

Equation (2.60) has been applied successfully to simulated luminescence signals, as well as to a wide variety of experimental data over the past 50 years [52, 69, 194].

The MOK equation is an intermediate between the first and second order expressions, in the sense that for $\alpha \cong 0$ it tends to first order, and for $\alpha \cong 1$, it tends to second order. Chen et al. [34] derived values of μ_g from the numerical values of T_1, T_m, and T_2 for given values of E, s, β, α and n_0. They found that μ_g is a relatively strong function of α, and a very weak function of E and s.

The symmetry factor μ_g as a function of α is found to change gradually from the value of 0.42 characteristic of first order when α equals zero, to a value of 0.52 of the second order when α equals one. Figure 2.19 shows the relationship between the mixed order kinetics parameter α, and the general order kinetics parameter b.

Chen et al. [34] pointed out that the MOK equation includes three free parameters, which are the activation energy E, the pre-exponential factor s, and the parameter α. As in the general order case, one disregards the additional parameter n_0 since it only influences the scale (or total intensity), rather than the shape of the peak. As far as the number of parameters is concerned, the MOK is thus similar to the GOK equation, and one may expect that both may be equally well able to

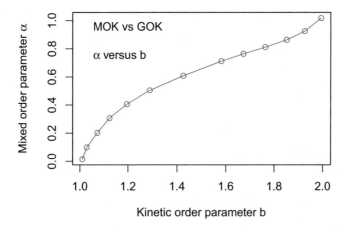

Fig. 2.19 The mixed order kinetics parameter α as a function of the general order kinetics parameter b

describe a general TL glow peak. Also, the two are similar in the sense that they tend to the first or second order cases when the appropriate values are chosen for the relevant parameters. However, the MOK parameter α has a physical meaning, while the kinetic order b in the GOK is a purely empirical parameter with no direct physical meaning.

An important advantage of the MOK approach over the GOK approach is that simple assumptions lead from the IMTS model differential equations, to the MOK equation. No similar derivation can result in the GOK equation, which therefore is entirely empirical.

The following code shows how to use the MOK model with the *tgcd* package, in order to carry out a CGCD analysis of typical TL data. The only parameter that needs to be changed from the corresponding code for the GOK model, is replacing "$g1$" in the model type with "$m1$".

In the example of Fig. 2.20, the value of the MOK parameter $\alpha = 10^{-16}$, indicating a first order TL peak.

2.17   Simulations Using the DELOC Functions in the R Package *RLumCarlo*

The previous sections in this chapter have been based on the use of differential equations to describe the luminescence signals associated with delocalized transition processes.

Code 2.14: MOK deconvolution of Glocanin TL (tgcd)

```
# Deconvolution of Reference GLOCANIN glow curve #1 with MOK
rm(list=ls())
library("tgcd")
data(Refglow)
# Load the data.
# Deconvolve data with 1 peak using the MOK expression
startingPars <-
  cbind(c(15.0),  c(1.0), c(520), c(0.1)) # Im, E, Tm, R
invisible(capture.output(TL1 <- tgcd(Refglow$x001, npeak=1,
model="m1",inisPAR=startingPars, nstart=10, edit.inis=FALSE)))
print.noquote("Best fit parameters")
TL1$pars

## [1] Best fit parameters
##              INTENS(Im) ENERGY(E) TEMPER(Tm)    aValue(a)
## 1th-Peak    10967.93  1.182694    490.341 0.000499707
```

Fig. 2.20 Deconvolution of Reference glow curve #1 in the GLOCANIN project, using Mixed Order Kinetics (MOK)

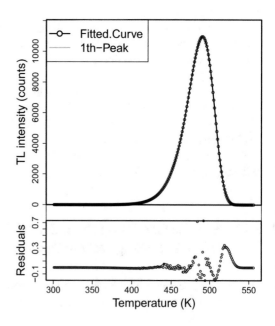

An alternative approach to the differential equations is the use of Monte Carlo methods. The MC approach is described in detail in Chaps. 8–10. In this section we give several examples of using the package *RLumCarlo* to simulate a variety of luminescence signals due to delocalized transitions. The following short table summarizes the available DELOC functions and their content. The details of these functions can be found in the documentation of the package *RLumCarlo* (see for example, Kreutzer et al. [87]).

RLumCarlo is a collection of energy band models that simulate luminescence signals using Monte Carlo (MC) methods for various stimulation modes. Each model is run by calling one of the R functions starting with *run_*. Currently, three different delocalized model types (TUN: tunneling, LOC: localized transition, DELOC: delocalized transition) are implemented in the package. Each of these model types is implemented for the stimulation types TL, CW-OSL or CW-IRSL, LM-OSL, and ISO (isothermal). It is important to note that each model has different input parameters and requirements.

Table 2.1 summarizes the available DELOC functions. Table 2.2 summarizes the parameters used in the implemented MC models, along with their physical meaning, units, and the range of realistic values. The ranges shown in this Table represent just a rough guideline, and might be exceeded for particular cases.

The following code simulates a ITL signal for three different temperatures $T = 220,\ 230,\ 240\,°C$ with thermal parameters $E = 1.4\,eV, s = 10^{12}\,s^{-1}$, retrapping ratio $R = 10^{-3}$. The code output in Fig. 2.21 simulates the ITL intensity, and the shaded area around the ITL curves indicates the standard deviation (or range) of the MC runs. As we increase the number of MC runs in this code, these shaded areas will become smaller.

The pipe operator ($\%>\%$) is used in several of the examples of this book, to transfer the results of the functions to the plotting function *plot_RLumCarlo*.

Table 2.1 Table of DELOCalized functions available in the package *RLumCarlo*

Function name	Description
plot_RLumCarlo	Plots "RLumCarlo" modeling results (the averaged signal or the number of remaining electrons), with modeling uncertainties.
run_MC_ISO_DELOC	Simulation of ITL signals using the one trap one recombination center (OTOR) model.
run_MC_LM_OSL_DELOC	Simulation of LM-OSL signals using the delocalized OTOR model.
run_MC_CW_OSL_DELOC	Simulation of CW-OSL signals using the OTOR model.
run_MC_TL_DELOC	Simulation of TL signals using the OTOR model.

Table 2.2 Table of input parameters for DELOC functions in *RLumCarlo*

Process	Symbol	Parameter in RLumCarlo function	Units	Typical values
Delocalized TL	E	Thermal activation energy of the trap	eV	0.5–3
	s	Frequency factor of the trap	1/s	1E8–1E16
	Times	Sequence of time steps for simulation (heating rate is 1 K/s)	s	0–700
	Clusters	Number of MC runs	1	1E1–1E4
	N_e	Total number of electron traps available	1	2–1E5
	n_filled	Number of filled electron traps at the beginning of the simulation	1	1–1E5
	R	Delocalized retrapping ratio	1	0–1
Delocalized CW-IRSL	A	Optical excitation rate from trap to conduction band	1/s	1E−3–1
	Times	Sequence of time steps for simulation	s	0–500
	Clusters	Number of MC runs	1	1E1–1E4
	N_e	Total number of electron traps available	1	2–1E5
	n_filled	Number of filled electron traps at the beginning of the simulation	1	1–1E5
	R	Delocalized retrapping ratio	1	0–1
Delocalized ISO	E	Thermal activation energy of the trap	eV	0.5–3
	s	Frequency factor of the trap	1/s	1E8–1E16
	T	Temperature of the isothermal process	°C	20–300
	Times	Sequence of time steps for simulation	s	0–1000
	Clusters	Number of MC runs	1	1E1–1E4
	N_e	Number of electrons	1	2–1E5
	n_filled	Number of filled electron traps at the beginning of the simulation	1	1–1E5
	R	Delocalized retrapping ratio	1	0–1
Delocalized LM-OSL	A	Optical excitation rate from trap to conduction band	1/s	1E−3–1
	Times	Sequence of time steps for simulation	s	0–3000
	Clusters	Number of MC runs	1	1E1–1E4
	N_e	Total number of electron traps available	1	2–1E5
	n_filled	Number of filled electron traps at the beginning of the simulation	1	1–1E5
	R	Delocalized retrapping ratio	1	0–1

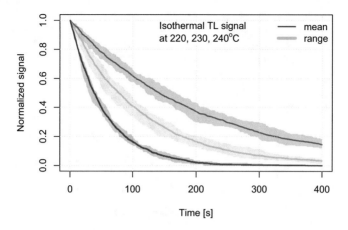

Fig. 2.21 Example of combining three plots for ITL signals from DELOCalized transitions, within the OTOR model

---

Code 2.15: Combine three plots for isothermal experiment

```
##=====================================================##
## COMBINE 3 PLOTS  FOR DELOCALIZED ITL
##=====================================================##
rm(list = ls(all=T))
suppressMessages(library(RLumCarlo))
## set time vector
times <- seq(0, 400)
run_MC_ISO_DELOC(T=220, E=1.4, s=1e12,R=1e-3, times = times) %>%
plot_RLumCarlo(norm = TRUE, col="red",legend = F)
run_MC_ISO_DELOC(T=230, E=1.4, s=1e12,R=1e-3, times = times) %>%
plot_RLumCarlo(norm = TRUE, col="green",add = TRUE)
run_MC_ISO_DELOC(T=240, E=1.4, s=1e12,R=1e-3, times = times) %>%
plot_RLumCarlo(norm = TRUE, col="blue",add = TRUE, times= times)
legend("top",bty="n",legend=c("Isothermal TL signal",
expression("at 220, 230, 240"^"o"*"C")))
```

The following example in Fig. 2.22 shows a simulation of a TL glow curve with the OTOR model, with the parameters $E = 1.45\,\mathrm{eV}$, $s = 3.5 \times 10^{12}\,\mathrm{s}^{-1}$, retrapping ratio $R = 0.1$.

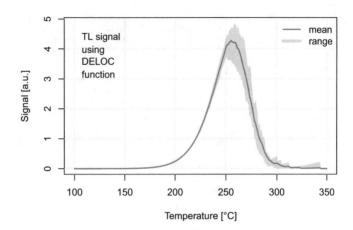

Fig. 2.22 MC simulation of TL glow curve from the OTOR delocalized transition model

---

Code 2.16: Single MC plot for delocalized TL

```
##====================================================##
## Example 1: Single MC Plot  for delocalized TL
##====================================================##
rm(list = ls(all=T))
library(RLumCarlo)
run_MC_TL_DELOC(
s = 3.5e12,
E = 1.45,
R = 0.1,
times = 100:350
) %>%
#Plot results of the MC simulation
plot_RLumCarlo
legend("topleft",bty="n",c("TL signal", "using
DELOC","function"))
```

Figure 2.23 shows a simulation of several scenarios at once, by varying many parameters in the model. A *for* loop is used to cycle through all the combinations of the values of the parameters *R*, *N_e* and *n_filled*.

Code 2.17: MC for delocalized TL: multiple parameters

```
##================================================##
##================================================##
## Plot multiple TL  curves with varying params
##================================================##
# define your parameters
rm(list = ls(all=T))
library(RLumCarlo)
times=seq(150,350,1)
s=rep(3.5e12,4)
E=rep(1.45,4)
R<-c(0.7e-6,1e-6,0.01,0.1)
clusters=100
N_e =c(400, 500, 700, 400)
n_filled =c(400, 500, 300, 70)
method="par"
output ="signal"
col=c(1,2,3,4) # different colors for the individual curves
plot_uncertainty <- c(TRUE,TRUE,TRUE,TRUE) # plot  uncertainty?
add_TF <- c(FALSE,rep(TRUE, (length(R)-1)))
for (u in 1:length(R)){
    results <-run_MC_TL_DELOC(times=times, s=s[u],E=E[u],
clusters =clusters, N_e = N_e[u],n_filled = n_filled[u],
R=R[u], method = method, output = output)
    plot_RLumCarlo(results,add=add_TF[u],legend = FALSE,
col=col[u], ylim=c(0,20))
}
legend("left",bty="n",c("TL","DELOC","many variables"))
legend("topright",bty="n",ncol=5,cex=0.55,title = "parameters" ,
        legend=c(paste0("E = ", E),paste0("s = ", s),
                paste0("n_filled = ", n_filled),
                paste0("N_e = ", N_e),
                paste0("R = ", R)), text.col=col)
```

2.18   Recommended Protocols for Analyzing TL Data

We recommend the following steps in studying the TL properties of a new dosimetric material.

1. *Initial rise method.* When isolated TL peaks are present, apply the initial rise method to obtain the activation energy E. As a rule of thumb, first order TL peaks with a maximum intensity located around 100 °C may have an energy value of about $E = 1$ eV, while peaks around 300 °C may have an energy of

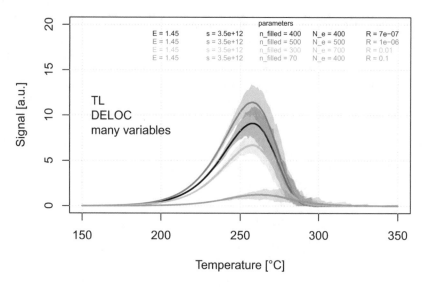

Fig. 2.23 Simulations of DELOCalized TL transitions, by varying many parameters in the model

about $E = 1.4\,\text{eV}$. If the initial rise method yields values much higher or much lower than these values, further data analysis is required.

2. *Measure the TL signal using several different heating rates and apply the variable heating rate method.* As was discussed in this chapter, analysis of TL data obtained with several different heating rates can be used to estimate both the activation energy E and the frequency factor s of each trap.

It is important to repeat the E, s analysis with at least four different heating rates, preferably low heating rates less than $10\,\text{K/s}$. Then compare these E, s values with the best fitting parameters obtained from other methods, and assess whether the results are consistent with each other.

3. *Partial thermal cleaning analysis.* This is also known as the $T_{\max} - T_{\text{STOP}}$ and $E - T_{\text{STOP}}$ method of analysis. In the simplest variation of this method, the TL glow curve is partially cleaned thermally by (a) heating the irradiated sample up to a temperature T_{STOP} lower than the temperature of maximum TL intensity $T_{\max}$, (b) measuring the TL glow curve, and (c) applying the initial rise method to this partial TL glow curve. This multi-step process must be repeated many times in order to obtain the $T_{\max} - T_{\text{STOP}}$ graph, and the corresponding $E - T_{\text{STOP}}$ plot. For a detailed discussion of this method and its variants, see the book by McKeever [111].

4. *TL dose response.* Measure the TL signal for several irradiation doses, plot the TL dose response $TL(D)$ both on a linear and on a log–log scale, to test for superlinear behavior.

If the dose response curve does not contain regions of superlinear behavior, fit the response curve with one of several possible fitting functions: Saturating exponential,

Double saturating exponential (DSE), and the Lambert equation (4.13) described in Chap. 4. If the dose response contains regions of superlinear behavior, then fit the response curve with the Pagonis–Kitis–Chen (PKC) Lambert equation (4.31) for superlinearity (Chap. 4).

5. *Analysis of kinetics of the peaks.* Measure the TL signal for several irradiation doses, and examine the shape of the individual peaks in the glow curve. If the shape does not change significantly with the irradiation dose, this is an indication of first order kinetics. Further evidence for first order kinetics is provided if the maxima of the individual peaks do not shift significantly with the irradiation dose. Perform a deconvolution of the TL glow curves using CGCD, and obtain the best fitting parameters.

It is important to repeat the CGCD procedure for more TL glow curves, which were measured at different doses. Then compare the best fitting parameters obtained from the different doses, and assess whether the results from different doses are consistent with each other.

6. *Isothermal decay experiments.* This is a valuable method of analysis, which can help identify individual isothermal components, which may correlate with specific peaks in the TL signal. This type of experiment can also help distinguish between delocalized energy transitions and transitions of a localized nature. For a detailed recent example of using isothermal experiments to obtain information on the luminescence process, see the experimental work by Kitis et al. [81].

Chapter 3
Analysis of Experimental OSL Data

3.1 Introduction

In this chapter we discuss luminescence signals produced by optical stimulation of the samples in the laboratory. We will look at both *continuous wave OSL* (CW-OSL) signals and *linearly modulated OSL* (LM-OSL) signals.

In Sects. 3.2–3.4 we discuss the first order kinetics for the CW-OSL luminescence process, and how it is described by simple exponential functions. Examples are given of computerized analysis of CW-OSL signals using exponential components, based on two different R packages, *Luminescence* and *numOSL*. Next, in Sects. 3.5–3.6 we present the GOT equation for CW-OSL signals derived from the OTOR model and show how this equation can be integrated numerically using R. This is followed by the analytical solution of the GOT equation for CW-OSL signals, which is based on the Lambert *W* function. Several examples of using the new package *RLumCarlo* are demonstrated in Sect. 3.7.

The presentation of the above topics is next extended to LM-OSL signals. Sections 3.8–3.12 discuss how to analyze these types of signals in order to extract the parameters characterizing the samples. Finally in Sect. 3.13 we show how to transform the shapeless CW-OSL signals into peak-shaped LM-OSL signals, and we conclude this chapter in Sect. 3.14 with a list of recommended protocols for analyzing OSL data in the laboratory.

3.2 First Order Kinetics in CW-OSL Experiments

During optical excitation in a luminescence experiment, the rate of change of the concentration of trapped electrons in the dosimetric trap (dn/dt) follows the differential equation:

© The Author(s), under exclusive license to Springer Nature Switzerland AG 2021
V. Pagonis, *Luminescence*, Use R!, https://doi.org/10.1007/978-3-030-67311-6_3

$$I = -\frac{dn}{dt} = n\,p(t) \tag{3.1}$$

where $p(t)$ (s^{-1}) is the rate of optical excitation. In the case of a CW-OSL experiment, the stimulation is optical in nature and occurs with a source of constant light intensity, and the function $p(t) = \sigma\,I = \lambda$, where $\sigma\,(cm^2)$ represents the optical cross section for the CW-OSL process, and $I\,(photons\,cm^{-2}\,s^{-1})$ represents the photon flux. The above equation for CW-OSL becomes

---

Differential equation for first order OSL kinetics

$$I_{OSL}(t) = -\frac{dn}{dt} = \lambda\,n \tag{3.2}$$

The solution of this differential equation is a simple exponential function:

$$n(t) = n_0\,e^{-\lambda t} \tag{3.3}$$

where n_0 is the initial concentration of trapped electrons at time $t = 0$. The corresponding luminescence intensity is

---

Analytical solution for first order OSL kinetics

$$I_{OSL}(T) = n_0\,\lambda e^{-\lambda t} \tag{3.4}$$

In many dosimetric materials, it is found that the CW-OSL signal contains several exponential components. For example, it is well established that in quartz there are several exponential components that are termed according to the magnitude of the optical stimulation cross section σ, as "*slow*," "*medium*," and "*fast*." It is then important to be able to analyze these composite CW-OSL signals into their constituent components. In the next section we describe how this is accomplished with the R package *numOSL*.

3.3   Fitting CW-OSL Data with the R Package *numOSL*

In this section we use the function *decomp* in the R package *numOSL* by Peng et al. [150]. This function allows decomposition of a CW-OSL or LM-OSL decay curve into a given number of first order components. The function uses a combination

of the differential evolution and Levenberg–Marquardt algorithms, as suggested by
Bluszcz et al. [17, 18] and Adamiec et al. [3].

The function *decomp* can analyze both CW-OSL and LM-OSL signals. Some of
the commonly used parameters in this function are *Sigdata* (a two-column matrix
with the stimulation time and photon count values), *ncomp* (an integer indicating the
number of exponential decay components), *constant* (a logical parameter indicating
whether a constant component should be subtracted from the decay curve), *typ*
(a character indicating the type of a decay curve "*cw*" or "*lm*"), and *weight*
(logical parameter indicating whether the fit should be performed using a weighted
procedure).

For a CW-OSL decay curve, the fitting model is (Bluszcz et al. [18])

$$I(t) = a_1 b_1 \exp(-b_1 t) + \cdots + a_k b_k \exp(-b_k t), \tag{3.5}$$

where $k = 1, 2, \ldots$ indicates the exponential components, $I(t)$ is the luminescence
intensity as a function of time, a_i is the number of trapped electrons for the i-
component, and b_i is the detrapping rate. A constant component c is added to
this equation, if one sets the parameter *constant=TRUE*, when calling the *decomp*
function.

Figure 3.1 shows how to analyze a CW-OSL curve into two exponential
components. The experimental data are for the CW-OSL signal from a freshly
irradiated aliquot of a feldspar sample KST4, measured at 50 °C, as described by
Pagonis et al. [146]. The total excitation time is taken to be 500 s.

The results of this analysis show two components with rate constants $b_1 = 0.17 \pm
0.02 \, \text{s}^{-1}$ and $b_2 = 0.0277 \pm 0.0005 \, \text{s}^{-1}$, and the value of $FOM = 7.7$ shows that
this fit is acceptable, but it could possibly be improved by using more components
to analyze the data.

In the next section we discuss how to analyze the same experimental CW-OSL
decay curve by using the R package *Luminescence*.

Code 3.1:   Fitting two-component CW-OSL signal (numosl)

```
### Fitting CW-OSL signal with package numOSL (2 components)
rm(list=ls())
library("numOSL")
data <- read.table("KST4ph300IR.txt")
data<-data.frame(data[2:500,1],data[2:500,2]) #data t=1-500 s
a<-decomp(data, ncomp=2)
```

```
print.noquote("Best fit parameters")
a$LMpars
cat("\nFOM=",a$FOM)

   ## [1] Best fit parameters
   ##              Ithn      seIthn        Lamda          seLamda
   ## Comp.1 58290.51  657.5422  0.17560609  0.0016782052
   ## Comp.2 83985.26  697.0981  0.02776157  0.0004630477
   ##
   ## FOM= 7.719679
```

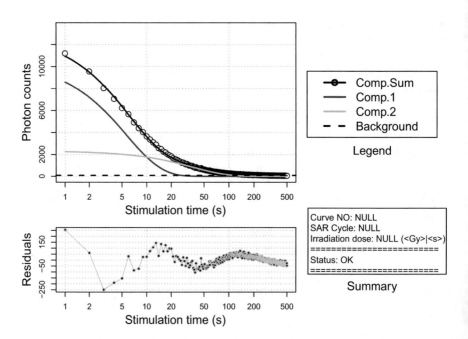

Fig. 3.1 Example of fitting the first 500 s of a CW-OSL signal with two exponential components, using the package *numOSL*. The CW-OSL data are from a freshly irradiated aliquot of feldspar sample KST4 (Pagonis et al. [146])

3.4 Fitting CW-OSL Data with the R package *Luminescence*

The code below uses the function *fit_CWCurve* within the R package *Luminescence*, to fit experimental CW-OSL data with three exponential components. It is assumed that the background has already been subtracted from the experimental CW-OSL data.

Typical parameters for the function *fit_CWCurve* are *values* (a data.frame containing the time and CW-OSL signal, e.g. in seconds and counts), *n.components.max* (a vector containing the maximum number of components to be used for fitting, with a maximum value of 7), and the parameter … (representing further arguments and graphical parameters passed to the plot). For the more detailed list of parameters, see the extensive documentation of the package *Luminescence* (Kreutzer et al. [88]).

The experimental data in Fig. 3.2 are for the first 800 s of the same CW-OSL signal analyzed in the previous section, for the feldspar sample KST4 (Pagonis et al. [146]). By using three components to analyze these data, we find $b_1 = 0.23\,\text{s}^{-1}$, $b_2 = 0.046\,\text{s}^{-1}$, and $b_3 = 0.0029\,\text{s}^{-1}$.

Code 3.2:   Fitting three-component CW-OSL signal (Luminescence)

```
# Fitting CW-OSL  with package Luminescence (three components)
rm(list=ls())
suppressMessages(library("Luminescence",warn.conflicts= FALSE))
##load and fit the data
data <- read.table("KST4ph300IR.txt")
data<-data.frame(data[1:800,1],data[1:800,2]) #use t=1-800 s
invisible(capture.output(fit <- fit_CWCurve(
   values = data,
   main = " ",
  n.components.max = 3)))
cat("\nBest fit parameters"," ")
cat("\nI01=",fit$data$I01," I02=",fit$data$I02,
   "  I03=",fit$data$I03)
cat("\nlambda1=",fit$data$lambda1," lambda2=",fit$data$lambda2)
cat("\nlambda3=",fit$data$lambda3)

   ##
   ## Best fit parameters
   ## I01= 40696.02   I02= 82084.83   I03= 91582.97
   ## lambda1= 0.2342178   lambda2= 0.04617059
   ## lambda3= 0.002939544
```

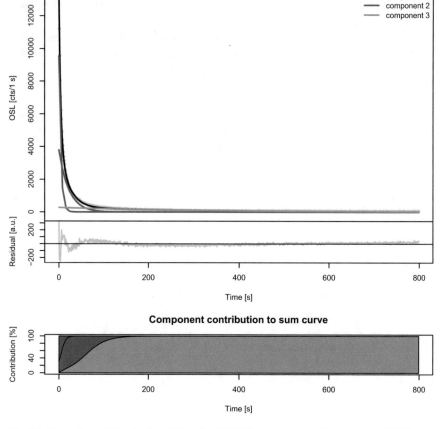

Fig. 3.2 Example of fitting the first 800 s of a CW-OSL signal from feldspar sample KST4, with three exponential components, using the package *Luminescence*. Experimental data from Pagonis et al. [146]

3.5   The GOT Equation for CW-OSL Signals

As we saw in the previous chapter, the GOT differential equation in the OTOR model where the stimulation is optical in nature and occurs with a source of constant light intensity is given by (Halperin and Braner [55])

$$I(t) = -\frac{dn}{dt} = \lambda \frac{n^2 A_m}{(N-n)A_n + n A_m} \tag{3.6}$$

where the function $p(t) = \sigma I = \lambda$, $\sigma \left(cm^2\right)$ represents the optical cross section for the CW-OSL process and $I \left(photons\ cm^{-2}\ s^{-1}\right)$ represents the photon flux.

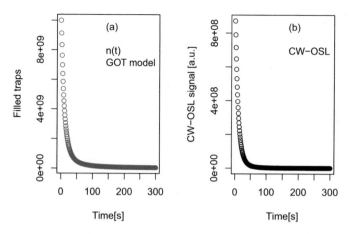

Fig. 3.3 Numerical solution of the GOT differential equation (3.6) for CW-OSL, using the R package *deSolve*. (**a**) Filled traps $n(t)$. (**b**) The corresponding CW-OSL signal

The following R code and Fig. 3.3 show an example of solving the GOT equation (3.6) for a CW-OSL process, with the numerical value $p(t) = \sigma I = 0.1$ (s^{-1}). The R package *deSolve* is used to solve the differential equation, as described previously for the TL process in Chap. 2.

3.6   Analytical Solution of the GOT Equation for CW-OSL

As discussed in Chap. 2, the analytical solution of Eq. (3.6) was found to be (Kitis and Vlachos [82])

Analytical solution of GOT with the Lambert W function

$$I(t) = \frac{N R}{(1 - R)^2} \frac{\lambda}{W[e^z] + W[e^z]^2} \qquad (3.7)$$

---

where $W[e^z]$ is the Lambert W function (Corless et al. [44, 45]). The concentration of trapped electrons $n(t)$ is equal to

$$n(t) = \frac{N R}{(1 - R)} \frac{1}{W[e^z]} \qquad (3.8)$$

Code 3.3: Solve the GOT equation for CW-OSL (deSolve)

```
# Numerically solve GOT equation for CW-OSL using package deSolve
rm(list=ls())
library("deSolve")
# Define Parameters
A_n <- 1e-10 # coefficient of retrapping
A_m <- 1e-8 # coefficient of recombination
Aopt <- 0.1 # Stimulation coefficient [1/s]
delta.t <- 1
t <- seq(0, 300, delta.t)
n.0 <- 1e10  # initial concentration of filled traps
N.traps <- 1e11   # total concentrations of available traps
# Calculate numerical ODE solution
ODE <- function(t, state, parameters){
  with(as.list(c(state, parameters)),{
      dn <-  -Aopt * n^2 * A_m / ((N.traps - n) * A_n + n * A_m)
      list(c(dn)) })}
parameters <- c(N.traps=N.traps,Aopt= Aopt,A_m = A_m,A_n = A_n)
state <- c(n = n.0)
num_ODE <- ode(y = state, times = t, func = ODE,
parms = parameters)
# Plot remaining filled traps  and TL signal
par(mfrow=c(1,2))
plot(num_ODE[,1],num_ODE[,2],xlab = "Time[s]",
ylab = "Filled traps",col = "red")
legend("topright",bty="n",c("(a)"," ","n(t)","GOT model"))
plot(num_ODE[-1,1],y = abs(diff(num_ODE[,2])),
xlab = "Time[s]", ylab = "CW-OSL signal [a.u.]")
legend("topright",bty="n",c("(b)"," ","CW-OSL"))
```

The following R code and Fig. 3.4 show a plot of the analytical solution of the GOT equation for CW-OSL and also show a plot of Eq. (3.7), using the function *lambertW0(x)* within the R package *lamW*. The parameters in the GOT/OTOR model are $\lambda = 0.01\,\text{s}^{-1}$, retrapping ratio $R = 0.01$, $N = 10^{10}\,\text{cm}^{-3}$, and $n_0 = 10^{10}\,\text{cm}^{-3}$.

Code 3.4: Plot of the Lambert W function solution for CW-OSL in the GOT model

```
## Plot the CW-OSL solution of GOT, using Lambert W function
rm(list=ls())
library(lamW)
t<-0:300
kB<-8.617E-5
no<-1E10
N<-1E10
R<-.01
c<-(no/N)*(1-R)/R
lambda<-.01
x<-unlist(t)
zCWOSL<-unlist((1/c)-log(c)+(lambda*no/(c*N*R))*t)
# plots
par(mfrow=c(1,2))
plot(x,(N*R/(1-R))/(lambertW0(exp(zCWOSL))),xlab="Time, s",
ylab="Filled traps n(T)",ylim=c(0,N),col="red")
legend("topright",bty="n",c("(a)   n(t)"," ","using","W0(x)"))
plot(x,(N*R/((1-R)^2))*lambda/(lambertW0(exp(zCWOSL))
        +lambertW0(exp(zCWOSL))^2),
    xlab="Time, s",ylab="CW-OSL signal")
legend("topright",bty="n",c("(b)",   "CW-OSL"))
```

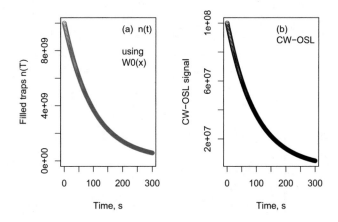

Fig. 3.4 Plot of the analytical solution of GOT differential equation for CW-OSL, using the Lambert W function solution from Eq. (3.7). (a) Filled traps $n(T)$ and (b) the corresponding CW-OSL signal

3.7 Simulations of CW-OSL Experiments Using the Package *RLumCarlo*

The following code and Fig. 3.5 show a simulation of a CW-OSL experiment using the function *run_MC_CW_OSL_DELOC* available with the package *RLumCarlo* (Kreutzer et al. [87]). The parameters in the GOT/OTOR model are $\lambda = 0.12\,\text{s}^{-1}$ and retrapping ratio $R = 0.1$. As in the TL examples in Chap. 2, the shaded area around the CW-OSL curve represents the range or standard deviation of the MC runs.

Code 3.5: Single plot MC for delocalized CW-OSL

```
##=================================================##
## MC simulations for delocalized CW-OSL
##=================================================##
rm(list = ls(all=T))
library(RLumCarlo)
run_MC_CW_OSL_DELOC(
    A = 0.12,
    R = 0.1,
    times = 0:100
) %>%
    #Plot results of the MC simulation
    plot_RLumCarlo(legend = F,ylab="CW-OSL [a.u.]")
legend("top",bty="n", legend=c("CW-OSL", "with function DELOC",
"Package RLumCarlo"))
```

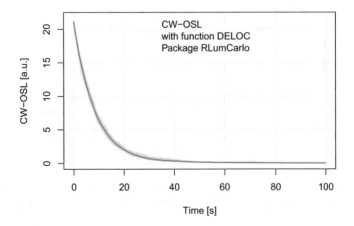

Fig. 3.5 Simplest example of MC simulation for CW-OSL DELOCalized transition process, within the OTOR model and based on the GOT equation (3.6). The parameters here are the excitation rate $\lambda = 0.12\,\text{s}^{-1}$ and retrapping ratio $R = 0.1$

The above example can be slightly varied, by specifying the parameter *output="remaining_e"* in the function *run_MC_CW_OSL_DELOC*. Instead of the luminescence signal, the code below will plot the number of remaining trapped electrons in the system. Note that inside the function *plot_RLumCarlo* we must specify the y-label with the argument
ylab="Filled traps n(t)", otherwise the default label of *"Average signal"* will be used on the y-axis in Fig. 3.6.

Code 3.6: Single plot MC for delocalized CW-OSL

```
##=====================================================##
## MC simulations for delocalized CW-OSL
##=====================================================##
rm(list = ls(all=T))
library(RLumCarlo)
run_MC_CW_OSL_DELOC(
n_filled=100,N_e=100,A = 0.12,
R = 0.1,
times = 0:100,
output="remaining_e"
) %>%
#Plot results of the MC simulation
plot_RLumCarlo(legend = F,xlab="Time (s)",
ylab="Filled Traps n(t)")
legend("topright",bty="n",legend=c("Remaining trapped electrons",
"during CW-OSL experiment"))
```

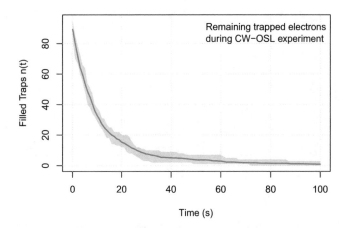

Fig. 3.6 Plotting the remaining electrons $n(t)$ during a DELOCalized CW-OSL simulation

The following example and Fig. 3.7 show how to combine two plots from several runs of the function *run_MC_CW_OSL_DELOC*, into one single plot. By using the logical arguments *norm = TRUE*, the two plots are normalized to the maximum value of the corresponding CW-OSL values, and the argument *add = TRUE* forces the two graphs on the same plot.

Code 3.7: Combine two plots from RlumCarlo

```
##==========================================================##
## Example: COMBINE TWO PLOTS  for delocalized CW-OSL
##==========================================================##
rm(list = ls(all=T))
library(RLumCarlo)
## set time vector
times <- seq(0, 50)
run_MC_CW_OSL_DELOC(A = 0.1, R = 0.01,n_filled=100,N_e=100,
                    times = times) %>%
plot_RLumCarlo(norm = TRUE, col="blue",
               legend = TRUE)
legend("top",bty="n",legend=c("CW-OSL signal",
expression("with A=0.1, 0.2 s"^"-1"*"")))
run_MC_CW_OSL_DELOC(A = 0.2, R = 0.01,n_filled=100,N_e=100,
                    times = times) %>%
plot_RLumCarlo(norm = TRUE, col="red", add = TRUE)
```

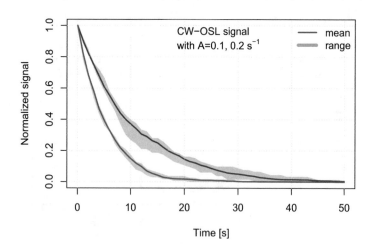

Fig. 3.7 Combine two plots for delocalized CW-OSL signals, with two different excitation rates $A = 0.1, \ 0.2 \, \mathrm{s}^{-1}$

In the last example of this section, we now simulate in Fig. 3.8 the delocalized CW-OSL process, by varying several of the parameters *A*, *R*, *n_filled, N_e*. The simulation is run 200 times (parameter *clusters* = 200).

A *for…loop* is used in this example, to cycle through all the combinations of the parameters in the model, with the parameter $u = 1 … 4$ determining how many times the loop is executed.

Code 3.8: MC for delocalized CW-OSL: multiple parameters

```
## Simulate CW-OSL DELOC with several parameter changes
##============================================================##
rm(list = ls(all=T))
library(RLumCarlo)
# define your parameters
A <- c(1,.3,.5,.1)
times <- seq(0,60,1)
R<-c(1e-7,1e-6,0.01,0.1) # sequence of different R values
clusters <- 200 # number of Monte Carlo simulations
N_e <- c(200, 500, 700, 400) # number of free electrons
n_filled <- c(200, 500, 100,
70) # number of filled traps
method <-"par"
output <- "signal"
col <- c(1,2,3,4) # different colors for the individual curves
plot_uncertainty <- c(T,F,T,F) # plot the uncertainty?
add_TF <- c(F,rep(T, (length(R)-1)))
for (u in 1:length(R)){
  results <-run_MC_CW_OSL_DELOC(A=A[u],times,clusters =clusters,
    N_e = N_e[u],   n_filled = n_filled[u],
R=R[u], method = method, output = output)
  plot_RLumCarlo(results,add=add_TF[u],legend = F, col=col[u])
}
legend("topright",ncol=4,bty="n",cex=.65,title = "CW-OSL
function DELOC" ,
        legend=c(paste0("A = ", A),
                paste0("n_filled = ", n_filled),
                paste0("N_e = ", N_e),
                paste0("R = ", R)), text.col=col)
```

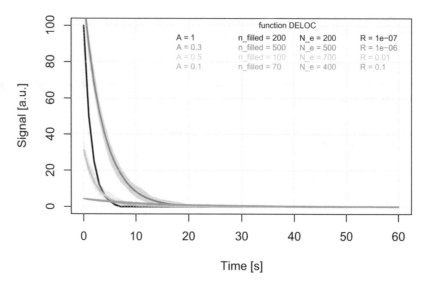

Fig. 3.8 Simulation of several CW-OSL curves using the DELOC functions of the *RLumCarlo* package, with several parameter changes

3.8 First Order Kinetics for LM-OSL Processes

An important variation of the OSL measurements under constant intensity of stimulating light has been first proposed by Bulur [26]. Since the OSL measured during constant excitation intensity (CW-OSL) or following a pulse of stimulating light is a rather featureless decaying function, Bulur suggested using linear modulation of the stimulating light, and the method used is termed LM-OSL. By using this technique, the OSL curves are observed in the form of peaks, and the peak maximum intensity, which is easily determined, can be utilized as the measured OSL signal. Thus, the resolution of several OSL signals is significantly easier, and the evaluation of the OSL parameters is usually much simpler.

As was discussed in Chap. 1, for LM-OSL processes the optical stimulation function is given by $p(t) = \sigma I t / P$, where σ, I represents again the optical cross section for the OSL process, $I \left(photons \, cm^{-2} \, s^{-1}\right)$ represents the photon flux, t is the elapsed time, and $P \, (s)$ is the total illumination time.

The differential equation for first order kinetics LM-OSL processes is:

First order kinetics differential equation for LM-OSL

$$\frac{dn(t)}{dt} = -\frac{\sigma I}{P} t \, n(t), \qquad (3.9)$$

---

where $n(t)$ is the concentration of trapped carriers, t is the stimulation time, and P is the total length of time of the LM-OSL measurement. The solution of this differential equation is:

First order kinetics analytical equation for LM-OSL

$$L(t) = n_0 \frac{\sigma I}{P} t \, \exp\left(-\frac{\sigma I}{2P} t^2\right), \qquad (3.10)$$

---

where $L(t)$ is the emitted LM-OSL intensity. This function $L(t)$ is the product of the increasing function t, and a peak-shaped Gaussian, and therefore $L(t)$ has a peak shape with a maximum.

Kitis and Pagonis [72] pointed out a major difference between TL and OSL deconvolution analysis. Once the time t_m of maximum LM-OSL intensity has been found in the experimental OSL curve, then the complete individual OSL curve can be identified within the composite LM-OSL curve. This simple property is very important for computerized curve deconvolution analysis (CCDA), since for every given time t_m, there is one and only one corresponding LM-OSL peak. This is in contrast to the respective property of TL peaks, in which every peak maximum temperature T_M corresponds to an infinite number of possible TL peaks with different pairs of kinetic parameters (E, s).

The usefulness of LM-OSL deconvolution techniques for analyzing LM-OSL curves from quartz in dating applications was examined by several researchers. For example, Singarayer and Bailey [167] reported on the use of LM-OSL in the study of OSL in quartz and found to contain typically five or six common LM-OSL components when stimulated at 470 nm. More work on LM-OSL in quartz, along with TL, has been reported by Kitis et al. [71]. Bulur and Yeltik [29] used LM-OSL in BeO ceramics and found two first order components.

3.9   General Order Kinetics and Transformed Equations for LM-OSL

Bulur and Göksu [28] amended the previous GOK equations by Bulur [26] and proposed the following equation for GOK LM-OSL signals:

General order kinetics differential equation for LM-OSL

$$L(t) = -dn/dt = (\sigma It/P) n^b / N^{b-1}, \qquad (3.11)$$

---

where the symbols were defined in the previous section and N is the total number of available traps. The solution of this differential equation is:

General order kinetics analytical equation for LM-OSL

$$L(t) = n_0^b(\sigma I)(t/N^{b-1}P)\left[1 + (b-1)(\sigma I n_0/2PN^{b-1})t^2\right]^{-b/(b-1)}$$

$$(3.12)$$

Kitis and Pagonis [72] studied the geometrical characteristics and symmetry factors of the peak-shaped LM-OSL curves and pointed out the similarities and differences between TL and OSL deconvolution analysis. These authors studied both first order and general order peaks and identified the presence of slowly varying pseudo-constants that can be used to develop expressions for the optical cross sections for OSL. In close analogy with the situation for TL, the following expressions were derived for first order and general order analysis:

Transformed equations for LM-OSL

$$I(t) = 1.64871 I_m \frac{t}{t_m} exp\left[-\frac{t^2}{2t_m^2}\right], \qquad \text{First Order} \qquad (3.13)$$

$$I(t) = I_m \frac{t}{t_m}\left[\frac{b-1}{2b}\frac{t^2}{t_m^2} + \frac{b+1}{2b}\right]^{b/(1-b)}. \qquad \text{General Order} \qquad (3.14)$$

In these expressions, t is the time (s), t_m is the time at which the maximum LM-OSL intensity I_m occurs, and b is the kinetic order of the OSL process.

3.10   The GOT Equation for LM-OSL Signals

As we saw in the previous chapter, the GOT differential equation for LM-OSL processes in the OTOR model is given by Halperin and Braner [55]:

$$I(t) = -\frac{dn}{dt} = p(t)\frac{n^2 A_m}{(N-n)A_n + n A_m} \qquad (3.15)$$

where the excitation rate is $p(t) = \frac{\sigma I}{P}t$ and σ, I represents again the optical cross section for the OSL process, $I\left(photons\ cm^{-2}\ s^{-1}\right)$ represents the photon flux, t is the elapsed time, and P (s) is the total illumination time. This gives:

GOT equation for LM-OSL processes

$$I(t) = -\frac{dn}{dt} = \frac{\sigma I}{P}t\frac{n^2 A_m}{(N-n)A_n + n A_m} \qquad (3.16)$$

The following R code gives an example of solving the GOT equation for LM-OSL processes, with the numerical value $p(t) = \frac{\sigma I}{P}t = 0.1t\ (s^{-1})$. As expected, the LM-OSL signal in Fig. 3.9 is peak shaped.

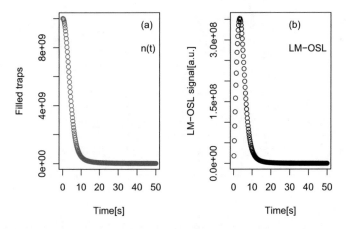

Fig. 3.9 Numerical solution of the GOT differential equation for LM-OSL. (a) Filled traps $n(t)$ and (b) the respective LM-OSL signal

---

Code 3.9: Solve the GOT equation for LM-OSL (deSolve)

```
# Numerically solve the GOT equation for LM-OSL
rm(list=ls())
library("deSolve")
# Define Parameters
A_n <- 1e-10 # coefficient of retrapping
A_m <- 1e-8 # coefficient of recombination
Aopt <- 0.1 # Stimulation coefficient [1/s]
delta.t <- .2
t <- seq(0, 50, delta.t)
n.0 <- 1e10   # initial concentration of filled traps
N.traps <- 1e11   # total concentrations of available traps
# Calculate numerical ODE solution
ODE <- function(t, state, parameters){
  with(as.list(c(state, parameters)),{
    dn <-  -Aopt *t*  n^2 * A_m / ((N.traps - n)* A_n+n* A_m)
    list(c(dn))
  })}
parameters <- c(N.traps =N.traps,Aopt = Aopt,A_m=A_m,A_n = A_n)
state <- c(n = n.0)
num_ODE <- ode(y = state, times = t, func = ODE,
parms = parameters)
# Plot remaining electrons and LM-OSL as a function of time
par(mfrow=c(1,2))
plot(x = num_ODE[,1],
     y = num_ODE[,2],xlab = "Time[s]", ylab = "Filled traps",
col = "red")
legend("topright",bty="n",c("(a)"," ","n(t)"))
plot(x = num_ODE[-1,1], y = abs(diff(num_ODE[,2])),
xlab = "Time[s]", ylab="LM-OSL signal[a.u.]")
legend("topright",bty="n",c("(b)"," ","LM-OSL"))
```

As discussed in Chap. 2, the analytical solution of Eq. (3.16) was found by Kitis and Vlachos to be [82]:

Analytical solution of GOT for LM-OSL with the Lambert W function

$$I(t) = \frac{N R}{(1 - R)^2} \frac{\sigma I}{P} t \frac{1}{W[e^z] + W[e^z]^2} \tag{3.17}$$

---

where $W[e^z]$ is the Lambert W function [44, 45]. The following code plots in Fig. 3.10 the analytical equation (3.17).

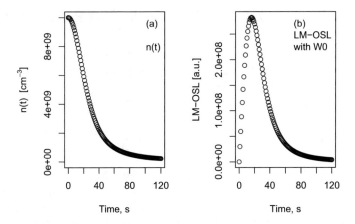

Fig. 3.10 Plots of the analytical solution of GOT Eq. (3.17) for LM-OSL, using the Lambert function $W[z]$. (**a**) Filled traps $n(t)$ and (**b**) the LM-OSL signal

---

Code 3.10: Plot W0(x) solution of GOT equation, for LM-OSL

```
# plot analytical solution of GOT equation for LM-OSL with W(x)
rm(list=ls())
library(lamW)
## Plot the analytical solution of GOT, using Lambert W function
t<-0:120
kB<-8.617E-5
no<-1E10
N<-1E10
R<-.46
c<-(no/N)*(1-R)/R
lambda<-.003
x<-unlist(t)
zLM<-unlist((1/c)-log(c)+(lambda*no/(c*N*R))*t^2/2)
# plots
par(mfrow=c(1,2))
plot(x,(N*R/(1-R))/(lambertW0(exp(zLM))),xlab="Time, s",
      ylab=expression("n(t)   [cm"^-3*"]"),ylim=c(0,N))
legend("topright",bty="n",c("(a)"," ","n(t)"))
plot(x,(N*R/((1-R)^2))*lambda*x/(lambertW0(exp(zLM))
  +lambertW0(exp(zLM))^2), xlab="Time, s",ylab="LM-OSL [a.u.]")
legend("topright",bty="n",c("(b)","LM-OSL","with W0"))
```

3.11   Fitting LM-OSL Signals Using the Package *numOSL*

The *decomp* function developed by Peng et al. [150] was previously discussed in this chapter for CW-OSL signals. The same function can also be used to analyze LM-OSL signals into first order components.

For an LM-OSL decay curve, the fitting model is (Bulur [27])

$$ I(t) = a_1 b_1 \frac{t}{P} e^{-b_1 t^2/(2P)} + \cdots + a_k b_k \frac{t}{P} e^{-b_k t^2/(2P)} \qquad (3.18) $$

where $k = 1, 2, \ldots, 7$ is the number of possible LM-OSL components in the function *decomp*, $I(t)$ is the luminescence intensity as a function of time, P is the total stimulation time, a_k is the concentration of initially trapped electrons for each component, and b_k is the detrapping rate. The constant component for the LM-OSL signal is taken as $c(t/P)$, if the parameter *constant* in this function is set to *constant=TRUE*. Parameters are initialized using the differential evolution method [2, 3, 18], then the Levenberg–Marquardt algorithm is used to optimize the parameters.

The following code analyzes typical LM-OSL experimental results in Fig. 3.11, for the important dosimetric material CaF_2:N (see Kitis et al. [78]).

Code 3.11:   Analysis of three-component LM-OSL signal (numOSL)

```
### Example: Analyze LM-OSL signal using package numOSL
rm(list=ls())
library("numOSL")
CaF2LMx <- read.table("CaF2LMx.txt")
CaF2LMy <- read.table("CaF2LMy.txt")
d<-data.frame(CaF2LMx,CaF2LMy)
a<-decomp(d,ncomp=3,typ="lm",log="",
          control.args=list(maxiter=10))
print.noquote("Best fit parameters")
a$LMpars
cat("\nFOM=",a$FOM)

   ## [1] Best fit parameters
   ##             Ithn     seIthn      Lamda      seLamda
   ## Comp.1   44763.64   1219.005 1.83464988 0.059252869
   ## Comp.2   86820.20  11296.986 0.09116347 0.011550623
   ## Comp.3 191063.67  11639.415 0.01385314 0.001722478
   ##
   ## FOM= 5.037462
```

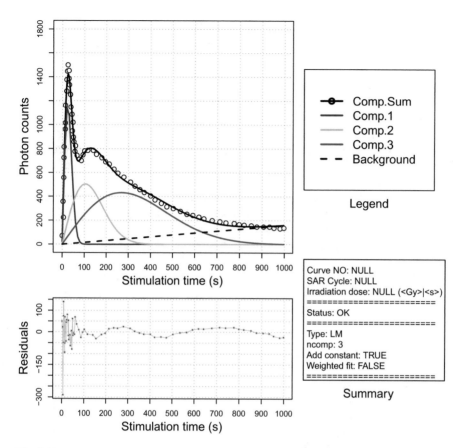

Fig. 3.11 Example of analyzing an LM-OSL signal from the dosimetric material CaF$_2$:N with three first order components, using the package *numOSL* (Kitis et al. [78])

3.12   Fitting Experimental LM-OSL Data with the R Package *Luminescence*

The fitting equation used for LM-OSL signals in the package *Luminescence* was developed by Kitis and Pagonis [72] :

$$I(t) = 1.684\, I_{m_1} \frac{t}{x_{m_1}} e^{-t^2/(2\,x_{m_1})} + \cdots + 1.684\, Im_i \frac{t}{x_{m_i}} e^{-t^2/(2\,x_{m_i})} \qquad (3.19)$$

where $i = 1 \ldots 7$ is the number of LM-OSL components in the signal, t_{m_i} is the time where the maximum of the i-component occurs, and I_{m_i} is the maximum height of the i-component. The quantities x_{m_i}, I_{m_i} in Eq. (3.19) are related by

$$x_{m_i} = \sqrt{\frac{t_{m_i}}{b_i}}$$

$$I_{m_i} = e^{-0.5 \frac{n_{0i}}{x_{m_i}}}$$

where b_i is the detrapping probability and n_{0i} is proportional to the initially trapped charge concentration for the i-component.

The following R code uses the package *Luminescence* to fit the previously discussed experimental LM-OSL data for CaF$_2$:N, with three first order components. It is assumed that the background has not already been subtracted from the experimental LM-OSL data. The code reads the LM-OSL data, and also creates an estimated linearly increasing background for the data from zero to a maximum of 10^3 cts, and stores it in the *data.frame* parameter *bgd*.

In this R code, the user specifies estimated starting values for the two kinetic parameters, the maximum of the peak-shaped LM-OSL signal (I_{m_i}), and for the location of these maxima on the horizontal time-axis (x_{m_i}). The results of the fitting process shown in Fig. 3.12 are stored in the parameter a and are accessed using *a$data*.

Notice that the output of the Luminescence package in this example is given in terms of the three optical cross sections σ of the OSL process, instead of the

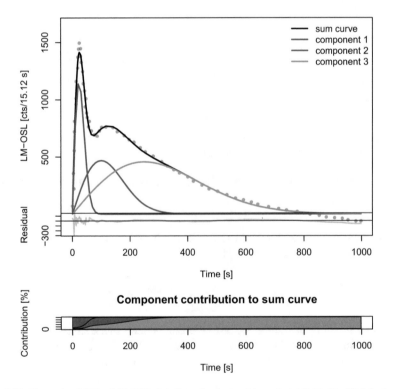

Fig. 3.12 Example of fitting LM-OSL data from the dosimetric material CaF$_2$:N with 3 first order components, using the package *Luminescence* (Kitis et al. [78])

excitation rates λ as in the example of the *tgcd* analysis in the previous section. This is because the *Luminescence* package assumes a value of the photon flux I and uses the relationship $\lambda = \sigma I$ to evaluate the optical cross section σ.

For more details, the readers are directed to the extensive documentation of the package *Luminescence*, and also to several papers by Kreutzer et al. [85, 86, 88].

Code 3.12: Fit LM-OSL data with three first order components (Luminescence)

```
# fit  LM-OSL  with three exponentials using Luminescence package
rm(list=ls())
suppressMessages(library("Luminescence",warn.conflicts = FALSE))
## fit LM data with background subtraction
CaF2LMx <- read.table("CaF2LMx.txt")
CaF2LMy <- read.table("CaF2LMy.txt")
d<-data.frame(CaF2LMx,CaF2LMy)
bgdx<-seq(from=15,to=1000,by=15)
bgdy<-seq(from=3,to=200,by=3)
bgd<-data.frame(bgdx,bgdy)
invisible(capture.output(a<-fit_LMCurve(d,bgd,
n.components = 3, main=" ",
start_values = data.frame(Im = c(1500,800,500),
xm = c(30,130,200)))))
cat("\nImax values: ",a$data$Im1,a$data$Im2,a$data$Im3)
cat("\ntmax values: ",a$data$xm1,a$data$xm2,a$data$xm3)
cat("\nn0 values: ",a$data$n01,a$data$n02,a$data$n03)
cat("\nCross section values (cm^2)")
cat("\n",a$data$cs1,a$data$cs2,a$data$cs3)
cat("\nR^2=",a$data$`pseudo-R^2`)
```

```
##
## Imax values:  1159.203 467.8042 456.2355
## tmax values:  23.24563 99.72249 248.6371
## n0 values:  44427.09 76913.83 187026.1
## Cross section values (cm^2)
##  2.168303e-17 1.178193e-18 1.895269e-19
## R^2= 0.9864
```

3.13 Transforming CW-OSL Into Pseudo-LM-OSL Signals

Bulur [27] suggested a method by which CW-OSL *or* CW-IRSL results measured under constant stimulating light intensity can be manipulated, so that the results are mathematically similar to the LM-OSL and LM-IRSL situation. The transformation described here is purely a mathematical change in the variables, which facilitates further analysis of the signals.

For the case of first order kinetics, Bulur wrote the decay function of luminescence as

$$I(t) = n_0 a \exp(-at), \tag{3.20}$$

where n_0 is the initial concentration of trapped electrons and a is the decay constant describing the CW luminescence curve. In order to convert the CW-OSL to a *pseudo-LM-OSL curve*, a new variable u is introduced and defined as

$$u = \sqrt{2tP} \tag{3.21}$$

or $t = u^2/2P$, where P is the measurement period and therefore variable u has dimensions of time.

In practice, one evaluates the new x-axis for the transformed data by using Eq. (3.21), and the new y-axis by using the expression:

$$I(u) = I(t)\, u/P \tag{3.22}$$

Using these two equations, one gets

$$I(u) = n_0 \frac{a}{P} u \exp\left[-\frac{a}{2P} u^2 \right]. \tag{3.23}$$

This has the same form as Eq. (3.10). The following R code and Fig. 3.13 show how to transform the shapeless CW-OSL or CW-IRSL signals into peak-shaped LM-OSL and LM-IRSL signals, respectively. The simple code reads a *.txt* file containing two

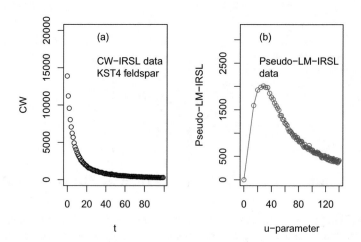

Fig. 3.13 Example of transforming a CW-IRSL signal into pseudo-LM-IRSL data, by reading a *.txt* file containing two columns, containing the time and the corresponding CW-intensities. The experimental data is discussed in Pagonis et al. [146]

columns, containing the time (in s) and CW-intensities (in cts/s, or some arbitrary units). The experimental data are a CW-IRSL curve for a freshly irradiated feldspar sample KST4, measured at 50 °C and for a total excitation time of 100 s, as described in Pagonis et al. [146].

Code 3.13: Transform CW into pseudo-LM data (.txt data file)

```
##Read CW-OSL data from .txt file and transform to pseudo-LM-
                                IRSL
rm(list=ls())
##produce x and y (time and count data for the data set)
par(mfrow=c(1,2))
CWdata <- read.table("KST4ph300IR.txt")
t<-CWdata[,1][1:100]
CW<-CWdata[,2][1:100]
u<-sqrt(2*t*max(t))
Iu<-u*CW/max(t)
plot(data.frame(t,CW),ylim=c(0,20000))
legend("topright",bty="n",legend=c("(a)"," ","CW-IRSL data",
  "KST4 feldspar"))
plot(u,Iu,xlab="u-parameter",typ="o",ylab="Pseudo-LM-IRSL",
    col="red",ylim=c(0,3200))
legend("topright",bty="n",legend=c("(b)"," ","Pseudo-LM-IRSL",
                                "data"))
```

The following code shows how to transform the same CW-IRSL signal into pseudo-LM-IRSL data, by using the package *Luminescence*. The result of this alternative code is of course the same as the plot shown in Fig. 3.13, apart from an arbitrary multiplication factor along the y-axis of the transformed data.

Code 3.14: CW-IRSL into pseudo-LM-IRSL data (Luminescence package)

```
##Read CW-IRSL data from .txt file and transform to pseudo-LM-
                                OSL
rm(list=ls())
suppressMessages(library("Luminescence",warn.conflicts =FALSE))
##produce x and y (time and count data for the data set)
par(mfrow=c(1,2))
values <- read.table("KST4ph300IR.txt")
```

```
t<-values[,1][1:100]
CW<-values[,2][1:100]
data<-data.frame(t,CW)
plot(data,xlab="Time [s]",ylab="Original CW-IRSL [cts/s]",
     ylim=c(0,20000))
legend("topright",bty="n",legend=c("(a)"," ","CW-IRSL data",
             "KST4 feldspar"))
##transform values
data.transformed <- CW2pLM(data)
plot(data.transformed,xlab="u-parameter",typ="o",
     ylab="Pseudo-LMIRSL",col="red",ylim=c(0,3000))
legend("topright",bty="n",legend=c("(b)"," ","Pseudo-LM-OSL",
                               "data"))
```

3.14 Recommended Protocols for Analyzing OSL Data

We recommend the following steps in studying the OSL properties of a new dosimetric material:

1. *Measure and fit the OSL signal using several different intensities of the stimulating light.* Analysis and fitting of OSL data obtained using variable light intensities can be used to obtain a better estimate for the optical cross section σ for each light sensitive trap.
2. *Measure and fit the OSL signal using several different irradiation doses.* It is important to repeat the analysis by irradiating the sample and measuring the OSL signal at different doses D. By comparing the best fitting parameters obtained from the different doses, one can assess whether the results from different doses are consistent with each other.
3. *Partial optical bleaching analysis.* This is also known as the $t_{max} - t_{STOP}$ method of analysis and is described in Kitis and Pagonis [72]. These authors suggested this experimental procedure as a method to separate composite LM-OSL curves into their constituent components.

The procedure is analogous to the well-known $T_{max} - T_{STOP}$ technique of analysis for composite TL glow curves; the sample is optically bleached for a time period t_{STOP}, then the complete LM-OSL signal is measured, and the time t_{max} of the maximum LM-OSL intensity is recorded. The process is repeated for a gradually increasing bleaching time t_{STOP}, and a graph of t_{max} vs. t_{STOP} is produced. For details, see Kitis and Pagonis [72].

4. *OSL dose response.* Measure the OSL signal for several irradiation doses D, plot the OSL dose response $OSL(D)$ both on a linear and on a log–log scale, to test for superlinear behavior. If the dose response curve does not contain regions of superlinear behavior, fit the response curve with one of several possible

fitting functions: saturating exponential, double saturating exponential, and the new Lambert equation (4.13) described in Chap. 4, derived recently by Pagonis et al. [135]. If the dose response contains regions of superlinear behavior, fit the response curve with the Pagonis–Kitis–Chen (PKC) Lambert equation (4.31) for superlinearity in Chap. 4, derived recently by Pagonis et al. [136].

5. *Analysis of kinetics of a single peak in LM-OSL data.* Measure the *LM-OSL* signal for several irradiation doses, and examine the shape of the signals by normalizing them to their maximum height. If the shape does not change significantly with the irradiation dose, this is a strong indication of possibly a single LM-OSL component, and also a strong suggestion of first order kinetics. Further evidence for first order kinetics is provided if the maxima of the individual peaks do not shift significantly with the irradiation dose.

6. *Isothermal decay experiments.* This is a valuable method of analysis, which can help identify individual isothermal components, which may correlate with specific components in the CW-OSL or LM-OSL signal. For examples of using isothermal experiments to obtain information on the luminescence process, see the recent experimental work by Kitis et al. [81] and Polymeris et al. [156].

Chapter 4
Dose Response of Dosimetric Materials

4.1 Introduction

In this chapter we discuss theoretical and experimental aspects of the dose response of dosimetric materials, and use analytical equations to fit experimental data.

Section 4.2 in this chapter provides a general discussion of the trap filling process in solids during irradiation, and how various nonlinear experimental dose responses can be described mathematically by using the superlinearity index $g(D)$, and the supralinearity index $f(D)$.

Section 4.3 contains a discussion and example of using the saturating exponential function, which is a common experimentally observed dose response in the laboratory. Section 4.4 presents the mathematical framework for describing the irradiation stage by using the OTOR model, and also discusses the analytical solution of this model using the Lambert W function. The analytical solution of the model is compared with the solution of the differential equations in the OTOR model.

Section 4.5 contains examples of fitting experimental ESR, TL, OSL data with a new analytical equation based on the Lambert function. Section 4.6 simulates experiments in which the sample temperature is variable during the irradiation process, and which may affect the dose response of the material. This is an important phenomenon which affects the luminescence signals, for both naturally irradiated and for laboratory irradiated samples.

Section 4.7 considers the model of Bowman and Chen [23], which describes superlinear dose response as being a result of competition between two electron traps during the irradiation of a sample. This is followed in Sect. 4.8 by an example of fitting experimental data, by using the new analytical PKC equation which describes superlinearity effects, and which was developed recently by Pagonis et al. [136].

This chapter will conclude in Sects. 4.9 and 4.10, with a discussion of the analytical dose response equations based on the Lambert function, and with a general overview of the importance of the Lambert function in the description of luminescence phenomena.

V. Pagonis, *Luminescence, Use R!*, https://doi.org/10.1007/978-3-030-67311-6_4

4.2   Nonlinear Dose Response of ESR, TL, and OSL Signals

The dose response of TL, OSL, and ESR signals is usually shown as a graph of the intensity of these signals as a function of the irradiation dose. Examples of this type of graph were given in Chap. 1. Experimentally researchers can measure two types of dose responses. The first type of experiment measures the trapped charge as a function of the irradiation dose *at the end of irradiation*, by using ESR or optical absorption techniques. In the second type of experiment, the trapped charge is measured *at the end of two combined stages,* i.e. of irradiation followed by thermal or optical stimulation of the sample.

The dose response graph is of fundamental importance in radiation dosimetry and luminescence dating. A desirable feature for these phenomena is that the dose dependence be linear in a broad range. However, dosimetric materials show many different types of dose responses.

Nonlinear dose responses are usually classified as *superlinearity* (or *supralinearity*), meaning a faster than linear dependence on the dose, and *sublinearity* meaning a lower than linear dose dependence.

Typical examples in dosimetric materials would be superlinear behavior at low doses, or linear behavior at low doses becoming superlinear at higher doses. Some materials show sublinear dose dependence from the lowest doses, and others show a linear and/or superlinear range followed by sublinear behavior. The dose response shows a saturation effect in many cases, where trapping states and recombination centers reach saturation, and this results in a plateau in the dose dependence. Finally, some materials might show a non-monotonic dose dependence of TL/OSL/ESR, in which the signal reaches a maximum at a certain dose and then decreases at higher doses.

If we consider the very simple OTOR model with only one trapping state and one kind of recombination center, the linear-to-saturation behavior is expected. This is discussed in Sect. 4.4. However, in real materials we have many trapping states and recombination centers, and competition effects become very important. The presence of competition leads to many different physical phenomena associated with TL/OSL/ESR, including different types of dose responses. This is discussed in Sect. 4.7.

In general, simulations and experiments have shown that competition during excitation can yield one kind of superlinear behavior, while competition during readout yields another kind of dependence. In addition, simulations have shown that the combination of the two types of competition can lead to different dose responses (Chen and Pagonis [38]).

Different definitions have been given to superlinearity. If, for example, the TL intensity is measured by the maximum intensity I_{max}, one way of presenting this property (see e.g. Halperin and Chen [56]) where the superlinearity starts from the lowest doses is by the expression:

$$I_{max} = a D^k \qquad (4.1)$$

where D is the applied dose, a is a proportionality factor, and k a constant. A value of $k > 1$ means that the dependence is superlinear whereas $k < 1$ means sublinearity, and obviously $k = 1$ means a linear dependence. When this behavior takes place in a certain dose range, a plot of I_{max} as a function of D on a log–log scale is expected to yield a straight line with a slope of k. For example, Halperin and Chen [56] reported on a strong superlinearity with $k \cong 3$, in UV irradiated semiconducting diamonds.

A more general definition was given by Chen and McKeever [36], who proposed a normalized quantity which evaluates numerically the amount of superlinearity, and which would also attain the value of k in the special case of Eq. (4.1). They defined the *superlinearity index* (or superlinearity factor) $g(D)$ as

$$g(D) = \frac{DS''(D)}{S'(D)} + 1 \qquad (4.2)$$

The condition $g(D) > 1$ indicates superlinearity, $g(D) = 1$ means a range of linearity, and $g(D) < 1$ signifies sublinearity. $g(D)$ is a dimensionless function, and for the expression in Eq. (4.1) or even for $S(D) = aD^k + b$, one gets $g(D) = k$.

A number of materials including $LiF : Mg, Ti$ have the property that at low doses the dose dependence is linear, followed by a nonlinear (i.e. superlinear) region before saturation effects set in (see e.g. Zimmerman [197]). Chen and McKeever[36] suggested the term *supralinearity* to describe this property of the measured quantity being above the continuation of the initial range. Some authors, e.g. Horowitz [58] and Mische and McKeever[113] quantified this property by defining the dimensionless quantity

$$f(D) = \frac{[S(D)/D]}{[S(D_1)/D_1]} \qquad (4.3)$$

termed by these authors the *supralinearity index* (or supralinearity factor, or supralinear dose response function). Here D_1 is a normalization dose in the initial linear range. This definition is obviously applicable only if such an initial range exists at the low dose regime. It should be noted that in the literature the terms superlinearity and supralinearity are used quite often interchangeably.

Analytical equations for the *supralinearity index* $f(D)$ are presented in Sect. 4.9.

4.3   The Saturating Exponential Function

A common behavior of many dosimetric materials is a linear dose dependence at low doses, followed by the approach to saturation. This type of behavior can be fitted many times with the well-known saturating exponential function:

$$n(t) = N \left[1 - \exp\left(-D/D_0\right)\right] \qquad (4.4)$$

where D_0 is the characteristic dose of the sample. As a rule of thumb, the dose response is usually fully saturated after a dose of about $5D_0$. The following code fits the dose response of experimental OSL data from Li et al. [97], for a quartz sample from Libya (Fig. 4.1).

Code 4.1:   Fit dose response data with saturating exponential

```
#Fit dose response data with Saturating Exponential
rm(list = ls(all=T))
options(warn=-1)
library("minpack.lm")
library("lamW")
## fit to saturation exponential ----
t = c( 0,50.7117,100.534,152.135,204.626,272.242)
y = c(0,33.144,42.205,43.1055,44.4157,43.7098)
fit_data <-data.frame( t ,y)
plot(fit_data,ylim=c(0,max(y)),xlab="Dose [Gy]",
ylab="OSL (L/T)")
fit <- minpack.lm::nlsLM(
  formula = y ~ N * (1-exp(-b* t)),
  data = fit_data,
  start = list(N= max(y),b = .01))
N_fit <- coef(fit)[1]
b_fit <- coef(fit)[2]
## plot analytical solution
t1<-0:300
lines(x = t1,  y = N_fit  * (1-exp(-b_fit* t1)),col = "blue")
legend("right",bty="n",legend=c("Quartz OSL"," ",
"Libyan quartz"))
## print results
cat("\nfitted N: ", N_fit)
cat("\nfitted Do: ", round(1/b_fit,2), "Gy")

  ##
  ## fitted N:   44.16116
  ## fitted Do:  35.83 Gy
```

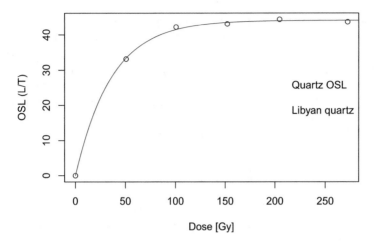

Fig. 4.1 Fitting dose response data with a saturating exponential. For more details see Pagonis et al. [135], original data from Li et al. [97]

4.4 Dose Response in the OTOR Model: The Lambert Analytical Solution

Figure 1.4 shows a schematic diagram of the OTOR model, and the relevant electronic transitions during the irradiation of a sample. The differential equations governing the traffic of electrons between the trapping level, the recombination center, and the conduction band in the OTOR model are [38]

$$\frac{dn}{dt} = A_n(N - n)n_c, \tag{4.5}$$

$$\frac{dm}{dt} = B(M - m)n_v - A_m m n_c, \tag{4.6}$$

$$\frac{dn_c}{dt} = X - A_n(N - n)n_c - A_m m n_c, \tag{4.7}$$

$$\frac{dn_v}{dt} = \frac{dn}{dt} + \frac{dn_c}{dt} - \frac{dm}{dt}. \tag{4.8}$$

$$m + n_v = n + n_c. \tag{4.9}$$

Here n (cm^{-3}) and m (cm^{-3}) are the concentrations of electrons in traps and of holes in recombination centers respectively, and N and M (cm^{-3}) are the total concentrations of trapping states and recombination centers. n_c (cm^{-3}) and n_v (cm^{-3}) are respectively the concentrations of free electrons and holes. A_n (cm^{3}s^{-1}) is the retrapping coefficient of electrons, A_m (cm^{3}s^{-1}) the recombination coefficient of electrons, and B $(cm^3 s^{-1})$ the trapping coefficient of holes in centers. X $(cm^{-3} s^{-1})$ is proportional to the dose rate of excitation, and actually denotes the rate of production of electron–hole pairs by the excitation irradiation per unit volume per second. Equation (4.9) is the charge neutrality condition.

The above set of equations cannot be solved analytically. Instead, one uses the quasi-equilibrium (QE) assumption to transform them into a single differential equation, as follows (Chen and Pagonis [38]). By setting $dn_c/dt = 0$ we find n_c, and the following single differential equation is obtained:

$$\frac{dn}{dt} = \frac{(N-n)A_n}{(N-n)A_n + n A_m} X \qquad (4.10)$$

This equation was derived and integrated previously by Lawless et al. [94], who showed that it can be easily integrated by separation of variables, to yield (Ref.[93], their Eq. 11)

$$Xt = n + \frac{A_m}{A_n}\left[-N \ln\left(1 - \frac{n}{N}\right) - n\right] \qquad (4.11)$$

The product $D = Xt$ represents the irradiation dose received by the sample, and by introducing the dimensionless retrapping ratio $R = A_n/A_m$ in this equation, we obtain

$$D = n + \frac{1}{R}\left[-N \ln\left(1 - \frac{n}{N}\right) - n\right] \qquad (4.12)$$

This is an equation for $D(n)$, which has been inverted numerically by previous researchers, in order to plot the inverse function $n(D)$. Lawless et al. [94] developed *approximate* solutions of this equation for low doses and high doses. Recently Pagonis et al. [135] developed the *exact* analytical solution $n(D)$ of this equation in terms of the Lambert W function:

New Lambert equation $n(D)$ in the OTOR model when $n(0) = 0$

$$\frac{n(D)}{N} = 1 + \frac{1}{1-R} W\left[(R-1)\exp\left(R - 1 - D/D_c\right)\right] \qquad (4.13)$$

---

where the constant D_c is defined as

$$D_c = N/R \qquad (4.14)$$

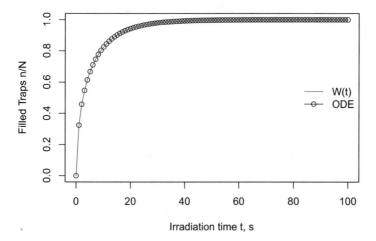

Fig. 4.2 Dose response in the OTOR model, using the Lambert analytical solution Eq. (4.13). The analytical solution (solid line) is compared with the numerical solution of the differential equation (4.10) shown as circles

Equation (4.13) is the new analytical expression for the irradiation stage of the OTOR model, which gives the trap filling ratio $n(D)/N$ as a function of the dose D, by using the Lambert W function. It shows that the function $n(D)/N$ depends only on two parameters, i.e. on the retrapping ratio R and on the constant $D_c = N/R$. The parameter D_c has the same units as the dose D and depends on the physical properties R, N of the material.

From a physical point of view, the retrapping ratio parameter R can have *any* positive real value, including values $R > 1$. The values $R = 0$ and $R = 1$ correspond to first and second order kinetics. Furthermore, under certain physical assumptions, values of R between 0 and 1 correspond to the empirical general order intermediate kinetic orders (see for example the discussion in Kitis et al. [78]).

As may be expected from a physical point of view, the approach to saturation and the shape of the $n(D)$ function depends on the amount of retrapping, i.e. on the value of the ratio R.

At the origin, the dose $D = 0$ and $W(0) = 0$, consistent with the initial condition of empty traps $n(0) = 0$.

The following code and Fig. 4.2 show a plot of the analytical equation (4.13) by using the implementation of the Lambert function in the R package *lamW*. The example also plots the numerical solution of the differential equation (4.10), by using the R package *deSolve*. As expected, the analytical equation (4.13) is identical with the numerical solution of the differential equations.

If the traps are not initially empty (i.e. $n(0) = n_0 \neq 0$), the solution of Eq. (4.13) is (Pagonis et al. [135])

Code 4.2: Irradiation: OTOR, Lambert analytical solution

```
# OTOR MODEL- IRRADIATION Lambert solution code
rm(list = ls(all=T))
options(warn=-1)
library(lamW)
library("deSolve")
## Plot  analytical solution of OTOR irradiation stage, using W
t<-0:100
N<-1E10
R<-.1
X<-1e10
k<-function(u){1+(1/(1-R))*lambertW0((R-1)*exp(-(R*X*u/N)+R-1))}
y1<-lapply(t,k)
x<-unlist(t)
y<-unlist(y1)
plot(x,y,xlab="Irradiation time t, s",ylab="Filled Traps n/N",
pch=1)
# Numerically Solve GOT equation for IRRADIATION  using  deSolve
n.0 <- 0  # initial concentration of filled traps
# Calculate numerical ODE solution
ODE <- function(t, state, parameters){
  with(as.list(c(state, parameters)),{
     dn <-    (N-n) * X*R / ((N - n) * R + n )
     list(c(dn)) })}
parameters <- c(N=N, X = X, R =R)
state <- c(n = n.0)
num_ODE <- ode(y = state, times =t,func=ODE,parms=parameters)
lines(x = num_ODE[,1],y = num_ODE[,2]/N,xlab ="Time[s]",
type="l",col = "red")
legend("right",bty="n",pch=c(NA,1),legend =c("W(t)", "ODE"),
       col = c("red","black"), lwd = 1)
```

Dose response n(D) in the OTOR model, when n(0) ≠ 0

$$\frac{n(D)}{N} = 1 + \frac{1}{1-R}W\left[(R-1)\exp\left(R-1-(D+D_{int})/D_c\right)\right] \qquad (4.15)$$

---

where the constant D_{int} has the same dimensions as the irradiation dose D and is defined by

$$D_{int} = D_c \left[n_0/N + \ln \left(1 - n_0/N \right) \right] \tag{4.16}$$

Equation (4.15) has the exact same mathematical form as Eq. (4.13), but is shifted along the horizontal dose D-axis by the amount D_{int} given by Eq. (4.16). In the next section, Eq. (4.13) is used to fit experimental data which start at the origin, and Eq. (4.15) is used to fit data which have a non-zero intercept on the y-axis.

4.5 Fitting Dose Response Data Using the Lambert W Function

The shape of the simulated dose response $n(D)/N$ from Eq. (4.13) depends strongly on the retrapping ratio R, and looks similar to a saturating exponential function (SE). The SE is often used to fit experimental dose responses in a variety of materials, and for a variety of luminescence signals, together with two more general equations, the SEL and DSE functions written as (Berger and Chen [15])

$$\frac{n(D)}{N} = 1 - \exp\left[-\frac{D}{D_0} \right] \qquad \text{SE} \tag{4.17}$$

$$\frac{n(D)}{N} = B_1 D + B_2 \left(1 - \exp\left[-\frac{D}{D_0} \right] \right) \qquad \text{SEL} \tag{4.18}$$

$$\frac{n(D)}{N} = B_3 \left(1 - \exp\left[-\frac{D}{D_{02}} \right] \right) + B_4 \left(1 - \exp\left[-\frac{D}{D_{01}} \right] \right) \qquad \text{DSE} \tag{4.19}$$

where D_0 is called the characteristic dose of the trap filling process. B_i ($i = 1 \ldots 4$) are constants, and D_{0i} are two constants characteristic of the sample with the dimensions of dose D. It is important to note that the SE, SEL, and DSE are considered more or less empirical analytical equations, and the constants B_i are not usually assigned a direct physical meaning.

In recent experimental work, the SEL (saturating plus linear) and DSE (double saturating exponential) functions have been used to fit experimental ESR data (Duval [46], Trompier et al. [181]), OSL data (Lowick [99], Timar-Gabor et al. [179, 180], Anechitei-Deacu et al. [6], Fuchs [50], Li et al. [97]), TL data (Berger and Chen [15], Berger [14], Bösken and Schmidt [22]), and ITL data (Vandenberghe et al. [183]).

Pagonis et al. [135] analyzed the relationship between the Lambert solution equation (4.13) and the saturating exponential function. When $R \cong 1$, the Lambert dose response equation (4.13) becomes identical with the SE in Eq. (4.17), while for small or large values of R (e.g. when $R = 0.001$ or $R = 3$), there is disagreement between the two expressions.

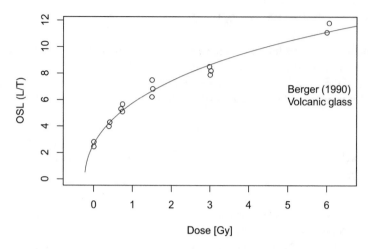

Fig. 4.3 Fit of experimental TL dose response data using the Lambert equation (4.15), with a non-zero intercept on the dose axis. For more details see Pagonis et al. [135], original data from Berger [14]

These authors chose experimental data from the literature, which could not be fitted with a SE function, and required instead the use of a SEL or a DSE fitting function. They found that Eq. (4.13) provides a satisfactory alternative to the empirical SEL and DSE regression models.

Berger and Chen [15] considered OSL signals, which were measured using the single-aliquot regenerative dose protocol (SAR) on fine grain sedimentary quartz.

The following code fits a set of TL data from Berger [14], their Fig. 1, using Eq. (4.13). This is a set of additive-dose data for purified volcanic glass, measured at the 321–330 °C temperature range of the glow curves, and preheated for 8 days to remove unstable TL. This type of additive-dose data often contains a non-zero y-intercept, therefore the fitting procedure introduces an extra fitting parameter. The data in Fig. 4.3 was fitted with the Lambert equation (4.15), with the x-intercept represents the equivalent dose D_E for this sample.

The following code analyzes ESR experimental data from Duval [46], by using the new dose response equation (4.13). These authors measured the dose response curves (DRCs) of the Al center from 15 sedimentary quartz samples from the Iberian Peninsula. The samples were irradiated in 11–14 dose steps up to a maximum dose of 23–40 kGy. It was found that the ESR signal grows almost linearly with the absorbed dose for doses above about 4 kGy. In this study it was concluded that *the ESR signal contains at least two components*, with the first component saturating at low doses and the second component showing no saturation even at these very high doses. The solid line in Fig. 4.4 is the least squares fit using Eq. (4.13); the observed

Code 4.3: Fit of experimental TL dose response data using W(x)

```
rm(list = ls(all=T))
options(warn=-1)
library("minpack.lm")
library("lamW")
## fit to Lambert equation   ----
t = c(0.00394056,0.00451523,0.39225,0.412035,0.703062,0.741318,
 0.742221,1.49553,1.49758,1.5158,2.98473,3.00304,3.02282,
 5.99852, 6.05755)
y = c(2.45103,2.80847,3.97964,4.28586,5.30465,5.10007,5.66177,
 6.21706,7.49364,6.82965,8.50226,7.88934,8.19555,11.0809,11.7953)
fit_data <-data.frame( t ,y)
plot(fit_data,ylim=c(0,max(y)),xlab="Dose [Gy]",
     ylab="OSL (L/T)",xlim=c(-.5,6.5))
fit <- minpack.lm::nlsLM(
  formula =y~N*(1+lambertW0((abs(R)-1)*exp(abs(R)-1-abs(b)*
  (t+abs(f))))/(1-abs(R))),
  data = fit_data,
  start = list(N= max(y),R=.9, b = .1,f=0.3))
N_fit <- coef(fit)[1]
R_fit <- abs(coef(fit)[2])
b_fit <- abs(coef(fit)[3])
f_fit <- abs(coef(fit)[4])
## plot analytical solution
t<- seq(from=-0.5,to=7,by=.02)
lines(
  x = t,
  y=N_fit*(1+lambertW0((R_fit-1)*exp(R_fit-1-b_fit*
                              (t+f_fit)))/(1-R_fit)),
  col = "blue")
legend("right",bty="n",legend=c("Berger (1990)",
                      "Volcanic glass"," "))
## print results
cat("\nfitted N: ", N_fit)
cat("\nfitted R: ", R_fit)
cat("\nfitted Dc: ", round(1/b_fit,2), "Gy")
cat("\nfitted f: ", f_fit)

    ##
    ## fitted N:  19.14201
    ## fitted R:  5.276899e-06
    ## fitted Dc:  21.6 Gy
    ## fitted f:  0.2279443
```

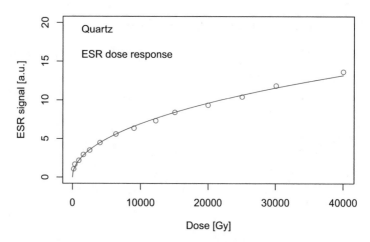

Fig. 4.4 Fit of experimental ESR dose response data using the Lambert equation (4.13). For more details see Pagonis et al. [135], original data from Duval [46]

good fit to the experimental data suggests that *this ESR signal may actually originate in a single trap instead*, which can be described by the new dose response function based on the Lambert function.

Pagonis et al. [135] also presented sets of experimental ITL data by Vandenberghe et al. [183], which were fitted using the Lambert equation (4.13).

The following code analyzes SAR-OSL experimental dose response curves by Timar-Gabor et al. [180] (their Figure 3), for fine grain and coarse grain quartz samples in a loess-palaeosol sequence. The data shown here are for samples MV 10 and MV 8, and were fitted by these authors using a DSE. The solid lines in Fig. 4.5 are the least squares fits using the new equation (4.13), showing that the Lambert function can be used to describe both types of dose responses in quartz with a smaller number of parameters than the DSE, even at the high doses involved in this experiment.

As an alternative to the Lambert function, we also recommend fitting a dose response curve using the R package *numOSL* by Peng et al. [150], using the Levenberg–Marquardt algorithm. When using function *fitGrowth*, five models are available: a linear model, a single saturation exponential model, a single saturation exponential plus linear model (SEL), a double saturation exponential model (DSE), and a general order kinetic model.

Code 4.4: Fit of experimental ESR dose response data using Lambert equation

```
rm(list=ls())
options(warn=-1)
library("minpack.lm")
library("lamW")
# Load the data
TLqzx<-c(174.13, 345.027, 931.847, 1603.74, 2524.39, 4031.12,
6372.18,9044.09,12217.5,15058.3,19981.5,25072.3,30082.3,40011.4)
TLqzy<-c(1.07478, 1.68389, 2.18591, 2.93875, 3.51271, 4.48122,
5.59377,6.3484,7.3184,8.39557,9.33133,10.4105,11.8478,13.6478)
plot(TLqzx,TLqzy,type="p",pch=1,col="red",
xlab=expression("Dose [Gy]"),ylab ="ESR signal [a.u.]",
ylim=c(0,20))
legend("topleft",bty = "n",
legend = c("Quartz"," ","ESR dose response"))
## fit to Lambert equation  ----
t <-TLqzx
y<-TLqzy
fit_data <-data.frame( t ,y)
#plot(fit_data,ylim=c(0,max(y)))
fit <- minpack.lm::nlsLM(
formula=y~N*(1+lambertW0((abs(R)-1)*exp(abs(R)-1-b*t))/
(1-abs(R))),
  data = fit_data,
   start = list(N= max(y),R=.9, b = .01)
)
N_fit <- coef(fit)[1]
R_fit <- abs(coef(fit)[2])
b_fit <- coef(fit)[3]
## plot analytical solution
t<- seq(from=0,to=40000,by=100)
lines( x = t,
y=N_fit*(1+lambertW0((R_fit-1)*exp(R_fit-1-b_fit*t))/(1-R_fit)),
  col = "blue")
## print results
cat("\nfitted N: ", N_fit)
cat("\nfitted R: ", R_fit)
cat("\nfitted Dc: ", 1/b_fit, " Gy")

   ##
   ## fitted N:   51.24355
   ## fitted R:   1.432573e-08
   ## fitted Dc:  995428.2  Gy
```

Code 4.5: Fit of experimental OSL dose response data using W(x)

```
rm(list = ls(all=T))
options(warn=-1)
library("minpack.lm")
library("lamW")
par(mfrow=c(1,2))
## fit to saturation exponential ----
t = c(-34.2466, 34.2466, 68.4932, 273.973, 1027.4,1986.3,3013.7,
      5000, 7979.45, 10000)
y = c(1.04664, 0.000978474, 2.76386, 7.24592,12.6008,14.5329,
      15.8956,   17.1905, 17.847, 18.0952)
fit_data <-data.frame( t ,y)
plot(fit_data,ylim=c(0,max(y)),xlab="Dose [Gy]",
ylab="OSL (L/T)")
fit <- minpack.lm::nlsLM(
formula=y~N*(1+lambertW0((abs(R)-1)*
exp(abs(R)-1-b*t))/(1-abs(R))),
  data = fit_data,
  start = list(N= max(y),R=.9, b = .01))
N_fit <- coef(fit)[1]
R_fit <- abs(coef(fit)[2])
b_fit <- coef(fit)[3]
## plot analytical solution
t<- seq(from=0,to=10000,by=100)
lines( x = t,
y=N_fit*(1+lambertW0((R_fit-1)*exp(R_fit-1-b_fit* t))/
(1-R_fit)),  col = "blue")
legend("right",bty="n",legend=c("(a)"," ","Fine grain",
"Quartz"," "))
## print results
cat("\nFine grain"," ")
cat("\nfitted N: ", N_fit)
cat("\nfitted R: ", R_fit)
cat("\nfitted Dc: ",round( 1/b_fit,2), "Gy")
## fit to Lambert equation ----
t = c(0, 3.5583, 44.1822, 258.718, 1051.62, 2044.98, 3003.94,
      5024.61, 7046.32, 9992.29)
y = c(0, 0.93512, 2.61108, 4.99104, 6.36704, 6.42148,
      6.43643, 6.46792, 6.77215, 6.97391)
fit_data <-data.frame( t ,y)
plot(fit_data,ylim=c(0,max(y)),xlab="Dose[Gy]",ylab="OSL (L/T)")
fit <- minpack.lm::nlsLM(
formula=y~N*(1+lambertW0((abs(R)-1)*
exp(abs(R)-1-b*t))/(1-abs(R))),
  data = fit_data,
  start = list(N= max(y),R=10, b = 1e-4))
N_fit <- coef(fit)[1]
R_fit <- abs(coef(fit)[2])
b_fit <- coef(fit)[3]
```

```
## plot analytical solution
t<- seq(from=0,to=10000,by=100)
lines( x = t,
y=N_fit*(1+lambertW0((R_fit-1)*exp(R_fit-1-b_fit*t))/(1-R_fit)),
   col = "blue")
legend("right",bty="n",legend=c("(b)"," ","Coarse grain",
"Quartz"," "))
## print results
cat("\nCoarse grain"," ")
cat("\nfitted N: ", N_fit)
cat("\nfitted R: ", R_fit)
cat("\nfitted Dc: ",round( 1/b_fit,2)," Gy")
```

```
##
## Fine grain
## fitted N:   17.57116
## fitted R:   0.3971654
## fitted Dc:  1325.83 Gy
## Coarse grain
## fitted N:   6.610193
## fitted R:   1.938374e-06
## fitted Dc:  403.02  Gy
```

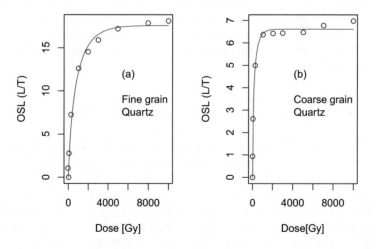

Fig. 4.5 Fit of experimental SAR-OSL experimental dose response data, for (**a**) fine grain and (**b**) coarse grain quartz samples, using the Lambert equation. For more details see Pagonis et al. [135], original data by Timar-Gabor et al. [180] (their Figure 3)

4.6 The Effect of Sample Temperature During the Irradiation Process

In this section we discuss the dose response of a thermally unstable dosimetric trap, when a material is irradiated in the laboratory or in nature. We consider again the OTOR model during the irradiation stage with a constant irradiation rate X, but now we add the possibility that electrons are thermally detrapped during the irradiation process. The equations in the model are (Chen et al. [35])

$$\frac{dn}{dt} = A_n(N - n)n_c - n\, p(t)\,, \tag{4.20}$$

$$\frac{dn_c}{dt} = X + n\, p(t) - A_n(N - n)n_c - A_m m n_c, \tag{4.21}$$

$$\frac{dm}{dt} = B\,(M - m)\,n_v - A_m m n_c, \tag{4.22}$$

$$\frac{dm}{dt} + \frac{dn_v}{dt} = \frac{dn}{dt} + \frac{dn_c}{dt}. \tag{4.23}$$

$$m + n_v = n + n_c \tag{4.24}$$

The term $p(t)$ represents the rate of thermal or optical excitation of trapped electrons, which in the case of isothermal TL processes is given by $p(t) = s\, \exp[-E/(kT_{\mathrm{ISO}})]$ where T_{ISO} is the temperature during the irradiation process.

Proceeding as before under the QE conditions, we set $dn_c/dt = 0$ and the following is obtained:

$$\frac{dn}{dt} = \frac{(N - n)A_n}{(N - n)A_n + n A_m} X - p(t)\frac{n^2 A_m}{(N - n)A_n + n A_m} \tag{4.25}$$

The following R code gives an example of numerically solving this differential equation for two different values of the temperature T_{ISO} during the irradiation process, by varying the thermal fading factor $p(t) = s\, \exp[-E/(kT_{\mathrm{ISO}})]$. At a low irradiation temperature $T_{\mathrm{ISO}} = 20\,°C$, all traps get filled at long times $t \sim 50\,s$. At a higher irradiation temperature $T_{\mathrm{ISO}} = 100\,°C$, only $\sim 80\%$ of the traps get filled at long times, and an equilibrium has been reached, between the charge loss due to thermal detrapping and the charge creation due to irradiation (Fig. 4.6).

Code 4.6: Irradiation of thermally unstable trap (OTOR)

```
# OTOR MODEL- IRRADIATION of thermally unstable trap (OTOR)
rm(list = ls(all=T))
options(warn=-1)
library(lamW)
library("deSolve")
## Plot the numerical solution of OTOR irradiation stage,
## at different irradiation temperatures
t<-0:100
N<-1E10
R<-.1
X<-1e10
E<-1
s<-1e12
kb<-8.617e-5
TISO<-20
p<-s*exp(-E/(kb*(273+TISO)))
# Numerically Solve GOT equation for IRRADIATION using deSolve
n.0 <- 0  # initial concentration of filled traps
# Calculate numerical ODE solution
ODE <- function(t, state, parameters){
  with(as.list(c(state, parameters)),{
     dn <-  ( (N-n) * X*R-(p*(n**2))) / ((N - n) * R + n )
       list(c(dn))  })}
parameters <- c(N=N, X = X, R =R, p=p)
state <- c(n = n.0)
num_ODE <- ode(y = state,times = t,func = ODE,parms=parameters)
plot(x = num_ODE[,1], y = num_ODE[,2]/N,pch=2,ylim=c(0,1.1),
xlab = "Time[s]", ylab = "Trap filling ratio, n/N",type="b",
      col = "red")
# Change the irradiation temperature TISO
TISO<-100
p<-s*exp(-E/(kb*(273+TISO)))
parameters <- c(N=N, X = X, R =R, p=p)
num_ODE2 <- ode(y = state, times = t, func=ODE,parms=parameters)
lines(x = num_ODE2[,1],    y = num_ODE2[,2]/N, pch=3,
col = "Black",type="b")
legend("right",bty="n",pch=c(2,3),expression("T=" ~ 20^o ~C ~"",
"T=" ~ 100^o ~C ~""),col=c("red","black"),lwd=2)
```

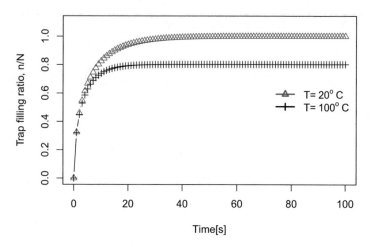

Fig. 4.6 Simulation of the effect of sample temperature during the irradiation process, within the OTOR model. For a more detailed study of this effect, see Chen et al. [35]

4.7   Competition During the Irradiation Stage: The Model of Bowman and Chen

In this section we consider the model of Bowman and Chen [23]. This model describes superlinear dose response as being a result of competition between two electron traps during the irradiation stage of a sample. Figure 4.7 shows a schematic diagram of the two trap one recombination center model (2T1C), and the relevant electronic transitions during the irradiation of a sample. The differential equations governing the traffic of electrons between the two trapping levels, the recombination center and the conduction band (CB) and valence band (VB) in the 2T1C model are (Chen and Pagonis [38])

$$\frac{dn_1}{dt} = A_1(N_1 - n_1)n_c, \tag{4.26}$$

$$\frac{dn_2}{dt} = A_2(N_2 - n_2)n_c, \tag{4.27}$$

$$\frac{dn_c}{dt} = X - A_1 n_c(N_1 - n_1) - A_2(N_2 - n_2)n_c - A_m m n_c, \tag{4.28}$$

$$\frac{dm}{dt} = -A_m m n_c + A_v (M - m) n_v, \tag{4.29}$$

$$m + n_v = n_1 + n_2 + n_c \tag{4.30}$$

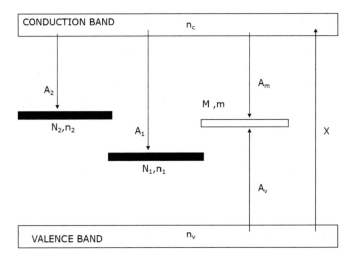

Fig. 4.7 The 2T1C model of Bowman and Chen [23], describing competition effects during the irradiation stage. Various electronic transitions are shown during the irradiation process. Transition X: Creation of electron–hole pairs by radiation. Transitions A_v and A_m: Trapping of holes and recombination of electrons at the recombination centers (RC). Transitions A_1, A_2: Trapping of electrons in the dosimetric and competitor trap. For more details, see Pagonis et al. [136]

Here n_1 (cm^{-3}), n_2 (cm^{-3}), and m (cm^{-3}) are the concentrations of electrons in the two traps and of holes in recombination centers respectively, and N_1, N_2, M (cm^{-3}) are the total concentrations of the two trapping states and recombination center. n_c (cm^{-3}) and n_v (cm^{-3}) are respectively the concentrations of free electrons and holes. A_1 $(\text{cm}^3\text{s}^{-1})$, A_2 $(\text{cm}^3\text{s}^{-1})$ are the trapping coefficient of electrons, A_m $(\text{cm}^3\text{s}^{-1})$ the recombination coefficient of electrons, and A_v $(\text{cm}^3\text{s}^{-1})$ is the trapping coefficient of holes in centers. X $(\text{cm}^{-3}\text{s}^{-1})$ is proportional to the dose rate of excitation, and actually denotes the rate of production of electron–hole pairs by the excitation irradiation per unit volume per second. If the irradiation time is t, the dose received by the sample is $D = Xt$ (cm^{-3}). Equations (4.26) and (4.27) express the trapping of electrons from the conduction band into the two traps. Equations (4.28) and (4.29) describe the rate of change of the concentration of electrons in the CB and of holes in the recombination center, respectively. Equation (4.30) is the charge neutrality condition, assuming initially empty traps and centers.

In this 2T1C model, the superlinear filling of the active trap N_1 can be described qualitatively as follows. At low doses the excitation fills both traps linearly, while at a critical dose the competing trap N_2 saturates, hence more electrons are made available for the trap of interest N_1. This causes a faster filling of this trap, with the transition region in the dose response curve above this critical dose appearing to be superlinear, before the dosimetric trap approaches saturation also.

The above set of differential equations cannot be solved analytically and one uses the quasi-equilibrium (QE) assumption, to transform them into a single differential equation. Specifically, Bowman and Chen [23] used the QE assumptions, and found the solutions $n_1(t)$ and $n_2(t)$ in terms of *implicit parametric* functions for $D(n_1)$ and $D(n_2)$, where D is the irradiation dose.

Recently Pagonis et al. [136] inverted these parametric equations, and brought them in the useful form $n_1(D)$ and $n_2(D)$, by carrying out a series expansion in terms of the parameter A_2/A_1. These authors obtained the following Pagonis–Kitis–Chen (PKC) equation describing the nonlinear dose response of a dosimetric trap:

PKC equation for nonlinear dose response

$$\frac{n}{N} = 1 - \left(\frac{1}{B} W \left[B \exp(B) \exp(-D/D_c) \right] \right)^{A_2/A_1} \qquad (4.31)$$

---

where the two constants B, D_c are

$$B = \frac{N_1 (A_1 - A_m)}{A_2 N_2 + A_m N_1}, \qquad (4.32)$$

$$D_c = \frac{A_2 N_2 + A_m N_1}{A_1} \qquad (4.33)$$

The dose response $n(D)/N$ in this rather simple equation (4.31) depends on only three parameters, the constants A_2/A_1, B, and D_c, which are given by Eqs. (4.32) and (4.33) in terms of the 5 parameters A_m, A_1, A_2, N_1, N_2 in the original model. The parameter D_c has the same dimensions at the irradiation dose D, so that the ratio D/D_c in Eq. (4.31) is dimensionless. As seen from the definition in Eq. (4.32), the parameter B is also dimensionless. The assumptions in deriving these analytical equations are the QE conditions, and the additional condition $A_2/A_1 < 1$, which allowed the series expansion for small values of A_2/A_1.

The overall dose response in this model will depend on the numerical values of the three parameters appearing in these equations: B, D_c, A_2/A_1. As the competitor trap approaches saturation, the dose response of the dosimetric trap n/N becomes superlinear. The initial short linear range in the curve n/N is followed by a range of superlinearity, which eventually becomes sublinear on its way to saturation.

The simulations of Pagonis et al. [136] showed that the behavior of the dose response $n(D)/N$ depends on the value of the parameters B and A_2/A_1. When *both* of the two conditions $B > 1$ and $A_2/A_1 < 1$ are satisfied, the dose response $n_2(D)$ of the dosimetric trap is superlinear. When only one of these conditions is satisfied, the dose response is similar to the behavior of the traps in the OTOR model, i.e. a linear-to-saturation behavior (Pagonis et al. [135]).

4.8 Fitting of Superlinear Experimental Data Using the Lambert Equation

Figure 1.18 shows ESR, TL and optical absorption data for fused silica, by Wieser et al. [189]. The E_1' center is observed in all forms of quartz and silica; however, these authors found an unexpected superlinear dose dependence of the E_1' center in fused silica. Previously a superlinear dose response in quartz had been observed with TL, but not with optical or ESR spectroscopy. They reported a superlinear increase of the E_1' center concentration by irradiation, and compared their results with the growth of the correlated optical E-band absorption, and with TL measurements which showed a correlation of the 415 °C glow to the E_1' center. The solid lines in Fig. 4.8 represent least squares fits using the analytical PKC equation (4.31). The Lambert-based equations in this chapter describe well all three types of dose responses for this material: ESR, TL, and OA.

As another example of fitted experimental data, Fig. 4.9 shows curve 2 from Nikiforov et al. [116], their Fig. 7. These authors measured the TL dose response characteristics of anion deficient aluminum oxide single crystals. Crystal 2 is a sample with low initial sensitivity, and the heating rate for these TL measurements was 6 K/s. Once more, the new analytical equation (4.31) describes well the variation of the superlinear dose response for the different samples of anion deficient aluminum oxide crystals.

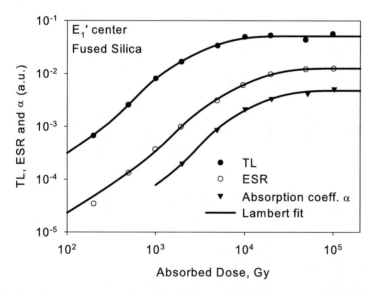

Fig. 4.8 Superlinear dose dependence of the E_1' center concentration (ESR), TL and OA signals, from a single sample of fused silica. Note the log scale in both axes. The solid lines are fitted using the PKC equation (4.31). For more details see Pagonis et al. [136], original data from Wieser [189]

For several additional examples of fitting TL,OSL, ESR data using the Lambert equations, see the papers by Pagonis et al. ([135, 136]).

Code 4.7: TL dose response of anion deficient aluminum oxide

```
rm(list = ls(all=T))
options(warn=-1)
library("minpack.lm")
library("lamW")
## fit to Pagonis-Kitis-Chen PKC superlinearity equation ----
t=c(0.0537568,0.103385,0.156481,0.211929,0.260776,
0.321015,0.36483,0.414625,0.478926,0.535569,0.589476,
0.638899,0.824344,0.951985,1.18991,1.63688,3.19332)
y=c(6694.89,15592.6,24767.6,39360.5,52176.2,78075.6,
101463,131855,189548,227223,272406,376197,469268,
634929,792121,1.16105e6,1.6992e6)
fit_data <-data.frame( t ,y)
fit <- minpack.lm::nlsLM(
formula=y~N*(1-(lambertW0(abs(B)*exp(abs(B)-lamda*t))/
abs(B))**beta),
  data = fit_data,
  start = list(N= max(y),B=5, lamda = 10,beta=0.1))
N_fit <- coef(fit)[1]
B_fit <- coef(fit)[2]
lamda_fit <- coef(fit)[3]
beta_fit <- coef(fit)[4]
## print results
cat("\nfitted N: ", N_fit)
cat("\nfitted B: ", B_fit)
cat("\nfitted Dc: ", round(1/lamda_fit,2)," Gy")
cat("\nfitted beta: ", beta_fit)
## plot analytical solution
t<- seq(from=0,to=4,by=.01)
plot(fit_data,log="xy",xlab="Dose [Gy]",ylab="TL")
legend("topleft",bty="n",legend=c("Nikiforov et~al. (2014)",
"Anion-deficient","Aluminum Oxide"))
lines( x = t,
y=N_fit*(1-(lambertW0(abs(B_fit)*exp(abs(B_fit)-lamda_fit*
t))/abs(B_fit))**beta_fit),
      col = "blue",log="xy")

  ##
  ## fitted N:  2095289
  ## fitted B:  6.870824
  ## fitted Dc:  0.05  Gy
  ## fitted beta:  0.02999771
```

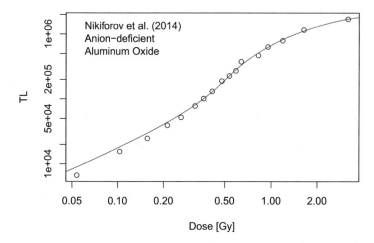

Fig. 4.9 TL dose response of an anion deficient aluminum oxide single crystal, which had low initial sensitivity to irradiation. The superlinear dose response is fitted with the PKC equation. For more details see Nikiforov et al. [116], see also the discussion in Pagonis et al. [136]

4.9 Analytical Equations for the Supralinearity Index f(D)

In this section we provide a general discussion of the nonlinear dose response of dosimetric materials, and present some analytical equations for the supralinearity index $f(D)$.

If we consider only the very simple OTOR model, a linear-saturation behavior is always expected. This case has been studied in detail in Pagonis et al. [135], who found the analytical equation for the trap filling ratio n/N:

$$\frac{n(D)}{N} = 1 - \frac{1}{R-1} \, W\left[(R-1) \exp\left(R - 1 - D/D_c \right) \right] \tag{4.34}$$

In this OTOR model, the function $n(D)/N$ depends only on two parameters, i.e. on the retrapping ratio R and on the dose scaling constant D_c. The parameter D_c has the same units as the dose D, and depends on the physical properties of the material. By using the definition of the supralinearity index $f(D)$ in Eq. (4.3), we obtain

$$f_{\text{OTOR}}(D) = \frac{1}{k_{\text{OTOR}} D} \left\{ 1 - \frac{1}{R-1} W\left[(R-1) \exp\left(R - 1 - D/D_c \right) \right] \right\} \tag{4.35}$$

where k_{OTOR} is a constant with dimensions of $1/D$ (usually Gy^{-1}), and represents the slope in the initial linear part of the dose response curve.

Equation (4.35) is the analytical equation for the supralinearity index $f_{\text{OTOR}}(D)$ in the OTOR model. In this model the index depends on only *two* parameters, the retrapping ratio R and the normalized characteristic dose D_c.

In real materials several trapping states and recombination centers exist, which take part in the irradiation and readout stages. The effects of competition can explain a wealth of different dose responses associated with TL, OSL, ITL, OA, and ESR signals. As far as superlinear dose dependence is concerned, it has been shown that competition during excitation can yield one kind of behavior, while competition during readout yields another kind of dependence. Simulations have also shown the combined effect of both kinds of competition (see for example Chen and Pagonis [38]).

In the 2T1C model, the superlinear dose response can be described by the analytical PKC equation (Pagonis et al. [136]):

$$\frac{n(D)}{N} = 1 - \left(\frac{1}{B}W\left[B\exp(B)\exp(-D/D_c)\right]\right)^{A_2/A_1} \tag{4.36}$$

In this 2T1C model, the function $n(D)/N$ depends on *three* parameters, i.e. on the dimensionless constant B, on the ratio of trapping coefficients A_2/A_1 of the two competing traps, and on the dose scaling constant D_c. The parameter D_c has the same units as the dose D, and the other two parameters B, A_2/A_1 are dimensionless.

From the definition of the supralinearity index $f(D)$ in Eq. (4.3), we obtain

$$f(D) = \frac{1}{kD}\left\{1 - \left(\frac{1}{B}W\left[B\exp(B)\exp(-D/D_c)\right]\right)^{A_2/A_1}\right\} \tag{4.37}$$

Here k is the slope in the initial linear part of the dose response. Equation (4.37) is the desired analytical expression for the dimensionless supralinearity index $f(D)$, in terms of the three parameters B, D_c, A_2/A_1 in the model.

The advantages of using these analytical equations to fit the $n(D)/N$ and $f(D)$ experimental data are

- The Lambert-based equations are simple,
- They contain physically meaningful parameters,
- They can be very easily adopted for various software, and
- They can be used to describe the superlinear behavior of ESR, TL, OSL, and OA signals.

In the following code and Fig. 4.10, we show how to fit experimental data for the supralinearity index $f(D)$, by using the analytical equation (4.37). The experimental data are taken from Edmund [47], who studied extensively the superlinear behavior of two Al_2O_3:C probes (probe A and B).

In the literature one also finds two alternative modeling approaches to analytical superlinearity equations. These are the track structure theory (TST) and the unified interaction model (UNIM). Other analytical expressions for $n(D)$ from models based on the irradiation stage were previously proposed by Waligórski et al. [186, 187], based on the track structure theory of Katz [65]. These previously used dose response curves have the form of a saturating exponential. Levy [96]

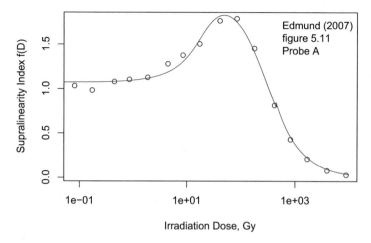

Fig. 4.10 Fit to the superlinear behavior of an $Al_2O_3:C$ probe using the PKC superlinearity equations. For more details of the analysis see Pagonis et al. [136], original experimental data from Edmund [47]

presented a phenomenological explanation of the origin of the SEL and DSE dose response functions, and how they can be used to describe the trap filling process. The proposed physical mechanism by Levy [96] is the creation of additional traps during the irradiation of the samples.

TST was introduced by Katz and collaborators in the 1960s [65, 186, 187], and is a model based on statistical considerations involving the cumulative Poisson distributions for a single and multiple hit interactions occurring during irradiation of a sample. The UNIM model is based on a combination of localized and delocalized recombination mechanisms, which are prevalent at different dose regimes (Horowitz et al. [57, 58]).

The existing models of nonlinear dose response $n(D)$ in dosimetric materials are usually classified into two major categories. Models in the first category describe the dose response $n(D)$, on the basis of mechanisms taking place during the irradiation stage. Such models provide an explanation for, among others, nonlinearity effects of ESR and OA measurements. Models in the second category describe nonlinearity effects as a result of competition mechanisms occurring during the readout stage, i.e. during the optical or thermal stimulation of the previously irradiated sample.

The analytical equation (4.13) belongs to the first group of models, and cannot provide an explanation for superlinearity effects, since there are no competing traps included in the OTOR model. The PKC equation (4.31) also belongs to this first group of models; however, the PKC equation can also describe the superlinearity dose response of materials.

A *third* analytical equation which includes superlinearity effects during both the irradiation and readout stages was developed by Pagonis et al. [136], based on the mixed order kinetics (MOK) model. This new analytical equation for the dose response of the luminescence intensity $I(D)$ is

$$I(D) = k \left\{ 1 - \left(\frac{1}{B} W[z] \right)^{A_2/A_1} \right\} \left(1 - \left(\frac{1}{B} W[z] \right)^{A_2/A_1} + \frac{N_1}{N_2} \right), \qquad (4.38)$$

$$z = B \exp(B) \exp(-D/D_c) \qquad (4.39)$$

where k is a constant, and the additional parameter in this analytical equation is the ratio of the total concentrations of traps N_1/N_2. The rest of the parameters have the same meaning as above.

Code 4.8: Fit to Supralinearity index f(D) using Lambert W

```
rm(list = ls(all=T))
options(warn=-1)
library("minpack.lm")
library("lamW")
## fit g(D) to Lambert equation ----
t = c(0.0811131, 0.171804,0.450923,0.857988,1.88341,
4.44069,8.44947,17.2683,40.7152,83.2104,176.246,
415.552,819.456,1674.74,3948.68,8983.32)
y = c(1.03326,0.983911,1.08074,1.10465,1.12844,1.28023,
1.37731,1.50481,1.76636,1.79021,1.45428, 0.813382,
0.428722,0.208669,0.0799713,0.0305689)
fit_data <-data.frame( t ,y)
fit <- minpack.lm::nlsLM(
   formula = y ~ N * (1-(lambertW0(abs(B)*exp(abs(B)- t/Dc))/
   abs(B))**beta)/t,
   data = fit_data,
   start = list(N= max(y),B=1.2, Dc = 5,beta=0.1)
)
N_fit <- coef(fit)[1]
B_fit <- abs(coef(fit)[2])
Dc_fit <- coef(fit)[3]
beta_fit <- coef(fit)[4]
## print results
cat("\nfitted N: ", N_fit)
cat("\nfitted B: ", B_fit)
cat("\nfitted Dc: ", round(Dc_fit,2)," Gy")
cat("\nfitted beta: ", beta_fit)
## plot analytical solution
t<- seq(from=0.01,to=10000,by=.2)
plot(fit_data,log="x",ylab="Supralinearity Index f(D)",
```

```
xlab="Irradiation Dose, Gy")
lines( x = t,
y=N_fit*(1-(lambertW0(abs(B_fit)*exp(abs(B_fit)-t/Dc_fit))/
abs(B_fit))**beta_fit)/t,        col = "blue",log="x")
legend("topright",bty="n",legend=c("Edmund (2007)",
"figure 5.11","Probe A"))

    ##
    ## fitted N:   375.0032
    ## fitted B:   1.237203
    ## fitted Dc:  5.07  Gy
    ## fitted beta:  0.03240109
```

4.10   On the Importance of the W Function in Describing Luminescence Phenomena

During the past decade, the Lambert function has been used to describe analytical equations for a variety of luminescence phenomena. In this section we summarize these applications in phenomenological luminescence models (TL, OSL, etc.), as they were reviewed by Kitis et al. [78]. For a general discussion of the properties of the W function and its many uses in science, the reader is referred to the original papers by Corless et al. [44, 45].

Description of the Readout Stage in the OTOR Model
The Lambert W function was first used by Kitis and Vlachos [82], to solve the OTOR model during the readout stage, in order to obtain an analytical expression for TL, OSL, etc. Even though the OTOR model has been studied for almost 50 years, this was the first time that an analytical solution was obtained. Later Singh and Gartia [169] also solved the OTOR model using the equivalent Wright ω function. In all modern software these functions are built-in functions like the common transcendental functions of sine, cosine, logarithm, etc.

Description of the Dose Response Stage in the OTOR Model
In this chapter we saw the new analytical Lambert equations developed by Pagonis et al. [135], which described the dose response $n(D)$ in the OTOR model. The analytical equations are characterized by the retrapping ratio R and the dose parameter D_c in the model. For values of $R \cong 1$ the new analytical equation can be approximated very well by a SE function, while for $R \leq 0.1$ or $R > 1$ the two functions diverge significantly.

The results of this chapter showed that the new dose response function in Eq. (4.13) is a more general function than the SE, SEL, and DSE equations. In addition, the proposed fitting function contains a smaller number of parameters than the SEL and DSE, and is based on physically meaningful parameters. Importantly, the new equation provides a simpler interpretation of the shape of the dose response curve, than the SEL and DSE. This is because by using the Lambert solution, the data is interpreted as the dose response of a *single trap*, instead of the two traps implied in the DSE and SEL functions.

Description of Superlinear Dose Response in the Model by Chen/Bowman
In this chapter we saw that Pagonis et al. [136] also obtained the PKC analytical solution of the dose response $n(D)$, within the model by Bowman and Chen [23].

The analytical PKC equation (4.31) is characterized by the dimensionless constant B, by the ratio of trapping coefficients A_2/A_1 of the two competing traps, and by the dose scaling constant D_c. The parameter D_c has the same units as the dose D, and the other two parameters B, A_2/A_1 are dimensionless.

Description of Localized Transitions in the LT Model
As we will see in Chap. 7, Kitis and Pagonis [75] obtained also the analytical solution of the LT model, in terms of the Lambert W function.

Chapter 5
Time-Resolved OSL Experiments

5.1 Introduction

The techniques of time-resolved optically and infrared stimulated luminescence (TR-OSL, TR-IRSL) are an important experimental tool for studying excitation and relaxation phenomena in a variety of materials. During the past decade, extensive TR measurements have been carried out using samples of both quartz and feldspars, due to the importance of these materials in dating and retrospective dosimetry applications. Time-resolved luminescence data provide a method of distinguishing between the different recombination routes in a variety of materials. The main advantage of TR-OSL over continuous wave optically stimulated luminescence (CW-OSL) measurements is that it allows study of recombination and/or relaxation pathways in the material, and therefore provides important information on the underlying luminescence mechanisms.

During TR-OSL measurements, the stimulation is carried out with a brief light pulse, and photons are recorded based on their arrival at the luminescence detector with respect to the light pulse. Summing signals from several pulses gives rise to a typical TR-OSL curve, which shows the buildup of luminescence during the pulse and the subsequent decrease when the optical stimulation is turned off.

The mathematical description of TR-OSL and TR-IRSL signals is an important subject, which can illuminate the underlying luminescence mechanisms. For example in quartz, the decaying luminescence signal immediately following a light pulse is commonly analyzed using the linear sum of exponential decays and can therefore be characterized using decay constants and characteristic luminescence lifetimes. By contrast, in feldspars, both TR-OSL and TR-IRSL have been the subject of several recent experimental studies, which have established that the signals *cannot* be described simply by the sum of exponentials.

Several researchers have studied the temperature dependence of luminescence lifetimes and luminescence intensity from TR experiments in quartz (see, e.g. the review paper by Chithambo et al. [41], and references therein). For example, luminescence lifetimes for unannealed sedimentary quartz are typically found to

V. Pagonis, *Luminescence, Use R!*, https://doi.org/10.1007/978-3-030-67311-6_5

remain constant at $\sim$42 µs for stimulation temperatures between 20 °C and 100 °C, and then to decrease continuously to $\sim$8 µs at 200 °C. These phenomena are usually described within the framework of *thermal quenching* of luminescence in quartz, which has been well known for several decades and is commonly described using the Mott–Seitz mechanism presented in Sect. 5.2.

In Sects. 5.2–5.3 we will study TR-OSL signals from materials in which luminescence is due to *delocalized* transitions involving the conduction and valence bands. In Sect. 5.4 we will study TR-IRSL signals from materials in which luminescence is due to *localized* transitions. This chapter will conclude in Sect. 5.5, with the presentation of a TR-photoluminescence (TR-PL) model for the important dosimetric material Al_2O_3:C.

5.2   Modeling of TR Signals and Thermal Quenching: The Mott–Seitz Mechanism

Thermal quenching has been observed in both TL and OSL experiments on quartz and is commonly described using the Mott–Seitz mechanism. This section presents the mathematical description of TR signals and thermal quenching in quartz.

Pagonis et al. [121] presented a numerical model for thermal quenching in quartz based on the Mott–Seitz mechanism. The model involves electronic transitions between energy states *within* the recombination center and was used to derive analytical expressions for the TR-OSL process in quartz. These authors presented analytical expressions for the luminescence intensity during and after the short pulses in a TR-OSL experiment and examined the relevance of the model for dosimetric applications. The expressions derived in Pagonis et al. [121] were shown to be equivalent to previously published analytical expressions, which were derived using a different physical approach by Chithambo[39, 40].

The Mott–Seitz mechanism is usually shown schematically using a configurational diagram as in Fig. 5.1 and consists of an excited state of the recombination center and the corresponding ground state. In this mechanism, electrons are captured into an excited state of the recombination center, from which they can undergo either one of the two competing transitions. The first transition is a direct radiative recombination route resulting in the emission of light and is shown as a vertical arrow in Fig. 5.1. The second route is an indirect thermally assisted non-radiative transition into the ground state of the recombination center; the activation energy W for this non-radiative process is also shown in Fig. 5.1.

The energy given up in the non-radiative recombination is absorbed by the crystal as heat, rather than being emitted as photons. One of the main assumptions of the Mott–Seitz mechanism is that the radiative and non-radiative processes compete within the confines of the recombination center; hence, they are referred to as localized transitions.

Figure 5.2 shows the energy level diagram corresponding to the model. The arrows in Fig. 5.2 indicate the electronic transitions that are likely to be taking place

Fig. 5.1 The configurational diagram for quartz, based on the Mott–Seitz mechanism of thermal quenching. From Pagonis et al. [121]

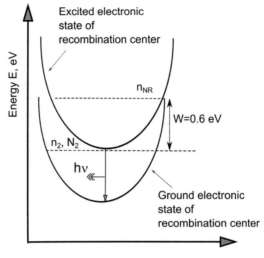

during a typical TR-OSL experiment. This simplified model consists of an optically sensitive electron trap referred to in this model as the dosimetric trap and shown as level 1, and several additional levels labeled 2–4 represent energy states within the recombination center. During the transition labeled 1, electrons from the dosimetric trap are raised by optical stimulation into the conduction band (CB), with some of these electrons being retrapped with a rate coefficient A_n as shown in transition 2. Transition 3 corresponds to an electronic transition from the CB into the excited state located below the conduction band, with rate coefficient A_{CB}.

Transition 5 indicates the direct radiative transition from the excited level into the ground electronic state with rate coefficient A_R, and transition 4 indicates the competing thermally assisted route. The probability for this competing thermally assisted process (transition 4) is given by a Boltzmann factor of the form $s \exp(-W/kT)$, where W represents the activation energy for this process and A_{NR} is a constant representing the non-radiative transition rate coefficient. Transition 6 denotes the non-radiative process into the ground state. The details of the non-radiative process in the model are not important in the model; what is critical in determining the thermal quenching effects is the ratio of the non-radiative and radiative rate coefficients A_R/A_{NR}, and the value of the thermal activation energy W.

Thermal quenching within this model is caused by the competing transitions 4 and 5 in Fig. 5.2. As the temperature of the sample is increased, electrons are removed from the excited state level 2 to the excited state level 3 according to the Boltzmann factor described above. This transition (level 2 → level 3) leads to both a decrease of the intensity of the luminescence signal, and to a simultaneous decrease of the luminescence lifetime with increasing stimulating temperature.

The parameters used in the model are defined as follows: N_1 is the total concentration of dosimetric traps (cm^{-3}), n_1 is the corresponding concentration of

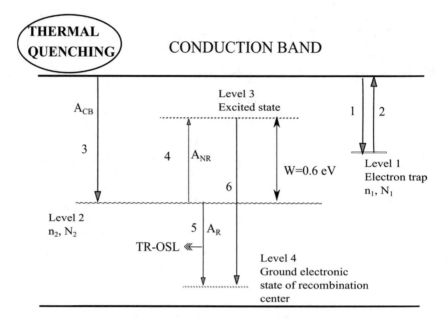

Fig. 5.2 The kinetic model of Pagonis et al. [121] for thermal quenching in quartz, based on the Mott–Seitz mechanism

trapped electrons (cm^{-3}), N_2 is the total concentration of luminescence centers, and $N_2 - n_2$ is the corresponding concentration of activated luminescence centers (cm^{-3}), respectively. A detailed discussion of the connection between the electronic concentrations n_2, $N_2 - n_2$, and N_2 and the concentrations of specific defects in quartz is presented in Pagonis et al. [121]. W is the activation energy for the thermally assisted process (eV), A_n is the conduction band to dosimetric electron trap transition coefficient (cm^3 s^{-1}); for quartz a typical value is $W = 0.64$ eV, while for Al$_2$O$_3$:C a typical value is $W = 1$ eV. A_R and A_{NR} are the radiative and non-radiative transition coefficients (s^{-1}), and A_{CB} is the transition coefficient (cm^3 s^{-1}) for the conduction band to excited state transition. The parameter n_c represents the instantaneous concentration of electrons in the conduction band (cm^{-3}), and P denotes the rate of optical excitation of electrons from the dosimetric trap (s^{-1}).

The equations used in the model are as follows:

$$\frac{dn_1}{dt} = n_c \left(N_1 - n_1 \right) A_n - n_1 P \tag{5.1}$$

$$\frac{dn_c}{dt} = -n_c \left(N_1 - n_1 \right) A_n + n_1 P - A_{CB} n_c \left(N_2 - n_2 \right) \tag{5.2}$$

$$\frac{dn_2}{dt} = A_{CB}n_c \left(N_2 - n_2\right) - A_R n_2 - n_2 A_{NR} \exp\left(-W/k_B T\right) \tag{5.3}$$

The instantaneous luminescence resulting from the radiative transition is defined as

$$I(t) = A_R n_2 \tag{5.4}$$

Transitions 4, 5, and 6 in Fig. 5.2 are of a localized nature, while transition 3 involves electrons in the CB and, hence, it is of a delocalized nature. The difference in the nature of these transitions can also be seen in their mathematical forms in Eqs. (5.1)–(5.4).

The values of the parameters used in the model of Pagonis et al. [121] are as follows: $A_n = 5 \times 10^{-14}\,\mathrm{cm^3 s^{-1}}$, $A_R = 1/(42\,\mu s) = 2.38 \times 10^4\,\mathrm{s^{-1}}$, $A_{CB} = 10^{-8}\,\mathrm{cm^3 s^{-1}}$, $P = 0.2\,\mathrm{s^{-1}}$, $N_1 = 10^{14}\,\mathrm{cm^{-3}}$, and $N_2 = 10^{14}\,\mathrm{cm^{-3}}$. The initial conditions for the different concentrations in the model are taken as: $n_1(0) = 9 \times 10^{13}\,\mathrm{cm^{-3}}$, $n_2(0) = 0$, $n_c(0) = 0$. The value of the delocalized transition coefficient A_{CB} was chosen in the model of Pagonis et al. [121], so that the conduction band empties quickly during the TR-OSL experiment after the optical stimulation is turned off, on a time scale of $\sim 1\,\mathrm{s}$. On the other hand, the value of the radiative transition probability was chosen so that it corresponds to the experimentally observed quartz luminescence lifetime of $\tau = 42\,\mu s$ at room temperature.

In later work, Pagonis et al. [142] derived the following analytical expressions for the luminescence intensity $I(t)$ measured during and after the TR-OSL pulse of duration $0 < t \le t_0$:

$$I(t) = A f \tau \left(1 - \exp\left[-t/\tau\right]\right) \quad \text{for } t \le t_0 \tag{5.5}$$

$$\tau(T) = \frac{1}{A_R + A_{NR} \exp\left(-W/kT\right)} \tag{5.6}$$

$$I(t) = B \exp\left[-\left(t - t_0\right)/\tau\right] \quad \text{for } t > t_0 \tag{5.7}$$

where A, B, f are constants and $\tau(T)$ represents the temperature dependent luminescence lifetime. Equations (5.5) and (5.7) are rising and decaying exponentials with the same decay constant $1/\tau$.

The following code fits TR-OSL curves for sedimentary quartz measured with a $60\,\mu s$ pulse and a dynamic range of $300\,\mu s$ and 10^6 sweeps, following a beta dose of 7 Gy. For more details on the experimental data in Fig. 5.3, see Chithambo et al. [41]. The rising part of the TR-OSL signal yields a lifetime of $38.1\,\mu s$, while the decaying part of the signal yields a luminescence lifetime of $38.5\,\mu s$.

Extensive experimental work has shown that in the case of most unannealed quartz samples, the luminescence lifetimes obtained from the exponential fits decrease smoothly from about $42\,\mu s$ to almost $2\,\mu s$ in the temperature range 20–

Code 5.1: Analysis of TR-OSL experimental data in quartz

```
rm(list=ls())
library("minpack.lm")
mydata <- read.table("chithamboTROSLqzdata.txt")
plot(mydata,pch=2,col="red",  ylab="TR-OSL [a.u.]",
    xlab=expression(paste("Time [",mu,"s]")))
legend("topright",bty="n",legend=c("Sedimentary","quartz"," ",
"470 nm","LEDs"))
# fit ON data
t<-mydata[,1][1:20]
y<-mydata[,2][1:20]
fit_data <-data.frame(t ,y)
fit <- minpack.lm::nlsLM(
formula = y ~ N * (1-exp(- t/b))+abs(bgd), data = fit_data,
   start = list(N= max(y),b = 40,bgd=10))
t1<-0:55
lines(x = t1,  y = coef(fit)[1]  *
(1-exp(- t1/coef(fit)[2]))+abs(coef(fit)[3]),col = "blue")
coef(fit)
# fit OFF data
t<-mydata[,1][21:length(mydata[,1])]-51.27
y<-mydata[,2][21:length(mydata[,1])]
fit_data <-data.frame(t ,y)
fit <- minpack.lm::nlsLM(
   formula = y ~ N * (exp(- t/b))+abs(bgd),data = fit_data,
   start = list(N= max(y),b = 40,bgd=10))
t1<-5:300
lines(x=t1+51.2,y=coef(fit)[1]*(exp(-(t1)/coef(fit)[2]))+
abs(coef(fit)[3]),col = "blue")
coef(fit)

##              N              b           bgd
## 254.2423690  38.0707228    0.8669149
##              N           b         bgd
## 221.49526   38.47338    2.67049
```

$300\,°C$. Additionally, the rising and decaying parts of the experimental TR-OSL signals in quartz in most cases yield the same luminescence lifetimes.

In the next section we discuss the effect of stimulation temperature on the luminescence lifetime and intensity, by using TR-OSL signals in the important dosimetric material Al_2O_3:C.

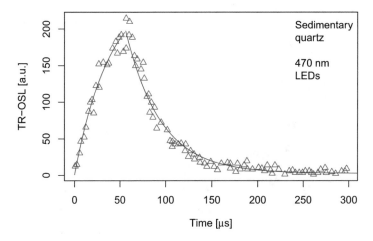

Fig. 5.3 Examples of TR-OSL curves for sedimentary quarts with 60 μs pulse. For more details, see Chithambo et al. [41]

5.3   The Effect of Stimulation Temperature on the Luminescence Lifetime and Luminescence Intensity

Pagonis et al. [119] studied TR-OSL signals in Al_2O_3:C using short pulses (0.5 s) of blue LEDs (470 nm), followed by relaxation period of 2.5 s, at different stimulation temperatures, as shown in Fig. 5.4a. These authors found that during the pulse excitation period, the integrated TR-OSL signal increases with the stimulation temperature between 50 and 150 °C, while the signal intensity decreases between 160 and 240 °C. This is shown in Fig. 5.4b. This behavior was interpreted to arise from the two competing effects of *thermal assistance* (activation energy, $E_{th} = 0.067 \pm 0.002$ eV) and *thermal quenching* (activation energy $W = (1.032 \pm 0.005)$ eV).

The effect of increasing the stimulating temperature T during the TR-OSL experiment is that the *integrated* TR-OSL intensity I_{total} varies with stimulation temperature according to the empirical equation:

$$I_{total}(T) = B \frac{\exp\left(-E_{th}/kT\right)}{1 + C \exp\left(-W/kT\right)} \tag{5.8}$$

where W is the activation energy for the thermal quenching process described in the previous section, C is a dimensionless constant, and E_{th} is a thermal activation energy associated with the luminescence intensity. Typical values of these parameters for Al_2O_3:C are $E_{th} = (0.067 \pm 0.002)$ eV and $W = (1.032 \pm 0.005)$ eV.

Similarly, as the stimulation temperature T increases, the luminescence lifetime τ decreases according to the empirical equation:

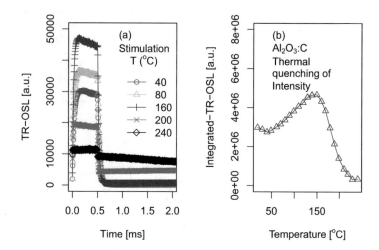

Fig. 5.4 Experimental dependence of (**a**) the TR-OSL luminescence signals, and (**b**) the integrated TR-OSL intensity, on the stimulation temperature for Al_2O_3:C sample. For more details, see Pagonis et al. [119]

$$\tau(T) = D\frac{1}{1 + C\exp\left(-W/kT\right)} \tag{5.9}$$

By comparing these empirical equations with the analytical equation (5.6) derived by Pagonis et al. [142], we see that $C = A_{NR}/A_R$, i.e. the constant C represents the ratio of radiative and radiative luminescence rates A_{NR}, A_R in the Mott–Seitz mechanism.

Pagonis et al. [119] analyzed the changes in the shape of the TR-OSL curves for Al_2O_3:C at different stimulation temperatures, by using analytical expressions available in the literature. The TR-OSL signals contain a slower temperature dependent phosphorescence signal, the "delayed-OSL" previously reported for this material. A typical result at a stimulation temperature of $100\,°C$ is shown in Fig. 5.5a, and fitting the signal with exponential functions yields a luminescence lifetime of $\tau = 37.4\,ms$ for both the rising and decaying parts of the signal.

In general, the temperature dependent luminescence lifetimes obtained from analysis of the optical stimulation period (LED ON) should be identical within experimental error to those obtained from the corresponding relaxation period (LED OFF). However, Pagonis et al. [119] found that the values of these luminescence lifetimes were systematically higher than previously reported values from time-resolved photoluminescence (TR-PL).

Figure 5.5b shows how the luminescence lifetime τ in the TR-OSL experiments for Al_2O_3:C depends on the stimulation temperature T. As the stimulation temperature increases, the value of τ increases initially up to $T = 30\,°C$ due to the thermal assistance process and then decreases to about $\tau = 2\,ms$ at a stimulation temperature of $220\,°C$ due to thermal quenching.

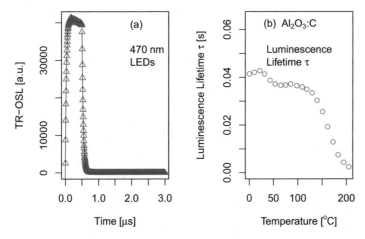

Fig. 5.5 (**a**) Analysis of the TR-OSL signal measured at a stimulation temperature of 100 °C for an Al_2O_3:C sample, using exponential functions shown as solid lines. The signal contains a slower temperature dependent phosphorescence signal, sometimes referred to as the "delayed-OSL" signal. (**b**) Experimental dependence of the luminescence lifetime τ on the stimulation temperature for the same sample. For more details, see Pagonis et al. [119]

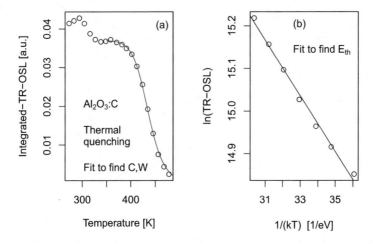

Fig. 5.6 Experimental determination of the thermal quenching parameters C, W, E_{th}. (**a**) The values of C, W are obtained by fitting the *decreasing* part of the data. The best fit is shown by the solid line. (**b**) The thermal activation energy E_{th} is obtained with an Arrhenius analysis of the *increasing* part of the data in (**a**). For more details, see Pagonis et al. [119]

The following code shows how to evaluate the thermal activation energy E_{th} and the thermal quenching activation energy W from the experimental TR-OSL data. The results of the least squares fit procedures shown in Fig. 5.6a, b are $W = 1.01$ eV, $C = 5.6 \times 10^{11}$, and $E_{th} = 0.065$ eV.

Code 5.2: Analysis of TR-OSL experimental data in alumina

```
rm(list=ls())
library("minpack.lm")
par(mfrow=c(1,2))
aluminatau470nmx<-unlist(read.table("aluminax1.txt"))
aluminatau470nmy<-unlist(read.table("aluminay1.txt"))
x<-aluminatau470nmx[8:20]+273
y<-aluminatau470nmy[8:20]
kB<-8.617*1e-5 # Boltzmann constant in eV/K
fit_data <-data.frame(x ,y)
plot(aluminatau470nmx+273,aluminatau470nmy,
xlab=expression("Temperature [K]"),
ylab="Integrated-TR-OSL [a.u.]")
legend("topright",bty="n","(a)")
legend("bottomleft",bty = "n", legend =
c(expression('Al'[2]*'O'[3]*':C',' ',
'Thermal','quenching',' ','Fit to find C,W')))
fit <- minpack.lm::nlsLM(
  formula = y ~ N /(1+c*exp(-W/(kB*x))),data = fit_data,
  start = list(N= max(y),c=1e6, W =1))
x1<-seq(from=350,to=500,by=1)
lines(x=x1,y=coef(fit)[1]/(1+coef(fit)[2]*
exp(-coef(fit)[3]/(kB*x1))),        col = "blue")
coef(fit)
al2o3risoOSLvsTempx<-unlist(read.table("aluminax2.txt"))
al2o3risoOSLvsTempy<-unlist(read.table("aluminay2.txt"))
x<-al2o3risoOSLvsTempx[4:10]
y<-al2o3risoOSLvsTempy[4:10]
y<-log(y)
x<-1/(kB*(x+273.15))
bestfit<-lm(y~x)
coefficients(bestfit)
plot(x, y, xlab = "1/(kT)   [1/eV]",ylab = "ln(TR-OSL)")
legend("topright",bty = "n", legend =c('(b)',expression(' ',
'Fit to find E'[th]*' ')))
abline(lm(y~x))

##                N            c            W
## 3.698208e-02 5.659581e+11 1.014901e+00
## (Intercept)            x
## 17.19471165 -0.06542585
```

5.4 TR-IRSL Experiments: Analytical Equations for Luminescence Intensity

As discussed in Chap. 1, TR experiments can provide crucial information about the luminescence mechanisms in a dosimetric material. Several experimental studies have identified five ranges of lifetimes in TR-IRSL signals, namely, 30–50 ns, 300–500 ns, 1–2 μs, 5 μs, and >10 μs (Sanderson and Clark [165]; Chithambo and Galloway [42]; Tsukamoto et al. [182]). Some of these lifetimes were interpreted as due to internal transitions within the recombination centers (Clark and Bailiff [43]). In several of these earlier papers, the TR-IRSL curves were fitted using a sum of exponentials.

Morthekai et al. [114] studied TR-IRSL from four feldspar mineral specimens, by assuming that detrapped electrons undergo random walk in the band-tail states before recombining by tunneling. The hopping time for the random walk was derived from the OFF-time data of TR-IRSL experiments, and the extracted parameters were shown to be consistent with the variable range hopping mechanism of the Mott kind. Pagonis et al. [144] also analyzed TR-IRSL data from the same four natural feldspars samples, in terms of the sum of an exponential and a stretched exponential function.

The technology of measuring such time-resolved signals has greatly improved during the past 10 years, and modeling advancements have contributed to a better understanding of the luminescence process. These recent developments in both experiments and models prompted researchers to re-examine the topic of the shape and mathematical characterization of TR-IRSL signals. During a TR-IRSL experiment, it is assumed that the excitation pulse is sufficiently short, so that only a small number of electrons are raised from the ground state into the excited state of the trap by IR excitation. Thus, during and after a TR-IRSL pulse, the concentration of electrons in the ground state does not change significantly.

Pagonis et al. [120] used the model of Jain et al. [62] to describe the shape of TR-IRSL signals during and following short infrared pulses on feldspars, in the microsecond time scale. They were able to fit experimental TR-IRSL data from feldspars, both during and after short excitation pulses, and examined different outcomes from the model, depending on the values of the various kinetic parameters. The model is described in detail in Chap. 6 and assumes a random distribution of trapped electrons and nearest neighbor recombination centers, and that emission of luminescence is the result of quantum tunneling recombination of a trapped electron from the excited state of the trap.

Pagonis et al. [120] developed the following analytical equations for the intensity of light during a short infrared pulse, with the assumption of a very weak de-excitation rate taking place from the excited state into the ground state.

> ### *Intensity of TR-IRSL signal during and after an infrared pulse*
>
> $$I_{ON}(t) = I_0 \left\{ 1 - \exp\left(-\rho' \ln [z \, s \, t]^3 \right) \right\} \qquad t < t_0 \qquad (5.10)$$
>
> $$I_{OFF}(t) = I_0 \left\{ \exp\left(-\rho' \ln [z \, s \, t]^3 \right) - \exp\left(-\rho' \ln [z \, s \, (t + t_0)]^3 \right) \right\} \qquad t > t_0$$
> $$(5.11)$$

The parameters in these equations are the saturation intensity I_0, the dimensionless positive charge density ρ', the excitation time t (s), the tunneling frequency s (s^{-1}), the duration of the IR pulse t_0 (s), and $z = 1.8$ is a constant in the model by Jain et al. [62]. These parameters are discussed in detail in Chap. 6.

The following R code fits the TR-IRSL data first published by Morthekai et al. [114] on alkali feldspar microcline FL1, a museum specimen of feldspar. For more details of the analysis, see Pagonis et al. [120]. Measurements were carried out on a Risø TL/OSL-20 reader equipped with an integrated pulsing option to control the IR LEDs, and a Photon Timer attachment to record the TR-IRSL pulses. The TR-IRSL signals in Fig. 5.7 were measured at a stimulation temperature of 50 °C, after the sample was given a dose of 61.8 Gy and was preheated to 280 °C for 60 s; the experiments used an ON-time of 50 μs and an OFF-time of 100 μs.

Code 5.3: Fit microcline TR-IRSL data with analytical equation

```
#Fit FL1 TR-IRSL data with analytical equation
rm(list = ls(all=T))
options(warn=-1)
library("minpack.lm")
par(mfrow=c(1,2))
## fit ON data to analytical TR-IRSL equation ----
mydata <- read.table("FL1ONdata.txt")
t<-1e-6*mydata[,1]
y<-mydata[,2]
mydata<-data.frame(t,y)
plot(t*1e6,y,xlab="Time [s]",ylab="TR-IRSL [Normalized]",
col="black",pch=1)
fit_data <-mydata
fit <- minpack.lm::nlsLM(
    formula=y~ imax*(1-exp (-rho*(log(1 + A*t)) ** 3.0))+bgd,
    data = fit_data,
    start = list(imax=3,A=1e6,rho=0.001,bgd=min(y)))
```

```r
imax_fit <- coef(fit)[1]
A_fit <- coef(fit)[2]
rho_fit <- coef(fit)[3]
bgd_fit <- coef(fit)[4]
## plot analytical solution
t1<-seq(from=1e-7,to=5e-5,by=1e-7)
lines(
  x = t1*1e6,
  y =imax_fit*(1-exp (-rho_fit*(log(1 + A_fit*
t1)) ** 3.0))+bgd_fit,    col = "red",lwd=2)
legend("topleft",bty="n","(a)")
legend("right",bty="n", pch=c(NA,NA,NA,1,NA),lwd=2,
       lty=c(NA,NA,NA,NA,"solid"),
       c(expression('TR-IRSL','Microcline',' ',
                    'Experiment','Analytical')),
col=c(NA,NA,NA,"black","red"))
## print results
cat("\nParameters from Least squares fit")
cat("\nImax=",formatC(imax_fit,format="e",digits=2)," cts/s",
       sep="    ","A=",round(A_fit,digits=2)," (s^-1)")
cat("\nrho=",round(rho_fit,digits=4),sep=" ",
       " bgd=",round(bgd_fit,digits=2)," cts/s")
## Use same parameters to fit the OFF data
mydata <- read.table("FL1OFFdata.txt")
t<-1e-6*mydata[,1]
y<-mydata[,2]
mydata<-data.frame(t,y)
plot(t*1e6,y,xlab="Time [s]",ylab="TR-IRSL [Normalized]",
col="black",pch=1)
## plot analytical solution
t1<-seq(from=5e-7,to=1e-4,by=1e-7)
lines(
  x = t1*1e6,
  y =.06+ imax_fit*(exp (-rho_fit*(log(1 + A_fit*t1)) ** 3.0)-
exp (-rho_fit*(log(1 + A_fit*(t1+5e-5)))) ** 3.0))+bgd_fit,
  col = "red",lwd=2)
legend("topleft",bty="n","(b)")
legend("right",bty="n", pch=c(NA,NA,NA,1,NA),lwd=2,
       lty=c(NA,NA,NA,NA,"solid"),
       c(expression('TR-IRSL','OFF data',' ',
'Experiment','Analytical')),col=c(NA,NA,NA,"black","red"))

##
## Parameters from Least squares fit
## Imax=    1.03e+00    cts/s    A=    6592795    (s^-1)
## rho= 0.0212   bgd= -0.04  cts/s
```

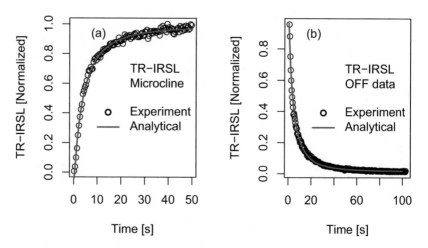

Fig. 5.7 Fit TR-IRSL data for microcline sample FL1, with the analytical Eqs. (5.10) and (5.11). (**a**) The IR excitation is ON. (**b**) The excitation is OFF. For more details, see Pagonis et al. [120]

The second sample analyzed by Pagonis et al. [120] was a potassium rich feldspar fraction (K-AlSi3O8) of a glacio-fluvial sediment from Jameson land, East Greenland (lab. code 951002FK). The TR-IRSL curves were measured at a stimulation temperature of 500 °C, after the sample was given a dose of 70 Gy, and then preheated to 250 °C for 60 s, using an ON-time of 500 μs, and an OFF-time of 300 μs. The natural luminescence signals were erased by exposing the samples to IR at an elevated temperature of 250 °C, followed by heating the sample from room temperature up to 500 °C. For sample 951002FK, the total stimulation time was 10 s, and the signals were detected in the UV, with the best fits to the data shown in Fig. 5.8.

An important conclusion from the analysis by Pagonis et al. [120] is that the duration of the excitation period clearly affects the rate of decay and the shape of the pulses after the LED has been turned off. This is because of the progressive shift in the peak of the distribution of distances of nearest neighbors with the duration of the excitation. This is also in contrast to the situation with quartz, where the rate of decay of the pulses is independent of the duration of the excitation period (Pagonis et al. [142]. From a practical point of view, there may be some discrepancies when trying to fit the two parts of the TR signal using the same set of model parameters. Such discrepancies between the fitting parameters most likely indicate the presence of a significant contribution from migration through the band-tail states, prior to recombination.

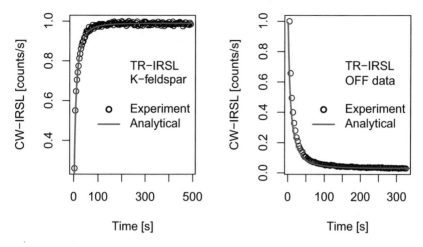

Fig. 5.8 Fit of K rich feldspar TR-IRSL data with analytical Eqs. (5.10) and (5.11). (**a**) The IR excitation is ON (**b**) The excitation is OFF. For more details, see Pagonis et al. [120]

5.5   A Model for TR-Photoluminescence (TR-PL) Experiments in Al₂O₃:C

Pagonis et al. [129] presented simulations of time-resolved photoluminescence (TR-PL) experiments in α-Al$_2$O$_3$:C. During TR-PL experiments, short pulses of UV-light are followed by relaxation periods of the charge carriers. The model used in these simulations was previously developed by Nikiforov et al. [117] in order to explain radioluminescence (RL), TL, and PL experiments for this material and is based on optical and thermal ionization of excited F-centers.

Pagonis et al. [129] modified the model of Nikiforov et al. [117], in order to describe thermal quenching in this material as a Mott–Seitz type of mechanism (similar to quartz), based on competition between radiative and radiationless electronic transitions occurring within the recombination center. They simulated a typical TR-PL experiment in α-Al$_2$O$_3$:C at different stimulation temperatures and compared the simulation results with the available experimental data. They also simulated TL and thermally stimulated conductivity (TSC), as a function of the heating rate used during such experiments.

In Fig. 5.9 the symbols are as follows: N denotes the main dosimetric trap, M_1, M_2 stand for deep electron traps, and ^1P and ^3P are the excited levels of the F-center. Upon optical excitation in the absorption band, the center is excited to the ^1P state from the ground state ^1S via the transition indicated by f. Luminescence from an F-center (centered at 420 nm) corresponds to the transition w_3, and thermal ionization of the F-center via the excited ^3P state corresponds to the transition indicated by P_F. The probability of thermal ionization of the excited ^3P state of the F-center is given by Boltzmann factor $P_F = C \exp(-W/k_B T)$, where W is the activation energy of the thermal quenching process and C is a frequency constant.

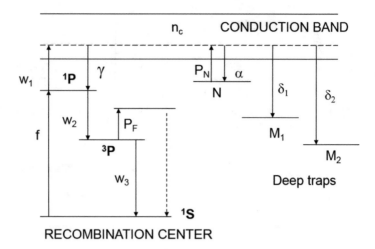

Fig. 5.9 The TR-PL model for time-resolved photoluminescence (TR-PL) experiments in α-Al$_2$O$_3$:C by Pagonis et al. [129], based on an earlier version of the model by Nikiforov et al. [117]

Thermal ionization leads to a decrease in the fraction of radiative transitions (w_3) taking place at the center and is believed to be the direct cause of thermal quenching of luminescence in this material. The optical transition denoted by w_1 results in the formation of an F^+ center according to $F^* - e = F^+$, while the F^+ center can change into an excited F^* center according to $F^+ + e = F^*$. Free electrons from the ionization of F-centers can be captured in the dosimetric trap, or in the deep traps M_1, M_2 (Yukihara et al. [196]).

The parameters used in the model are defined as follows: N is the total concentration of dosimetric traps (cm^{-3}); n is the corresponding concentration of trapped electrons (cm^{-3}); M_1 and M_2 are the total concentrations of deep traps; n, m_1, m_2, n_{1P}, and n_{3P} (cm^{-3}) stand for the concentration of occupied levels in $N, M_1, M_2,^1 P$, and ^3P, respectively; n_{F+} (cm^{-3}) is the concentration of F^+ centers; n_c (cm^{-3}) is the concentration of electrons in the conduction band; $\alpha, \delta_1, \delta_2$ and γ (cm^3s^{-1}) are the capture coefficients of carriers at the corresponding levels shown in Fig. 5.9; w_1, w_2, and w_3 (s^{-1}) denote the transition coefficients; f (cm^3s^{-1}) is the excitation intensity. The emptying probability for dosimetric traps is described by the Arrhenius expression $P_N = ns \exp(-E/k_B T)$, where E is the trap depth of the main dosimetric trap and s is the corresponding frequency factor. W is the activation energy for the thermally assisted process (eV) in Al$_2$O$_3$:C, with a typical value of $W = 1$ eV.

The equations used in the model are as follows:

$$\frac{dn}{dt} = n_c (N - n) \alpha - nP \tag{5.12}$$

$$\frac{dm_1}{dt} = \delta_1 \left(M_1 - m_1\right) n_c \tag{5.13}$$

$$\frac{dm_2}{dt} = \delta_2 \left(M_2 - m_2\right) n_c \tag{5.14}$$

$$\frac{dn_c}{dt} = -n_c \left(N_1 - n_1\right) \alpha + n P - \delta_1 \left(M_1 - m_1\right) - \delta_2 \left(M_2 - m_2\right) - \gamma n_{F+} n_c + w_1 n_{1P} \tag{5.15}$$

$$\frac{dn_{F+}}{dt} = -\gamma n_{F+} n_c + w_1 n_{1P} \tag{5.16}$$

$$\frac{dn_{3P}}{dt} = w_2 n_{1P} - P_F n_{3P} - w_3 n_{3P} \tag{5.17}$$

$$\frac{dn_{1P}}{dt} = f + \gamma n_{F+} n_c - w_1 n_{1P} - w_2 n_{1P} \tag{5.18}$$

The instantaneous luminescence resulting from the radiative transition is defined as

$$I(t) = w_3 n_{3P} \tag{5.19}$$

Transitions corresponding to the parameters f, w_2, w_3, and P_F in Fig. 5.9 are of a *localized* nature, while the rest of the transitions involve electrons in the CB and hence are of a *delocalized* nature. The difference in the nature of these transitions can also be seen in their mathematical forms in Eqs. (5.12)–(5.19). The term in Eqs. (5.15) and (5.16) expresses the fact that there are empty electronic states available for electrons from the CB; these are excited states of the recombination center, in agreement with the general assumptions of the Mott–Seitz mechanism of thermal quenching.

The values of the parameters used in the model of Pagonis et al. [129] are as follows: $E = 1.3$ eV, $s = 10^{13}$ s^{-1}, $\alpha = 10^{-14}$ (cm^{3}s^{-1}), $\delta_1 = 10^{-12}$ (cm^{3}s^{-1}), $\delta_2 = 10^{-14}$ (cm^{3}s^{-1}), $\gamma = 10^{-11}$ (cm^{3}s^{-1}), $N = 10^{13}$ cm^{-3}, $M_1 = 10^{15}$ cm^{-3}, $M_2 = 10^{14}$ cm^{-3}, $C = 10^{13}$ s^{-1}, $W = 1$ eV, $w_1 = 1$ s^{-1}, $w_2 = 3 \times 10^3$ s^{-1}, and $w_3 = 29$ s^{-1}, $f = 10^{10}$ cm^3 s^{-1}. The initial conditions for the different concentrations in the model are taken as: $m_1(0) = 10^{14}$ cm^{-3}, $n_{F+}(0) = 10^{14}$ cm^{-3}, and all other concentrations are taken to be 0.

The following R code solves the system of differential equations in this model, and the results are shown in Fig. 5.10 for a simulation of time-resolved experiments for Al$_2$O$_3$:C, while the light excitation is ON. Figure 5.10a shows the TR-OSL intensity at three different stimulation temperatures $T = 120, 180, 210\,^\circ$C, while Fig. 5.10b plots the area under the TR-OSL curves, as a function of the stimulation temperature T. As T increases, both the lifetime in (a) and the areas in (b) decrease, illustrating the effect of thermal quenching.

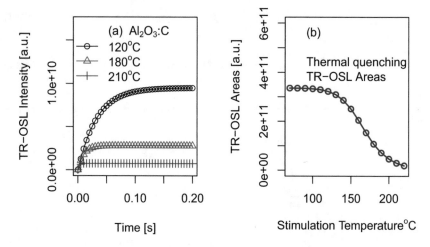

Fig. 5.10 Simulation of TR-PL experiments for Al$_2$O$_3$:C, while the light excitation is ON. (**a**) The TR-OSL intensity at three different stimulation temperatures T =120, 180, 210 °C. (**b**) The area under the TR-OSL curves in (**a**) is plotted as a function of the stimulation temperature and shows the effect of thermal quenching. For more details, see Pagonis et al. [129]

A *for* loop is used to solve the system of differential equations for each stimulation temperature from 70 °C to 220 °C, in steps of 10 °C. The function *TempVar* is called within the *for* loop, for each stimulation temperature.

Code 5.4: Simulation of TR-PL experiments in Al2O3:C

```
#Simulate TR-PL experiments with Nikiforov/Pagonis model
rm(list = ls(all=T))
library("deSolve")
PagonisAlumina <- function(t, x, parms) {
  with(as.list(c(parms, x)), {
    dn<- -s*exp(-E/(kb*(273+T)))*n+alpha*(N-n)*nc
    dm1<- delta1*(M1-m1)*nc
    dm2<- delta2*(M2-m2)*nc
    dnc<- s*n*exp(-E/(kb*(273+T)))-delta1*(M1-m1)*nc-
    delta2*(M2-m2)*nc-Gamma*nF*nc-alpha*(N-n)*nc+w1*n1P
    dnF<- -Gamma*nF*nc+w1*n1P
    dn3P<-w2*n1P-C*exp(-W/(kb*(273+T)))*n3P-w3*n3P
    dn1P<-f+Gamma*nF*nc-w1*n1P-w2*n1P
    res <- c(dn,dm1,dm2,dnc,dnF,dn3P,dn1P)
    list(res)   })}
```

```r
TempVar<-function(T){
parms<- c(E=1.3, s=1e13,alpha=1e-14, delta1=1e-12, delta2=1e-14,
  Gamma=1e-11,N=1e13,M1=1e15,T=T,M2=1e14,C=1e13,W=1,w1=1,w2=3e3,
  w3=w3,f=1e10,kb=8.617e-5)
  y <- xstart <- c(n = 0, m1=1e14,m2=0,nc=0,nF=1e14,n3P=0,n1P=0)
  out <-  ode(xstart, times, PagonisAlumina, parms)  }
w3<-29
times  <- seq(0, .2,by=.005)
Temps<-seq(70,220,10)
Lc<-temps<-matrix(NA,nrow=length(times),ncol=length(Temps))
areaLc<-vector(length=length(Temps))
for (i in 1:length(Temps)){
  T<-Temps[i]
  a<-TempVar(T)
  Lc[,i]<- w3*a[,"n3P"]
  areaLc[i]<-sum(Lc[,i],rm.NA=TRUE)}
par(mfrow=c(1,2))
plot(times,Lc[,6],typ="o",col="black",xlim=c(0,.2),
ylim=c(0,1.7e10),pch=1, xlab="Time [s]",
ylab="TR-OSL Intensity [a.u.]")
lines(times,Lc[,12],typ="o",col="red",pch=2)
lines(times,Lc[,15],typ="o",col="blue",pch=3)
legend("topleft",bty="n",pch=c(NA,1,2,3),
lty=c(NA,rep("solid",3)),col=c(NA,"black",
"red","blue"),legend=c(expression("(a)  Al"[2]*"O"[3]*":C",
"120"^o*"C","180"^o*"C","210"^o*"C")))
plot(Temps,areaLc,typ="o",col="blue",lwd=2,pch=1,ylim=c(0,6e11),
xlab=expression("Stimulation Temperature"^o*"C"),
ylab="TR-OSL Areas [a.u.]")
legend("topleft",bty="n",legend=c("(b)"," ","Thermal quenching",
  "TR-OSL Areas"))
```

Part II
Luminescence Signals from Localized Transitions

Chapter 6
Localized Transitions and Quantum Tunneling

6.1 Introduction to the TLT Models

As discussed in Chap. 1, there are three types of localized luminescence models commonly used in the literature: the tunneling localized transition (TLT) model, the localized transition (LT) model, and the semilocalized transition (SLT) model. In this chapter we will consider the TLT models, and in the next chapter we will consider the LT and SLT models.

Specifically, in this chapter we will look at four different types of TLT models: ground state tunneling (GST) model, irradiation ground state tunneling (IGST) model, excited state tunneling (EST) model, and thermally assisted excited state tunneling (TA-EST) model. We will present the physical principles behind each model and discuss approximate analytical solutions to the differential equations describing each model. A more complete presentation of these models, which is not based on these approximate analytical equations, will be presented in Chap. 12.

In Sect. 6.2 we will consider the geometrical concept of the nearest neighbor distribution in a random distribution of defects in a solid. This is an important concept that is found in all four tunneling models presented in this chapter. In Sect. 6.3 we introduce the fundamental ideas of quantum tunneling in a solid and present a brief historical overview of the four TLT luminescence models. This is followed in Sects. 6.4 and 6.5 by a discussion of ground state tunneling phenomena and an example of analyzing experimental data for the associated phenomenon of anomalous fading (AF). An approximate analytical equation is presented to describe the AF phenomenon and is used to obtain the well-known decade g-factor. Section 6.6 presents a model describing simultaneous irradiation and GST, and an approximate analytical solution for the dose response of samples under these conditions.

Section 6.7 introduces models used to describe excited state tunneling phenomena (EST), and Sect. 6.8 presents the approximate analytical solutions developed by Kitis and Pagonis [73] for the general TLT model developed by Jain et al. [62].

Sections 6.9 and 6.10 provide the R codes for analyzing TL and CW-IRSL signals for freshly irradiated samples exhibiting quantum tunneling phenomena. In Chap. 12 we will see how these models can be used to describe luminescence signals from feldspar samples that underwent either an optical or thermal treatment after irradiation.

Section 6.11 presents the thermally assisted excited state model (TA-EST) by Brown et al. [24]. Finally, Sect. 6.12 shows how quantum tunneling phenomena can be simulated using the TUN functions in the package *RLumCarlo*.

6.2   Quantum Tunneling and the Distribution of Nearest Neighbors

During the past decade, significant progress has been made both experimentally and theoretically in understanding the behavior of luminescence signals from feldspars, apatites, and other natural materials. Quantum mechanical tunneling and the associated phenomenon of "anomalous fading" of these luminescence signals are now well established as dominant mechanisms in these materials (see, e.g. Pagonis et al. [127]).

In a TLT model, it is customary to consider a random distribution of electrons and positive charges in a crystal and to introduce the concept of the distribution of nearest neighbor distances. The positive charges in this random distribution can be ions, or holes, or more generally acceptors.

Two of the common assumptions of quantum tunneling models based on random distributions of electrons and positive charges are: (a) an electron tunnels from a donor to the nearest positive charge and (b) the concentration of electrons is much lower than that of positive charges at all times during the tunneling process. Because of these assumptions, the concentration ρ (acceptors/m^3) of positive charges would remain practically constant during the tunneling process, and it can be used to characterize the system.

In the TLT model, the quantum tunneling mechanism takes place in this random distribution of holes and electrons, directly from the ground state of the system. The probability per unit time $P(r)$ (in s^{-1}) that a trapped electron will tunnel to a positive charge is given by the exponential decay behavior of the wavefunction (Tachiya and Mozumder [175])

$$P(r) = s \exp\left(-r/a\right) \tag{6.1}$$

where the frequency parameter s (s^{-1}) characterizes the tunneling process for the trap, and the length a (m) represents the attenuation length of the ground state of a wavefunction. The inverse of the tunneling length a is called the potential barrier penetration constant $\alpha = 1/a$ (m^{-1}).

One introduces now two dimensionless quantities r' and ρ', instead of r and ρ. First, a dimensionless distance parameter r' is defined by

$$r' = (4\pi\rho/3)^{1/3} r \qquad (6.2)$$

where ρ (acceptors/m^{-3}) represents the charge density of recombination centers in the material per unit volume, and r is the actual donor–acceptor distance. Notice that r' is directly proportional to the distance r, simply scaled by a factor $(4\pi\rho/3)^{1/3}$ that depends on the acceptor density ρ. Secondly, one also introduces a dimensionless acceptor density parameter ρ' by

$$\rho' = (4\pi\rho/3)\,a^3 \qquad (6.3)$$

where a is the tunneling length. Notice that ρ' is directly proportional to the acceptor density ρ, simply scaled by a "unit tunneling volume" $4\pi a^3/3$.

As a concrete example, let us consider a cube with side $d = 100$ nm, containing 50 electrons and 300 recombination centers (acceptors), as shown in Fig. 6.1a. The physics in this example would be determined, for example, by specifying the tunneling distance $a = 0.11$ nm, and its inverse would represent the tunneling parameter $\alpha = 1/a = 9 \times 10^9\,\mathrm{m}^{-1}$. The density of acceptors in this cube is $\rho = 300/d^3 = 3 \times 10^{23}$ (acceptors/m^{-3}). The dimensionless acceptor density is then $\rho' = (4\pi\rho/3)\,a^3 = 1.67 \times 10^{-6}$.

For random distributions of defects, the distribution of nearest neighbors is given by the Poisson probability of finding no neighbors in a sphere of radius r (see, e.g. Jain et al. [62])

$$g(r) = 4\pi\rho r^2 \exp\left[-4\pi\rho/3\,(r)^3\right] \qquad (6.4)$$

By using the dimensionless length r' and dimensionless acceptor density ρ', this distribution becomes

---

Nearest neighbor distribution

$$g(r') = 3\,(r')^2 \exp\left[-(r')^3\right] \qquad (6.5)$$

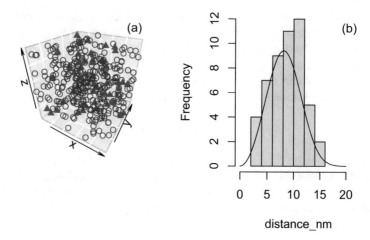

Fig. 6.1 (**a**) A cube with side $d = 100$ nm contains 50 electrons (triangles) and 300 recombination centers (circles). (**b**) Histogram of the nearest neighbor distances of electron–acceptor pairs from the cube in (**a**). The solid line in (**b**) represents the analytical equation for the distribution of nearest neighbors equation (6.5). For more information, see Pagonis and Kulp [140]

The following code evaluates the nearest neighbor distribution of distances between electrons and acceptors in the cube discussed above. The code uses the R packages *plot3D* to plot the random locations of the electrons (triangles) and acceptors (circles) in Fig. 6.1a. The R package *FNN* is used to evaluate the nearest neighbor distance between each pair of electron and positive charge, and the code produces the histogram in Fig. 6.1b for the distribution of nearest neighbor in the system.

It must be noted that the distribution $g(r')$ in Eq. (6.5) is a purely *geometrical* concept, and there is no specific physical process associated with it. The *physics* of the system is contained in the tunneling length a and will be considered in the next section.

6.3   Overview of Four TLT Models

Several types of models have been investigated in order to explain the luminescence signals in feldspars, and these are shown schematically in Fig. 6.2. The first type of quantum mechanical tunneling is considered to take place directly from the ground state of the trap, as shown in Fig. 6.2a. In this book, we will refer to this type of model as ground state tunneling (GST) model, and initially it was developed by Tachiya and Mozumder [175]), within the context of the kinetics of chemical reactions. Several decades later this model was shown to be associated with a power-law type of decay of the luminescence signal (Huntley [59]).

Code 6.1: The nearest neighbors distribution

```
# Fig 1 in Pagonis and Kulp paper
# Original Mathematica Program written by V Pagonis
# R  version written by Johannes Friedrich
rm(list = ls(all = TRUE)) # empties the environment
library("plot3D")
library("FNN")
## Define Parameters ----
sideX <- 100e-9 # lenght of quader in m
sideX_nm <- sideX*1e9 # length of quader in nm
N_pts <- 50
alpha <- 9e9
N_centers <- 300
rho <- N_centers/sideX^3
r_prime <- function(r) (4*pi*rho/3)^(1/3) * r * 1e-9
xyz_traps <- data.frame(
   x = sample(1:sideX_nm, N_pts, replace = TRUE),
   y = sample(1:sideX_nm, N_pts, replace = TRUE),
   z = sample(1:sideX_nm, N_pts, replace = TRUE))
xyz_centers <- data.frame(
   x = sample(1:sideX_nm, N_centers, replace = TRUE),
   y = sample(1:sideX_nm, N_centers, replace = TRUE),
   z = sample(1:sideX_nm, N_centers, replace = TRUE))
par(mfrow=c(1,2))
   plot3D::scatter3D(xyz_centers$x, #plot centers (blue)
                 xyz_centers$y,cex=1,
                 xyz_centers$z, bty = "g", pch = 1, theta = 30,
                 phi = 30, col = "blue")
   plot3D::scatter3D(xyz_traps$x, # add traps (red)
                 xyz_traps$y, cex=1,
                 xyz_traps$z, bty = "g", pch = 17,theta = 30,
                 phi = 30, col = "red",add = TRUE)
legend("topright",bty="n","(a)")
    ## find nearest neighbour
    dist <- FNN::get.knnx(data = as.matrix(xyz_centers),
                       query = as.matrix(xyz_traps),k = 1)
    ## plot histogram
    distance_nm<-as.vector(dist$nn.dist)
    hist(distance_nm,xlim=c(0,22),main=" ")
    ## calc analytical solution
    r <- seq(0, 20, 0.1)
    distr_ana <- 3 * 8 * r_prime(r)^2 * exp(-(r_prime(r)^3))
    ## plot analytical solution
    lines(x = r,  y = distr_ana)
legend("topright",bty="n","(b)")
```

Huntley and Lian [61] suggested an extension of the GST model, and this is shown schematically in Fig. 6.2b. These authors suggested a differential equation to describe simultaneous natural irradiation and anomalous fading effects on the luminescence of feldspars, which was examined in detail by Li and Li [98]. We will refer to this model as the irradiation ground state tunneling (IGST) model. Kars et al. [64] applied the model of Huntley and Lian [61] to construct unfaded and natural dose response curves (DRCs) of IRSL signals.

The third type of quantum mechanical tunneling is considered to take place via the excited state of the system of electron–hole pairs, as shown in Fig. 6.2c. Historically this type of model was first considered by Thioulouse et al. [177] and Chang and Thioulouse [30]. Almost 30 years later, Jain et al. [62] developed a general kinetic model that quantifies localized recombination within randomly distributed donor–acceptor pairs. We will refer to this type of model as an excited state tunneling (EST) model. Note that we will refer to electron–hole pairs or to donor–acceptor pairs for simplicity.

The fourth type of model was developed recently by Brown et al.[24] for the purposes of low temperature thermochronology, and is shown schematically in Fig. 6.2d. These authors extended the original model by Jain et al. [62], to include irradiation processes. This model also uses a differential equation to describe simultaneous irradiation, quantum tunneling, and thermal excitation effects. We will refer to this model as a thermally assisted excited state tunneling (TA-EST) model.

In the rest of this chapter we will summarize the mathematical description of these four models, point out their similarities and differences, and show the available approximate analytical equations for each model. In Chap. 12 we will develop appropriate R functions for these four models, which can simulate a wide variety of processes in feldspars.

6.4   Ground State Tunneling: The Anomalous Fading Effect

Extensive experimental and modeling studies have revealed a time dependent localized tunneling/recombination probability. Of major practical interest is the "anomalous fading" of luminescence signals observed mainly in feldspar samples because of their importance in luminescence dating studies (Lamothe et al. [91], and references therein). This anomalous fading effect is now believed to be caused by quantum mechanical tunneling from the ground state of the trap to the luminescence center (Visocekas et al. [184, 185]).

In the GST model, the tunneling mechanism takes place in a random distribution of holes and electrons in the crystal, and transitions take place directly from the ground state of the system. As discussed previously in this chapter, the probability per unit time that a trapped electron will tunnel to a positive charge is given by (Tachiya and Mozumder [175])

$$P(r) = s \exp(-r/a) \tag{6.6}$$

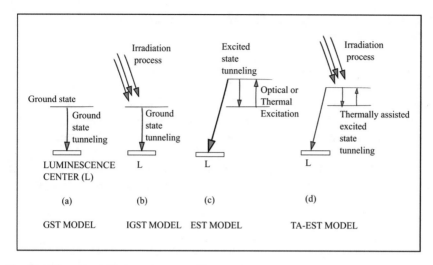

Fig. 6.2 Schematic depiction of several TLT models: (**a**) The ground state tunneling (GST) model (Tachiya and Mozumder [175], Huntley [59]). (**b**) The more general irradiation and ground state tunneling (IGST) model studied by Li and Li [98], in which anomalous fading and natural irradiation are taking place simultaneously. (**c**) The excited state tunneling (EST) model (Jain et al. [62]). (**d**) Simultaneous irradiation and thermally assisted excited state tunneling (TA-EST) model by Brown et al. [24]

The tunneling lifetime $\tau(r') = 1/P(r')$ (in s) in this model is the inverse of $P(r')$. Substituting the dimensionless quantities r' and ρ' in this equation, we obtain

$$\tau(r') = s^{-1} \exp\left[\left(\rho'\right)^{-1/3} r'\right] \tag{6.7}$$

The tunneling lifetime τ (s) depends on the distance parameter r'. The instantaneous concentration of electrons in the ground state is denoted by $n(r', t)$ and depends both on the elapsed time t and on the distance parameter r'. The physical picture here is that as time t progresses, electrons will first tunnel quickly to the nearest neighbors. As the number of electrons and holes in the system decreases, the electrons have to tunnel farther away in the crystal, so that the tunneling process becomes slower and the tunneling lifetime increases.

The distribution of electrons in the ground state $n(r', t)$ varies both with r' and with t, according to the differential equation (Tachiya and Mozumder [175]):

$$\frac{\partial n(r', t)}{\partial t} = -n(r', t)\, s \exp\left[-\left(\rho'\right)^{-1/3} r'\right] \tag{6.8}$$

We assume that at time $t = 0$ the distribution of distances is given by the nearest neighbor distribution $g(r') = 3\left(r'\right)^2 \exp\left[-\left(r'\right)^3\right]$ in Eq. (6.5). The solution of this first order differential equation, for a *constant distance parameter* r', is the simple exponential function:

$$n\left(r',t\right) = n_0 \, 3 \left(r'\right)^2 \, \exp\left[-\left(r'\right)^3\right] \exp\left[-s\,t\,\exp\left[-\left(\rho'\right)^{-1/3} r'\right]\right] \qquad (6.9)$$

where n_0 is the initial concentration of filled traps in the material at time $t = 0$. The value of n_0 is treated as a scaling factor in all four models discussed in this chapter, since we are interested in the trap filling ratio $n(t)/n_0$, instead of the actual value of n_0.

The *total* concentration of trapped electrons at time t is evaluated numerically by integrating $n\left(r',t\right)$ over all possible values of the variable r' (see Huntley [59]):

$$n\left(t\right) = \int_0^\infty n\left(r',t\right) dr' \qquad (6.10)$$

$$n\left(t\right) = n_0 \int_0^\infty 3 \left(r'\right)^2 \, \exp\left[-\left(r'\right)^3\right]\right) \exp\left\{-s \exp\left[-\left(\rho'\right)^{-1/3} r'\right] t\right\} dr' \qquad (6.11)$$

The following code shows an example of the distribution obtained using Eq. (6.9) as a function of the dimensionless distance r', and at times $t = 0, 10^2, 10^4, 10^6$ years after the start of the tunneling process, with the results shown in Fig. 6.3. The solid line indicates the initial peak-shaped symmetric distribution $g(r')$ at time $t = 0$. The values of the parameters used in Fig. 6.3 are typical for ground state tunneling in feldspars, $\rho' = 1 \times 10^{-6}$ and $s = 3 \times 10^{15}\,\text{s}^{-1}$. The sharply rising dashed lines in Fig. 6.3 represent the "moving tunneling fronts" in the tunneling process (Huntley [59]). The characteristic shape of this tunneling front is the product of the two functions appearing in Eq. (6.9), namely of the sharply rising double exponential function $\exp\left[-s\,t\,\exp\left[-\left(\rho'\right)^{-1/3} r'\right]\right]$ and of the symmetric nearest neighbor distribution $3\left(r'\right)^2 \exp\left[-\left(r'\right)^3\right]$ (see the detailed discussion in Pagonis and Kitis [134]).

Code 6.2: Time evolution of the nearest neighbors distribution

```
rm(list=ls())
s<-3e15                      # frequency factor
rho<-1e-6                    # rho-prime values 0.005-0.02
rc<-0.0                      # for freshly irradiated samples, rc=0
times<-3.154e7*c(0,1e2,1e4,1e6)      # times in seconds
rprimes<-seq(from=rc,to=2.2,by=0.002)    # rprime=0-2.2

##### function to find distribution of distances ###
fingDistr<-function(tim){3*(rprimes**2.0)*exp(-(rprimes**3.0))*
    exp(-exp(-(rho**(-1/3))*rprimes)*s*tim)}
######

distribs<-sapply(times,fingDistr)
# Plots
cols=c(NA,NA,NA,1:4)
matplot(rprimes,distribs,xlab="Dimensionless distance r'",
ylab="Nearest neighbor Distribution g(r')",type="l",lwd=4)
legend("topright",bty="n", lty=c(NA,NA,NA,1:4), lwd=4,
col=cols,legend = c("Elapsed", "time"  ," ","t=0 s",
  expression("10"^"2"*" years"),
expression("10"^"4"*" years"),expression("10"^"6"*" years")))
```

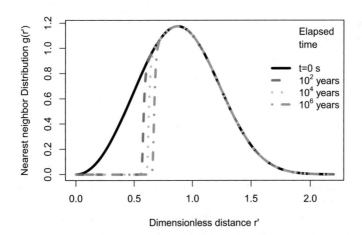

Fig. 6.3 Examples of the nearest neighbor distribution at different times $t = 0, 10^2, 10^4, 10^6$ years, based on Eq. (6.9). The solid black line represents the nearly symmetric distribution $g(r')$ at time $t = 0$. As time increases, the "tunneling front" is the almost vertical line that moves to the right, as more and more electrons are recombining at larger distances r'

In the rest of this section, we will discuss the *approximate* analytical version of the GST model (Tachiya and Mozumder [175], Huntley [59]). Later in Chap. 12 we will see how to evaluate more accurately the total concentration $n(t)$ of the remaining electrons after time t, by developing appropriate R functions.

In the approximate version of the GST model, one introduces a critical lifetime τ_c and a corresponding critical radius r'_c, which describe the behavior of the physical system. From a physical point of view, this approximation means that at time $t = \tau_c$ all electrons within the critical radius $r = r'_c$ have recombined, while all electrons within distance $r' > r'_c$ are still present in the system. Geometrically, this approximation corresponds to replacing the dashed lines in Fig. 6.3 by a vertical line. Mathematically the value of the critical distance is

$$r'_c = \left(\rho'\right)^{1/3} \ln\left(zs\, t\right) \tag{6.12}$$

where $z = 1.8$ is a correction factor that was introduced arbitrarily by Huntley [59] but was explained mathematically later by Pagonis and Kitis [134]. In the example of Fig. 6.3 we see that after 100 years, the almost vertical line representing the tunneling front has moved to a critical radius $r'_c \simeq 0.6$. We can verify this value using Eq. (6.12), by substituting the values of the parameters in this example:

$$r'_c = \left(\rho'\right)^{1/3} \ln\left(zs\, t\right) = \left(10^{-6}\right)^{1/3} \ln\left[1.8(3 \times 10^{15})\,(100 \times 3.154 \times 10^7\right] = 0.58$$

in agreement with Fig. 6.3. In this approximate model of ground state tunneling, the following analytical equation expresses the concentration of charge carriers in the ground state $n(t)$ during geological time scales (Tachiya and Mozumder [175], Huntley [59]):

Loss of charge due to ground state tunneling

$$n(t) = n_0 \exp\left(-\rho'\, \ln[z\, s\, t]^3\right) \tag{6.13}$$

The following R code plots this equation for three different dimensionless acceptor densities $\rho' = 10^{-6}$, 5×10^{-6}, 10^{-5}. As ρ' increases in Fig. 6.4, the rate of charge loss increases. In addition, the initial fast loss of charge is followed by a "long tailed" decay, which is characteristic of quantum tunneling phenomena.

Code 6.3: Ground state tunneling: Remaining electrons n(t)

```
# Simulate Loss of charge due to ground state tunneling (GST)
rm(list=ls())
z<-1.8
s<-3e15                    # frequency factor
rho1<-1e-6                 # rho-prime values
rho2<-5e-6
rho3<-1e-5
years<-1000
elapsedt<-3.154e7*years
t<-seq(from=1,to=elapsedt,by=1e6)
gr1<-100*exp(-rho1*(log(z*s*t)**3.0))
gr2<-100*exp(-rho2*(log(z*s*t)**3.0))
gr3<-100*exp(-rho3*(log(z*s*t)**3.0))
plot(unlist(t)/(3.154e7),unlist(gr1),type="l",lty=1,lwd=2,
ylim=c(0,130),pch=1,ylab="Remaining electrons [%]",
xlab="Elapsed time t [years]")
lines(unlist(t)/(3.154e7),unlist(gr2),lwd=2,col="red",lty=2)
lines(unlist(t)/(3.154e7),unlist(gr3),lwd=3,col="green",lty=2)
legend("topright",bty="n",lwd=2, lty=c(NA,1,2,3),
legend = c("Acceptor Density"  ,expression("10"^"-6"*""),
expression("5x10"^"-6"*""),expression("10"^"-5"*"")),
 col=c(NA,"black","red","green"))
```

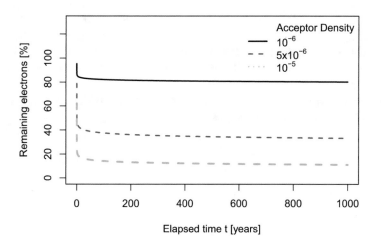

Fig. 6.4 Examples of the loss of charge due to ground state tunneling, for different dimensionless acceptor densities $\rho' = 10^{-6}, 5 \times 10^{-6}, 10^{-5}$, based on Eq. (6.13). As ρ' increases, the rate of charge loss increases. As the elapsed time t increases, the initial fast loss of charge is followed by a "long tailed" decay, characteristic of quantum tunneling phenomena

6.5 Example of Analyzing Experimental Data for the Anomalous Fading Effect

Pagonis and Kitis [134] discussed the traditional description of anomalous fading effects, in which the decay of luminescence signal after the end of irradiation is written as (Li and Li [98], Polymeris et al. [157], Lamothe et al. [91])

$$\frac{L}{L_c} = 1 - \frac{g}{100} \log_{10}\left(\frac{t}{t_c}\right) \tag{6.14}$$

or, in terms of the natural logarithm function instead of the base-10 logarithm (Polymeris et al. [157]):

$$\frac{L}{L_c} = 1 - K \ln\left(\frac{t}{t_c}\right) \tag{6.15}$$

$$g = 230.2\, K \tag{6.16}$$

where the dimensionless g-factor describes the percentage loss of luminescence signal per decade of time, and t_c is the time when the first measurement L_c takes place. In practical situations one plots the remnant luminescence signal L as a function of $\log_{10}(t)$, and the slope of the linear part of this graph is the g-factor. According to this definition, the g-value depends on the exact point in time (t_c) in which the first measurement is carried out, as well as on the choice of the linear part of the experimental data (see Li and Li [98], for a more detailed discussion).

An example of how the g-factor is evaluated from experimental data is shown in the following code, using the experimental data by Polymeris et al. [157], on Durango apatite. Figure 6.5a shows the remnant TL signal measured at two different times. The results of the calculation are shown in Fig. 6.5b. In this method one finds the slope K of the graph of L/L_c vs. $\ln(t)$, and then Eq. (6.16) is used to find the value of g.

In this example we obtain $g = 20.5\%$ per decade, which is a high g-value characteristic of several types of apatites.

Code 6.4:   Anomalous fading (AF) and the *g*-factor

```r
rm(list=ls())
# Load the data of Anomalous fading (AF) and calculate the g-factor
par(mfrow=c(1,2))
mydata <- read.table("durnago0sok.txt")
T<-mydata[,1]
TL<-mydata[,2]/1e5
plot(T,TL,type="o",pch=1,col="red",xlim=c(50,450),ylim=c(0,1.6),
  xlab=expression("Temperature ["^"o"*"C]"),ylab ="TL (a.u.)")
mydata2 <- read.table("durango10daysok.txt")
T2<-mydata2[,1]
TL2<-mydata2[,2]/1e5
lines(T2,TL2,type="o",pch=2,col="blue")
legend("left",bty="n","(a)")
legend("topleft",bty="n",lwd=2, lty=c(NA,NA,1,2,3),
pch=c(NA,NA,1,2),
legend=c(expression('Anomalous', 'fading ', 't=0 s',
't=10 days')),col=c(NA,NA,"red","blue"))
mydata3 <- read.table("DurangoAFdataok.txt")
t<-mydata3[,1]
RTL<-mydata3[,2]
#plot(t,RTL,xlab="ln(t/to)",ylab="Remnat TL [%]")
y<-RTL
x<-t
bestfit<-lm(y~x)
coefficients(bestfit)
plot(x, y, xlab=expression("ln(t/to)"),ylab = "RTL [%]")
abline(lm(y~x))
slope<-abs(coefficients(bestfit)[[2]])
g<-230.2*slope
paste0("g-factor=",round(g,digits=2)," % per decade")
legend("topright",bty="n",
legend=c(expression('Durango apatite',
'g-factor=20.5% ',)))
legend("left",bty="n","(b)")

## (Intercept)           x
##  0.99615757 -0.08918921
## [1] "g-factor=20.53 % per decade"
```

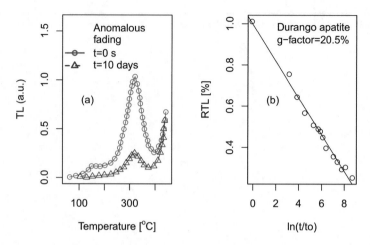

Fig. 6.5 Anomalous fading effect in Durango apatite. (**a**) The TL signal is measured immediately after irradiation, and after 10 days have elapsed at room temperatures. (**b**) Analysis of the remnant TL signal from (**a**), to obtain the g-factor for this material. For more details, see Polymeris et al. [157]

Pagonis and Kitis [134] derived a relationship between the g_{50} values at the 50% reduction point of the luminescence signal, as a function of the concentration ρ' of charge carriers in the material.

$$g_{50} = 2.7035 \left(\rho'\right)^{1/3} \tag{6.17}$$

Using a value of $\rho' = 3 \times 10^{-6}$, this equation gives $g = 0.039 = 3.9\%$ per decade, in agreement with experimental data (Huntley and Lian [61], Huntley and Lamothe [60]).

By contrast, using a value of $\rho' = 5 \times 10^{-3}$, this equation gives a very high factor $g = 0.462 = 46.2\%$ per decade, in agreement with experimental data for excited state tunneling for various apatites. Pagonis and Kitis [134] tested this new analytical equation for g_{50}, by fitting experimental data for several types of apatites that are known to exhibit high anomalous fading effects.

Experimental work on feldspars has shown the existence of two different ranges of values for the dimensionless acceptor density ρ', which characterizes feldspars. One observes either a low value of the density $\rho' \sim 10^{-6}$ that explains anomalous fading phenomena (see for example Li and Li [98]), or a much higher value $\rho' \sim 10^{-3}$ that explains excited state tunneling phenomena, about three orders of magnitude larger. For example, see the detailed experimental study of the CW-IRSL signals in 23 feldspar samples by Pagonis et al. [132]), which established a range of $\rho' = 0.002-0.01$.

Pagonis et al. [141] discussed the possible physical basis of these two ranges of values of the dimensionless density ρ'. Low values of $\rho' = 10^{-6}$ are most likely associated with a small value of the tunneling constants of the order of $a = 0.2$ nm in Eq. (6.1). Similarly, high values of $\rho' = 10^{-3}$ may be associated with much larger tunneling constants a in Eq. (6.1), of the order of $a = 2$ nm. For a discussion of the crystal structure and of the corresponding wavefunctions in feldspars, the reader is referred to the two excellent papers by Poolton [158, 159].

6.6   Simultaneous Ground State Tunneling and Irradiation of the Sample

Li and Li [98] applied the model of Huntley and Lian [61] in an extensive experimental and modeling study of both laboratory-irradiated and naturally irradiated feldspars. The modeling results of Li and Li [98] were expressed in terms of integral equations that require numerical integration over the donor–acceptor distances in the model. Later Kars et al. [64] applied the model by Huntley and Lian [61] to construct unfaded and natural dose response curves (DRCs) of IRSL signals. On the basis of their modeling and experimental results, Kars et al. [64] proposed a new correction method for anomalous fading effects, and their model predicted the anomalous fading rate for samples that had reached field saturation.

Li and Li [98] investigated the simultaneous effects of irradiation and tunneling by using the differential equation:

$$\frac{\partial n\left(r',t\right)}{\partial t} = \frac{\dot{D}}{D_0}\left[N\left(r'\right) - n\left(r',t\right)\right] - n\left(r',t\right)s\exp\left[-\left(\rho'\right)^{-1/3}r'\right] \tag{6.18}$$

where s (s^{-1}) is the frequency characterizing the ground state tunneling process, $\dot{D}$ (Gy/s) is the dose rate, and D_0 (Gy) is the characteristic dose of the sample. This equation is valid for samples irradiated in nature with a very slow dose rate $\dot{D}$ (Gy/s) of the order of 10^{-11} Gy/s, but also for samples irradiated with the much higher dose rates of about 0.1 Gy/s used in the laboratory. The parameter $N(r') = N g(r')$ represents the total concentration of traps corresponding to a distance parameter r', and N is the total number of traps in the sample. The rest of the parameters in this equation have the same meaning as in the previous section.

The first term in the right hand side of Eq. (6.18) represents the rate of increase of the concentration $n(r',t)$ due to irradiation, while the second term represents the decrease in the concentration $n(r',t)$ due to the effect of ground state tunneling. The solution of the first order differential equation (6.18) when one starts with initially empty traps $n\left(r',0\right) = 0$ is the following saturating exponential function:

$$n\left(r',t\right) = N g\left(r'\right)\frac{\dot{D}}{D_0 s_{eff}\left(r'\right) + \dot{D}}\left[1 - \exp\left\{-\left[\frac{\dot{D}}{D_0} + s_{eff}\left(r'\right)\right]t\right\}\right] \tag{6.19}$$

where for simplicity of presentation we defined the effective frequency $s_{eff}\left(r'\right)$ for the process as

$$s_{eff}\left(r'\right) = s \exp\left[-\left(\rho'\right)^{-1/3} r'\right] \tag{6.20}$$

The total concentration $n(t)$ of trapped electrons at time t is evaluated numerically by integrating $n\left(r', t\right)$ over all possible values of the variable r':

$$n(t) = \int_0^\infty n\left(r', t\right) dr' \tag{6.21}$$

$$n(t) = \int_0^\infty Ng(r') \frac{\dot{D}}{D_0 s_{eff}\left(r'\right) + \dot{D}} \left[1 - \exp\left\{-\left[\frac{\dot{D}}{D_0} + s_{eff}\left(r'\right)\right]t\right\}\right] dr' \tag{6.22}$$

The modeling results of Li and Li [98] were expressed in terms of these types of *integral equations*, which require numerical integration over the distances r' in the model. As mentioned in the previous section, several R functions will be presented later in Chap. 12 and will circumvent the need for these numerical integrations, by replacing them with finite sums over the distances r'.

In the rest of this section we will study the approximate analytical equation developed by Pagonis and Kitis [134], for the dose response of feldspar samples. These authors showed that the integral equations for DRCs in the model of Li and Li [98] can be replaced with the following analytical equation:

Simultaneous irradiation and ground state tunneling

$$L_{\text{FADED}}(D) = M\left(1 - \exp\left[-D/D_0\right]\right) \exp\left[-\rho' \ln\left(\frac{D_0 s}{X}\right)^3\right] \tag{6.23}$$

---

where L_{FADED} is the luminescence signal measured at the end of the anomalous fading process, X is the natural irradiation rate, D_0 is the characteristic dose of the unfaded sample, M is the total number of trapping sites, and the term $M\left(1 - \exp\left[-D/D_0\right]\right)$ represents the luminescence signal L_{UNFADED} that would have been obtained in the absence of any fading. The ratio of the faded over the unfaded signal will depend on the parameters s, ρ', D_0, and X and is equal to

$$\frac{L_{\text{FADED}}(D)}{L_{\text{UNFADED}}(D)} = \exp\left[-\rho' \ln\left(\frac{D_0 s}{X}\right)^3\right] \tag{6.24}$$

Figure 6.6 shows examples of Eq. (6.23) for the loss of charge $n(t)$ during simultaneous irradiation and fading in nature, for three different dimensionless acceptor densities $\rho' = 10^{-6}$, 2×10^{-6}, 3×10^{-6}. The loss is due to ground state tunneling. As ρ' increases, the rate of charge loss increases. The tunneling frequency is $s = 3 \times 10^{15}\,\text{s}^{-1}$, the characteristic dose is $D_0 = 502\,\text{Gy}$, and natural irradiation rate is $X = 3\,\text{Gy/ka}$.

By using the values $\rho' = 3 \times 10^{-6}$, $s = 3 \times 10^{15}\,\text{s}^{-1}$, $D_0 = 538\,\text{Gy}$, and $X = 3\,\text{Gy/ka}$ in Eq. (6.24), we obtain the fading ratio $= 0.438$, i.e. 43.8% of the original signal remaining in the sample. This is in agreement with the results in Fig. 6.6.

Code 6.5: Simultaneous irradiation and anomalous fading in nature

```
# Simultaneous irradiation and anomalous fading in nature
rm(list=ls())
s<-3e15                   # frequency factor
rho1<-1e-6               # rho-prime values 0.005-0.02
rho2<-2e-6
rho3<-3e-6
Do<-538
X<-3/(1000*365*3600*24)        #natural dose rate X=3 Gy/Ka
curve((1-exp(-x/Do))*exp(-rho1*(log(Do*s/X)**3.0)),1,3500, 200,
lty=1,lwd=3,
ylab=expression('L'[FADED]*'  [a.u.]'),xlab="Natural Dose [Gy]",
col=1,ylim=c(0,1.2))
curve((1-exp(-x/Do))*exp(-rho2*(log(Do*s/X)**3.0)),1,3500,200,
col=2,lty=2, lwd=3,  add=TRUE)
curve((1-exp(-x/Do))*exp(-rho3*(log(Do*s/X)**3.0)),1,3500,200,
col=3,   lwd=3,   lty=3,add=TRUE)
legend("topright",bty="n",lwd=3, lty=c(NA,1,2,3),legend =
c("Acceptor Density"  ,expression("10"^"-6"*""),
expression("2x10"^"-6"*""),expression("3x10"^"-6"*"")),
col=c(NA,1:3))
legend('topleft',bty="n",c("Simultaneous",
"Irradiation and fading") )
```

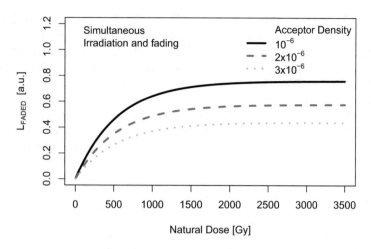

Fig. 6.6 Examples of the accumulated charge $n(t)$ during simultaneous irradiation and fading in nature, due to ground state tunneling, for three different dimensionless acceptor densities $\rho' = 10^{-6}, 2 \times 10^{-6}, 3 \times 10^{-6}$. As ρ' increases, the rate of charge accumulation and the corresponding saturation signal *decrease*. For more details, see Pagonis and Kitis [134]

6.7 Excited State Tunneling Phenomena (EST Model)

Historically the model was first considered by Thioulouse et al. [177], and Chang and Thioulouse [30]. Almost 30 years later, Jain et al. [62] developed a general kinetic model that quantifies localized recombination within randomly distributed donor–acceptor pairs. The energy band model is shown in Fig. 6.7. The recombination takes place via the excited stage of the donor, with the nearest neighbor acceptor. Jain et al. [62, 63] presented the model in two forms. In the first exact version of the model, the concentrations of trapped electrons $n(r', t)$ evolve in both space and time, while in the second approximate version of the model the concentrations $n(t)$ evolve only in time.

The more general version of the model requires multiple numerical integrations of the differential equations, over the distances r' and over the time variable t.

The equations for the model by Jain et al. [62] are

$$\frac{\partial n_g \left(r', t \right)}{dt} = -p(t) \, n_g \left(r', t \right) + B \, n_e \left(r', t \right) \tag{6.25}$$

$$\frac{\partial n_e}{dt} = p(t) \, n_g \left(r', t \right) - B \, n_e \left(r', t \right) - \frac{n_e \left(r', t \right)}{\tau (r')} \tag{6.26}$$

$$L(t) = -\frac{dm}{dt} = \int_{r'=0}^{\infty} \frac{n_e \left(r', t \right)}{\tau (r')} dr' \tag{6.27}$$

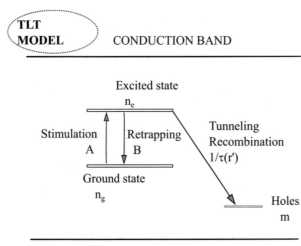

Fig. 6.7   The energy band model for the EST model (Jain et al. [62]). Electrons are excited from the ground state (n_g) into the excited state (n_e) of the trap, at a rate A. From the excited state, they can either tunnel to the nearest recombination center at the rate $1/\tau$ or relax back to the ground state of the trap at the rate B. The tunneling lifetime τ is a function of the donor–acceptor separation distance. The concentrations n_g and n_e are a function of both distance and time in the model. From Pagonis et al. [131]

$$\tau(r') = (s_{tun})^{-1} \exp\left[(\rho')^{-1/3} r' \right] \tag{6.28}$$

where n_g (cm^{-3}) is the instantaneous concentration of electrons (donors) in the ground state, which is a function of time t (s) and of the distance parameter r', defined previously by $r' = (4\pi\rho/3)^{1/3} r$ with ρ (cm^{-3}) representing the density of recombination centers in the material and r (m) the donor–acceptor distance. As in the previous section, ρ' is the dimensionless number density of recombination centers (acceptors), defined by $\rho' = (4\pi\rho/3)\,\alpha^{-3}$, where $\alpha(m^{-1})$ is the potential barrier penetration constant (Jain et al. [62]). The term $p(t)$ describes the experimental thermal or optical excitation rates, as in the previous chapters. s_{tun} (s^{-1}) is the frequency factor characterizing the tunneling process taking place from the excited state of the system, $A(t)$ (s^{-1}) is the excitation rate from the ground to the excited state (transition A in Fig. 6.7), and B (s^{-1}) is the rate of de-excitation from the excited state back to the ground state (transition B in Fig. 6.7). From a physical point of view, there is no direct relationship between the three different frequency factors A, B, and s_{tun}. The tunneling lifetime $\tau = (s_{tun})^{-1} \exp\left[(\rho')^{-1/3} r' \right]$ (in s) depends on the distance parameter r', and $L(t)$ is the instantaneous photon emission rate resulting from tunneling recombination via the excited state (recombination transition $1/\tau$ in Fig. 6.7).

Pagonis et al. [145] discussed the EST model of Jain et al. [62] and noted that the above system of equations can be approximated using QE conditions, with the following single differential equation:

$$\frac{\partial n\left(r',t\right)}{\partial t} = -n\left(r',t\right)\frac{A\left(t\right)s_{tun}}{B}\exp\left[-\left(\rho'\right)^{-1/3}r'\right] \tag{6.29}$$

Equation (6.29) is a first order differential equation *for a fixed distance r'*, and its solution is

$$n\left(r',t\right) = Ng\left(r'\right)\exp\left\{-\frac{s_{tun}}{B}\exp\left[-\left(\rho'\right)^{-1/3}r'\right]\int_0^t A\left(t\right)dt\right\} \tag{6.30}$$

The excitation rate $A(t)$ is different for various types of experiments (TL, CW-IRSL, etc.) and is discussed in the next two sections.

The *total* concentration of trapped electrons at time t is found once more by integrating over all distances r':

$$n(t) = \int_0^\infty n\left(r',t\right)dr' \tag{6.31}$$

$$n(t) = \int_0^\infty Ng\left(r'\right)\exp\left\{-\frac{s_{tun}}{B}\exp\left[-\left(\rho'\right)^{-1/3}r'\right]\int_0^t A\left(t\right)dt\right\}dr' \tag{6.32}$$

and the time dependent luminescence intensity $L(t)$ is evaluated numerically by integrating the derivative $\partial n(r',t)/\partial t$ over all possible values of the variable r', and using Eq. (6.29):

$$L\left(t\right) = -\int_0^\infty \frac{\partial n\left(r',t\right)}{\partial t}dr' = \int_0^\infty n\left(r',t\right)\frac{A\left(t\right)s_{tun}}{B}\exp\left[-\left(\rho'\right)^{-1/3}r'\right]dr'$$

$$\tag{6.33}$$

We will encounter these equations again in Chap. 12, when we develop R functions to replace these numerical integrations with finite sums.

In the next section, we present an approximate analytical solution for Eq. (6.29).

6.8 The Kitis–Pagonis Analytical Solution of the EST Model

Kitis and Pagonis [73] derived an analytical equation solution for the EST model, by considering quasi-equilibrium conditions (QE). One of the assumptions in the model is that the concentration of electrons $n_e(t)$ in the excited state at any time is several orders of magnitude smaller than the concentration of electrons $n_g(t)$ in the ground state of the system. Furthermore, the relaxation of the excited states is much faster than the time scales of TL and OSL experiments. Therefore, one can model the excited state in the quasi-steady approximation. Specifically, the time scale for electronic relaxation processes involving the excited states is of order of ns or ms. On the other hand, the time scale for TL/OSL processes is of the order of s.

Based on these observations, Kitis and Pagonis [73] carried out extensive algebra and obtained the following analytical solutions for the remaining concentration of electrons $n(t)$:

Loss of charge due to excited state tunneling

$$n(t) = n_0 \exp\left(-\rho' \, \ln\left[1 + \frac{z \, s_{tun}}{B} \int_0^t A(t) \, dt\right]^3\right) \qquad (6.34)$$

Notice that this equation for the loss of charge $n(t)$ due to *excited state tunneling* in the EST model is almost identical to the corresponding equation (6.13) for the loss of charge due to *ground state tunneling* in the GST model. As in the original paper by Jain et al. [62], Kitis and Pagonis [73] assumed that $s_{tun} = B$ in order to simplify the notation in the model. This assumption does not alter the results from the model because the equations contain the combination $A(t)s_{tun}/B$, rather than the variables $A(t)$, s_{tun}, and B separately.

The corresponding luminescence intensity $I(t)$ is found from the derivative $-dn/dt$ of Eq. (6.34), and we will refer to this analytical equation as the *general Kitis–Pagonis (KP) equation*:

The general KP equation for luminescence intensity

$$I(t) = 3 \, n_0 \, \rho' \, z \, A(t) \, F(t)^2 \, e^{-F(t)} \, e^{-\rho'(F(t))^3} \qquad (6.35)$$

$$F(t) = \ln\left(1 + \frac{z \, s_{tun}}{B} \int_0^t A(t) \, dt\right) \qquad (6.36)$$

Equation (6.35), along with Eq. (6.36), has been termed the *fifth master equation* in the review paper by Kitis et al. [78]. These analytical expressions can characterize TL, IRSL, and ITL signals in the EST model, as long as one is dealing with freshly irradiated samples, i.e. samples that have not undergone any thermal or optical treatments after irradiation.

The function $F(t)$ takes the following form for the various stimulated luminescence phenomena.

In the case of CW-IRSL decay curves, we have $A(t) = \sigma I$, where σ, I are the optical cross section and photon flux, so that the function $F(t)$ is

$$F_{\text{CW-IRSL}}(t) = \ln\left(1 + \frac{z\,\sigma I\,s_{tun}}{B}\,t\right) = \ln\left(1 + z\,A\,t\right) \tag{6.37}$$

where $A = \sigma I\,s_{tun}/B$ (s^{-1}) is the effective rate constant characterizing the CW-IRSL process in the EST model.

In the case of TL, we have $A(t) = s_{th}\exp(-E/k_B T)$, where E, s_{th} are the thermal parameters for the trap, and $F(t)$ has the form:

$$F_{\text{TL}}(t) = \ln\left(1 + \frac{z\,s_{th}\,s_{tun}}{B}\,\frac{1}{\beta}\int_{T_0}^{T} e^{-\frac{E}{k_B T'}}\,dT'\right) \tag{6.38}$$

$$F_{\text{TL}}(t) = \ln\left(1 + \frac{z\,s}{\beta}\int_{T_0}^{T} e^{-\frac{E}{k_B T'}}\,dT'\right) \tag{6.39}$$

where $s = s_{th}\,s_{tun}/B$ is the effective rate constant (or effective frequency factor) characterizing the TL process in the EST model.

Similarly, for LM-IRSL peaks, we have $A(t) = \sigma It/P$, where P is the total excitation time, and the function $F(t)$ takes the form:

$$F_{\text{LM-IRSL}}(t) = \ln\left(1 + \frac{z\,\sigma I\,s_{tun}}{B}\,\frac{t^2}{2P}\right) = \ln\left(1 + z\,\frac{A\,t^2}{2P}\right) \tag{6.40}$$

where $A = \sigma I\,s_{tun}/B$ (s^{-1}) is the effective rate constant characterizing the LM-IRSL process in the EST model.

Finally, for ITL decay curves, we have $A(t) = s_{th}\exp(-E/k_B T_{ISO})$, where E, s_{th} are the thermal parameters for the trap and T_{ISO} is the temperature of the isothermal process, and $F(t)$ has the form:

$$F_{\text{ITL}}(t) = \ln\left(1 + \frac{z\,s_{th}\,s_{tun}}{B}\exp(-E/k_B T_{ISO})\,t\right) = \ln\left(1 + z\,A\,t\right) \tag{6.41}$$

where $A = s_{th}\,s_{tun}\exp(-E/k_B T_{ISO})/B$ (s^{-1}) is the effective rate constant characterizing the ITL process in the EST model.

The fifth master equation Eq. (6.35) was tested by Kitis and Pagonis [73], by comparing it with the numerical solution of the differential equations in the EST model (Kitis and Pagonis [73, 74], Pagonis and Kitis [134]). This equation has also been tested extensively during the past decade, by comparing it with many different types of experimental signals, from different types of natural and artificial dosimetric materials (Sfampa et al. [166], Sahiner et al. [163], Kitis et al. [81], Polymeris et al. [155]).

In the next two sections we give specific examples of using the general KP equations (6.35), to analyze experimental CW-IRSL and TL data, by using appropriate R codes.

6.9 Fitting CW-IRSL Data Using the KP-CW Equation

In the previous section, we saw that CW-IRSL signals can be described with the analytical equations:

$$I_{CW-IRSL}(t) = 3\, n_0\, \rho'\, z\, p(t)\, F(t)^2\, e^{-F(t)}\, e^{-\rho'(F(t))^3} \tag{6.42}$$

where $F_{CW-IRSL}(t)$ is given above in Eq. (6.37).

For practical work fitting experimental data, we use this equation in the following form, which will be hereafter named the *KP-CW equation*:

The KP-CW equation

$$I_{CW-IRSL}(t) = \frac{I_0\, F(t)^2\, e^{-\rho'(F(t))^3}}{1 + zAt} + bgd \tag{6.43}$$

$$F_{CW-IRSL}(t) = \ln(1 + zAt) \tag{6.44}$$

---

where $z = 1.8$ and the fitting parameters are the effective rate constant A (in s^{-1}) for the CW-IRSL process, the scaling constant I_0, the dimensionless acceptor density ρ', and a constant experimental background bgd (if necessary). Several experimental studies have found that typical values of the infrared stimulation rate A are $1-10\,s^{-1}$ and that for feldspars in the EST model, the typical values of the dimensionless density are in the range $\rho' = 0.003-0.02$ (see, e.g. the comprehensive study of 23 feldspar samples by Pagonis et al. [132]).

The following code fits a CW-IRSL experimental curve from the feldspar sample KST4 studied by Kitis et al. [80], and the least squares fitting result is shown in Fig. 6.8. The best fitting values in this example are $A = 7.07\,s^{-1}$, $I_0 = 2.72 \times 10^4$ cts/s, $\rho' = 0.0073$, and $bgd = 47.05$ cts/s.

Note the following practical detail: the analytical KP-CW equation (6.43) gives a non-physical answer of zero for the intensity at time $t = 0$, i.e. $I_{CW-IRSL}(0) = 0$. For this reason, it is necessary that the file that contains the experimental data starts at some value $t > 0$ (e.g. at time $t = 1$ s), and *not* at time $t = 0$. This is a necessary condition for the fitting process to work correctly.

Code 6.6: CW-IRSL data fitted with the KP-CW equation

```
#Fit CW-IRSL data with KP-CW equation
rm(list = ls(all=T))
options(warn=-1)
library("minpack.lm")
## fit to analytical KP-CW equation for CW-IRSL (TLT model)-
mydata <- read.table("ph300s0IR.asc")
t<-as.numeric(gsub(",", ".", gsub("\\.", "", mydata[,1])))
y<-as.numeric(gsub(",", ".", gsub("\\.", "", mydata[,3])))
mydata<-data.frame(t,y)
plot(t,y,log="xy",
xlab="Time [s]",ylab="CW-IRSL [counts/s]",col="black",pch=1)
fit_data <-mydata
fit <- minpack.lm::nlsLM(
 formula=y~ imax*exp (-rho*(log(1 + A*t)) ** 3.0)*
    (log(1 + A*t) ** 2.0)/(1 +  t*A)+bgd,
  data = fit_data,
  start = list(imax=3,A=1.1,rho=0.03,bgd=min(y)))
imax_fit <- coef(fit)[1]
A_fit <- coef(fit)[2]
rho_fit <- coef(fit)[3]
bgd_fit <- coef(fit)[4]
## plot analytical solution
lines(
 x = t,
 y =imax_fit*exp (-rho_fit*(log(1 + A_fit*t)) ** 3.0)*
    (log(1 + A_fit*t) ** 2.0)/(1 +  t*A_fit)+bgd_fit,
  col = "red",lwd=2,log="xy")
legend("topright",bty="n", pch=c(NA,NA,1,NA),lwd=2,
lty=c(NA,NA,NA,"solid"),
c(expression('KST4 feldspar',' ',
'Experiment','KP-CW equation')),col=c(NA,NA,"black","red"))
## print results
cat("Parameters from Least squares fit"," ")
cat("\nImax=",formatC(imax_fit,format="e",digits=2)," cts/s",
sep="   ","A=",round(A_fit,digits=2)," (s^-1)")
cat("\nrho=",round(rho_fit,digits=4),sep=" ",
" bgd=",round(bgd_fit,digits=2)," cts/s")

   ## Parameters from Least squares fit
   ## Imax=    2.72e+04     cts/s    A=    7.07     (s^-1)
   ## rho= 0.0073   bgd= 47.05  cts/s
```

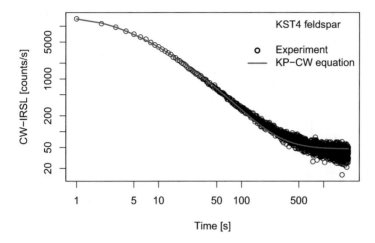

Fig. 6.8 Experimental CW-IRSL glow curves from freshly irradiated KST4 feldspar sample, fitted using the KP-CW analytical equation (6.43). For more details and examples, see Pagonis et al. [131]

6.10 Fitting TL Data from Freshly Irradiated Samples Using the KP-TL Equation

For freshly irradiated samples, the TL signals are analyzed using the following KP equation for the intensity of a TL signal (see Eqs. 29–30 in Kitis and Pagonis [73]):

$$I_{\text{TL}}(t) = 3\, n_0\, \rho'\, z\, p(t)\, F(t)^2\, e^{-F(t)}\, e^{-\rho'(F(t))^3} \tag{6.45}$$

where $F_{\text{TL}}(t)$ is given above in Eq. (6.38). For practical work fitting experimental data, we use this equation in the following form, which will be hereafter named as the *KP-TL equation*:

The KP-TL equation

$$I_{\text{TL}}(t) = \frac{I_0\, F(t)^2\, e^{-\rho'(F(t))^3}\,\left(E^2 - 6k_B^2 T^2\right)}{E k_B s T^2 z - 2k_B^2 s T^3 z + \exp\left(E/k_B T\right) E\beta} + bgd \tag{6.46}$$

$$F_{\text{TL}}(t) = \ln\left(1 + \frac{z\, s k_B T^2}{\beta E}\, e^{-\frac{E}{k_B T}}\left(1 - \frac{2k_B T}{E}\right)\right) \tag{6.47}$$

The fitting parameters are the scaling constant I_0, thermal activation energy E (eV), effective frequency factor s (s^{-1}) for the TL process, dimensionless acceptor density

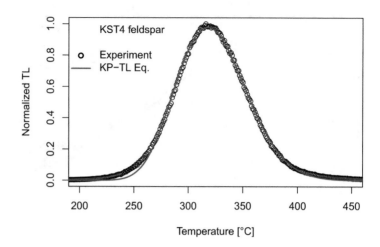

Fig. 6.9 Experimental TL glow curves from freshly irradiated KST4 feldspar sample, fitted using the KP-TL analytical equation (6.46). Note that the solid line does not describe the experimental data very accurately at low temperatures, due to the approximations involved in the KP equations. For more details and MC examples, see Pagonis et al. [131]

ρ', and a constant background bgd (if necessary). Here β is the heating rate, and in the following code, we use $\beta = 1$ K/s. Even though the KP-TL equation looks complex, it is rather easy and straightforward to code it.

In order to obtain reliable results with the KP-TL equation, one *must* constraint the code using experimental data, because there are infinite combinations of the parameters E, s, ρ' that will fit the data. It is best to use the known value of E, which can be obtained from separate initial rise and $T_{max} - T_{stop}$ experiments. In the following code we use $E = 1.45$ eV, a value obtained from a separate analysis using the initial rise method, and the least squares fitting parameters are I_0, s, and ρ'. The experimental data are from a feldspar sample KST4 that was irradiated and heated up to 300 °C, to thermally remove the lower temperature peaks in the TL glow curve, before measuring the TL signal. It is assumed that heating the sample to 300 °C does not affect the trap responsible for the TL signal at 320 °C.

Notice that in Fig. 6.9, the KP-TL equation fails to describe the TL glow curve very well at low temperatures. This is due to the approximations made during derivation of the KP-TL equation (see the detailed discussion in Kitis and Pagonis [74]).

For more details on the experimental data of Fig. 6.9, see Pagonis et al. [131], and Polymeris et al. [155]. It is also important to remember that the physical meaning of the parameter s in the KP-TL equation is an *effective* frequency factor for the TL process (equal to $s = s_{th} \, s_{tun}/B$) and *not* the frequency factor s_{th} for the thermal activation process.

Code 6.7:   Kitis–Pagonis analytical equation for TL (KP-TL)

```
rm(list = ls(all=T))
options(warn=-1)
library("minpack.lm")
library(expint)
## Least squares fit to TL using the KP-TL eqt (TLT model)-
mydata <- read.table("ph300s0.asc")
t<-as.numeric(gsub(",", ".", gsub("\\.", "", mydata[,1])))
y<-as.numeric(gsub(",", ".", gsub("\\.", "", mydata[,3])))
y<-y/max(y)
mydata<-data.frame(t,y)
plot(t,y,xlab="Temperature [\u00B0C]",ylab="Normalized TL",
col="black",pch=1,xlim=c(200,450))
kb<-8.617e-5
z<-1.8
fit_data <-mydata
kB<-8.617E-5
En<-1.45
T<-t+273
fit <- minpack.lm::nlsLM(
  formula=y~imax* exp(-rho*( (log(1+z*s*kB*((T**2.0)/
abs(En))*exp(-En/(kB*T))*(1-2*kB*T/En)))**3.0))*
(En**2.0-6*(kB**2.0)*(T**2.0))*( (log(1+z*s*kB*((T**2.0)/
abs(En))*exp(-En/(kB*T))*(1-2*kB*T/En)))**2.0)/
(En*kB*s*(T**2)*z-2*(kB**2.0)*s*z*(T**3.0)+exp(En/(kB*T))*En),
  data = fit_data,
start = list(imax=1e12,s=1e11,rho=.009),upper=c(1e20,1e13,.02),
  lower=c(1e11,1e11,.008))
# Obtain parameters from best fit
imax_fit <- coef(fit)[1]
s_fit <- coef(fit)[2]
rho_fit <- coef(fit)[3]
En_fit <- En
## plot analytical solution
lines(
x = t,imax_fit* exp(-rho_fit*( (log(1+z*s_fit*kB*((T**2.0)/
abs(En_fit))*exp(-En_fit/(kB*T))*(1-2*kB*T/En_fit)))**3.0))*
(En_fit**2.0-6*(kB**2.0)*(T**2.0))*((log(1+z*s_fit*kB*((T**2.0)/
abs(En_fit))*exp(-En_fit/(kB*T))*(1-2*kB*T/En_fit)))**2.0)/
(En_fit*kB*s_fit*(T**2)*z-2*(kB**2.0)*s_fit*z*(T**3.0)+
exp(En_fit/(kB*T))*En_fit),col="red",lwd=2)
legend("topleft",bty="n", pch=c(NA,NA,1,NA),lwd=2,
lty=c(NA,NA,NA,"solid"),
c(expression('KST4 feldspar',' ',
'Experiment','KP-TL Eq.')),col=c(NA,NA,"black","red"))
```

```
## print results
cat("\nBest fit parameters"," ")
cat("\nImax=", formatC(imax_fit, format = "e", digits = 2),
" s=",formatC(s_fit, format = "e", digits = 2)," (s^-1) ")
cat("\nrho=",round(rho_fit,digits=4),sep=" ","E=",En_fit," eV")

##
## Best fit parameters
## Imax= 1.40e+13  s= 3.50e+12   (s^-1)
## rho= 0.0096 E= 1.45   eV
```

6.11   The Low Temperature Thermochronology Model by Brown et al.

Brown et al. [24] investigated the simultaneous effects of irradiation and quantum tunneling on the TL glow curves in materials with a random distribution of defects. This model is shown schematically in Fig. 6.2d and will be referred to as the thermally assisted EST (TA-EST) model.

These authors extended the original model by Jain et al. [62, 63] and Li and Li [98], to include all three processes: irradiation, quantum tunneling, and thermally assisted processes. This model was developed for low temperature thermochronology; however, it is completely general and can be used for any thermally active dosimetric trap. The model is based on the following differential equation:

$$\frac{\partial n\left(r',t\right)}{\partial t} = \frac{\dot{D}}{D_0}\left[N\left(r'\right) - n\left(r',t\right)\right] - n\left(r',t\right)\exp\left(-\frac{E}{k_B T}\right)\frac{P\left(r'\right)s}{P\left(r'\right)+s} \tag{6.48}$$

where $P(r')$ is the rate of excited state tunneling given by

$$P(r') = P_0 \exp\left[-\left(\rho'\right)^{-1/3} r'\right] \tag{6.49}$$

The parameters in this model are the tunneling frequency P_0 (s^{-1}), T represents the temperature of the sample, k_B is the Boltzmann constant, s (s^{-1}) is the trap frequency factor, and E (eV) is the thermal activation energy of the trap from the ground state to a state of higher energy. In order to simplify the notation, we can now define an effective tunneling probability $P_{eff}\left(r'\right)$ that depends also on the sample temperature T

$$P_{eff}\left(r'\right) = \frac{P\left(r'\right)s}{P\left(r'\right)+s}\exp\left[-E/\left(k_b T\right)\right] \tag{6.50}$$

The solution of the first order differential equation (6.48) *for a constant value of the distance r'*, and for initially empty traps $n(r', 0) = 0$, is the following saturating exponential function:

$$n(r', t) = N g(r') \frac{\dot{D}}{D_0 P_{eff}(r') + \dot{D}} \left[1 - \exp\left\{ -\left[\frac{\dot{D}}{D_0} + P_{eff}(r') \right] t \right\} \right]$$
(6.51)

The total concentration of trapped electrons at time t is evaluated numerically by integrating $n(r', t)$ over all possible values of the variable r':

$$n(t) = \int_0^\infty n(r', t) \, dr'$$
(6.52)

$$n(t) = \int_0^\infty N g(r') \frac{\dot{D}}{D_0 P_{eff}(r') + \dot{D}} \left[1 - \exp\left\{ -\left[\frac{\dot{D}}{D_0} + P_{eff}(r') \right] t \right\} \right] dr'$$
(6.53)

In Chap. 12 we develop R functions that calculate $n(t)$, by replacing this integral with finite sums over the distances r'.

It must be noted that Brown et al. [24] used a high value of $\rho' = 0.00132$ for their TA-EST model, which is three orders of magnitude larger than the GST model value of $\rho' = 10^{-6}$. It is also noted that this model contains two frequency factors s and P_0, while Eq. (6.29) introduced previously in this chapter for the EST model contains a single effective frequency factor $s_{eff}(r') = A(t)s_{tun}/B$. The input parameters in this TA-EST model are ρ', P_0, E, s, D_0, $\dot{D}$, T_{irr}, and t_{irr}. It is noted that there are no approximate analytical solutions for this model.

We also mention two other important recent papers on thermochronology applications of feldspars, the model by King et al. [67] that includes the effect of band-tail states, and the model by Biswas et al. [16]. For a review of thermochronometry applications, see King et al. [66].

6.12 Simulations Using the TUN Functions in the R Package *RLumCarlo*

The previous sections in this chapter have been based on the use of differential equations to describe the luminescence signals associated with quantum tunneling processes.

An alternative approach to the differential equations is the use of Monte Carlo methods. The MC approach is described in detail in Chap. 9. In this section we give several examples of using the package *RLumCarlo* to simulate a variety of luminescence signals due to tunneling from the excited state of the trapped electrons

Table 6.1 Table of TUNneling functions available in the package *RLumCarlo*

Function name	Description
plot_RLumCarlo	Plots "RLumCarlo" modeling results (the averaged signal or the number of remaining electrons), with modeling uncertainties
run_MC_CW_IRSL_TUN	Simulation of CW-IRSL signals due to tunneling transitions from the excited state of the trap into a recombination center (RC) for the EST model
run_MC_ISO_TUN	Simulation of ITL signals due to tunneling from the excited state of the trapped charge into the RC (EST)
run_MC_LM_OSL_TUN	Simulation of LM-IRSL signals due to tunneling from the excited state of the trapped charge into the RC (EST)
run_MC_TL_TUN	Simulation of TL signals due to tunneling from the excited state of the trapped charge into the RC (EST)

(Kreutzer et al. [87]). The following short table summarizes the available TUN functions and their content (Table 6.1).

The following code simulates a CW-IRSL signal for a sample with acceptor density $\rho' = 0.005$, an IR excitation rate $A = 5\,\text{s}^{-1}$, and a critical radius $rc = 0.5$. The code runs parallel evaluations if available, and the output in Fig. 6.10 is the CW-IRSL intensity. Table 6.2 shows typical numerical values for the input parameters.

Code 6.8: Single plot MC simulations for tunneling CW-IRSL

```
##===================================================##
## Example:Single Plot for  tunneling CW-IRSL
##===================================================##
rm(list = ls(all=T))
suppressMessages(library("RLumCarlo"))
run_MC_CW_IRSL_TUN(
A = 5,
rho = 5e-3,
times = 0:500,
r_c = 0.5,
delta.r = 1e-2,
method = "par",
output = "signal"
) %>%
#Plot results of the MC simulation
plot_RLumCarlo(norm = F, legend = F)
legend("top",bty="n",legend=c("CW-IRSL","TUN function"))
```

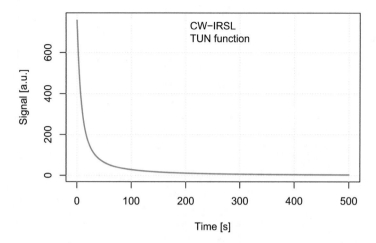

Fig. 6.10 Simulation of CW-IRSL signal using the function *run_MC_CW_IRSL_TUN* in the package *RLumCarlo*

Table 6.2 Table of input parameters for TUN functions in *RLumCarlo*

Process	Parameter	Description	Units	Typical values
TL with tunneling	E	Thermal activation energy	eV	0.5–3
	s	Effective frequency factor	1/s	1E8–1E16
	rho	Dimensionless density of recombination centers ρ'	1	1E-6–1E-2
	r_c	Critical distance (>0) for thermally/optically pretreated samples	1	0.1–0.8
	Times	Sequence of time steps (heating rate 1 K/s)	s	0–700
	Clusters	Number of MC runs	1	1E1–1E4
	N_e	Total number of electron traps available	1	2–1E5
	delta.r	Increments of the distance r'	1	0.01–0.1
CW-IRSL with tunneling	A	Effective optical excitation rate	1/s	1E-3–1
	rho	Dimensionless density of recombination centers ρ'	1	1E-6–1E-2
	Times	Sequence of time steps	s	0–500
	Clusters	Number of MC runs	1	1E1–1E4
	N_e	Total number of electron traps available	1	2–1E5
	r_c	Critical distance (>0) for thermally/optically pretreated sample	1	0.1–0.8
	delta.r	Increments of the distance r'	1	0.01–0.1

(continued)

Table 6.2 (continued)

Process	Parameter	Description	Units	Typical values
ISO with tunneling	E	Thermal activation energy	eV	0.5–3
	s	Effective frequency factor	1/s	1E8–1E16
	T	Temperature of the isothermal process	°C	20–300
	rho	Dimensionless density of recombination centers ρ'	1	1E-6–1E-2
	Times	Sequence of time steps for simulation	s	0–1000
	Clusters	Number of MC runs	1	1E1–1E4
	N_e	Total number of electron traps available	1	2–1E5
	r_c	Critical distance (>0) for thermally/optically pretreated sample	1	0.1–0.8
	delta.r	Increments of the distance r'	1	0.01–0.1
LM-OSL with tunneling	A	Effective optical excitation rate	1/s	1E-3–1
	rho	Dimensionless density of recombination centers ρ'	1	1E-6–1E-2
	Times	Sequence of time steps	s	0–3000
	Clusters	Number of MC runs	1	1E1–1E4
	N_e	Total number of electron traps available	1	2–1E5
	r_c	Critical distance (>0) for thermally/optically pretreated sample	1	0.1–0.8
	delta.r	Increments of the distance r'	1	0.01–0.1

The following code simulates and plots together two CW-IRSL signals for a sample with acceptor density $\rho' = 0.003$, and two different excitation rates $A = 3,\ 6\,\mathrm{s}^{-1}$. The code combines the two plots as shown in Fig. 6.11.

Code 6.9: Combining two plots in CW-IRSL experiment

```
rm(list = ls(all=T))
library(RLumCarlo)
times <- seq(0, 200)
run_MC_CW_IRSL_TUN(A = 3, rho = 0.003, times = times) %>%
  plot_RLumCarlo(norm = TRUE, lty=1,legend = TRUE)
run_MC_CW_IRSL_TUN(A = 6, rho = 0.003, times = times) %>%
  plot_RLumCarlo(norm = TRUE, lty=2,col="blue",add = TRUE)
legend("top",bty="n",legend=c(expression("CW-IRSL","TUN function",
  "A=3.0 s"^"-1*" ","A=6.0 s"^"-1*" ")),lty=c(NA,NA,1:2))
```

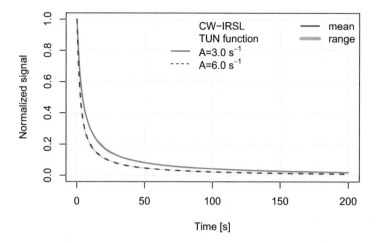

Fig. 6.11 Simulation of two CW-IRSL signals using the function *run_MC_CW_IRSL_TUN* in the package *RLumCarlo*, for two different excitation rates A

The following example simulates the number of remaining electrons $n(t)$ and the resulting ITL signal during an ITL experiment, using the function *run_MC_ISO_TUN* in the package *RLumCarlo*. The simulated sample has an acceptor density $\rho' = 0.01$ and thermal trap parameters $E = 1.2\,\text{eV}$, $s = 10^{12}\,\text{s}^{-1}$. The ITL experiment in Fig. 6.12a, b takes place at a temperature of $T = 200\,°\text{C}$.

Code 6.10: MC for tunneling ITL: remaining electrons

```
# MC Model for remaining charges and ITL signal
rm(list = ls(all=T))
library(RLumCarlo)
par(mfrow=c(1,2))
results <- run_MC_ISO_TUN(
    E = 1.0,
    s = 1e12,
    T = 250,
    rho = 0.01,
    clusters=100,
    times = seq(0, 200),
    output = "remaining_e"
) %T>%
plot_RLumCarlo(
    legend = FALSE,
```

```
ylab = "Remaining electrons" )
    legend("topright",bty="n",legend=c(" (a)"," ","n(t)",
"TUN function"))

results <- run_MC_ISO_TUN(
  E = 1.2,
  s = 1e12,
  T = 250,
  rho = 0.01,
  times = seq(0, 200)
) %T>%
  plot_RLumCarlo(norm = FALSE, legend = FALSE)
legend("topright",bty="n",legend=c(" (b)"," ","ITL signal"))
```

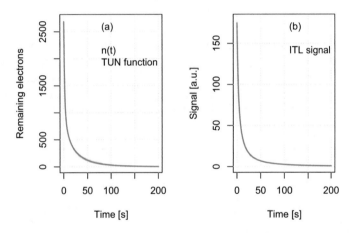

Fig. 6.12 Simulation of (**a**) remaining electrons $n(t)$ during an ITL experiment, and (**b**) of the resulting ITL signal, using the function *run_MC_ISO_TUN* in the package *RLumCarlo*

The following code and Fig. 6.13 simulate the LM-IRSL signal using the function *run_MC_LM_OSL_TUN*, for a sample with acceptor density $\rho' = 0.01$, and an excitation rate of $A = 3.0\,\mathrm{s}^{-1}$.

Code 6.11:   Single plot MC simulations for tunneling LM-OSL

```
##=======================================================##
## Example 1: MC simulations of tunneling LM_IRSL
##=======================================================##
rm(list = ls(all=T))
library(RLumCarlo)
run_MC_LM_OSL_TUN(
A =3.0,
rho = 1e-2,
times = 0:200,
clusters = 100,
N_e = 20,
r_c = 0.001,
delta.r = 1e-1,
method = "par",
output = "signal"
) %>%
# Plot results of the MC simulation
plot_RLumCarlo(norm = F,legend=F)
legend("topright",bty="n",legend=c("LM-IRSL signal",
"TUN function",""))
```

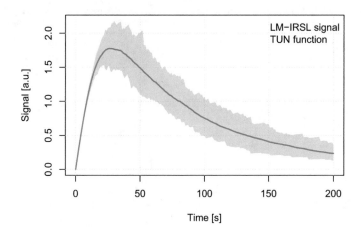

Fig. 6.13 Simulation of LM-IRSL signal using the function *run_MC_LM_OSL_TUN* in the package *RLumCarlo*

The following code and Fig. 6.14 simulate the TL signal using the function *run_MC_TL_TUN,* for a sample with acceptor density $\rho' = 0.01$, and thermal trap parameters $E = 1.3$ eV, $s = 10^{12}\,\text{s}^{-1}$. The sample has been thermally treated after irradiation, and the effect of the thermal treatment is described by the critical radius parameter $r_c' = 0,\ 0.8,\ 1.0$. The R function *lapply* is applied to the list *r_c*, that contains the r_c' values. The value of $r_c' = 0$ corresponds to the freshly irradiated sample, which has not undergone any treatment after irradiation. The default heating rate is 1 K/s.

Code 6.12: Simulation of TL from pretreated sample

```
##==================================================
## Simulate TL from pretreated sample
##==================================================
rm(list = ls(all=T))
library(RLumCarlo)
s <- 3.5e12
rho <- 0.015
E <- 1.45
r_c <- c(0,0.8,1.0)
times <- seq(100, 450) # time = temperature
results <- lapply(r_c, function(x) {
  run_MC_TL_TUN(
    s = s,
    E = E,
    rho = rho,
    r_c = x,
    times = times
  )})%>%
plot_RLumCarlo(norm = FALSE, legend = TRUE)
legend("topleft",bty="n",legend=c(expression('Critical Radius',
  ' ','r'[c]*'=0, 0.8, 1.0')) )
```

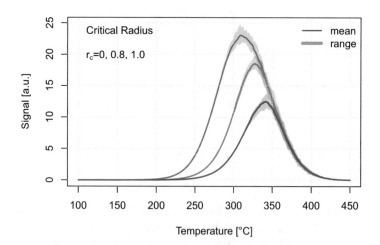

Fig. 6.14 Simulation of TL signal using the function *run_MC_TL_TUN* in the package *RLum-Carlo*. The sample has been thermally treated after irradiation, and the effect of the thermal treatment is described by the variable critical radius parameter $r_c' = 0, 0.8, 1.0$

Chapter 7
Localized Transitions: The LT and SLT Model

7.1 Localized Models with Constant Recombination Coefficients

In the luminescence literature, we find several types of localized transition models, the localized transition model (LT), the TLT model, and the semilocalized transitions model (SLT). In this chapter we study the LT and SLT models.

The astute reader will notice that, with the exception of the anomalous heating rate effect, there are no experimental data presented in this chapter, in connection with the LT model. The reason is that there are no luminescence materials whose behavior is unambiguously described by the LT model.

Section 7.2 presents a short overview of the LT model in the literature. Section 7.3 contains the differential equations for the LT model and their numerical and analytical solutions. This is followed in Sect. 7.4 by several examples of using the R package *RLumCarlo* to simulate these types of localized transition phenomena.

Section 7.5 examines the SLT model developed by Mandowski [101], which is a hybrid model with elements from both localized and delocalized transitions. In Sect. 7.6 we will discuss the anomalous heating rate effect which is observed in several dosimetric materials, and will provide description of this effect based on the Mandowski SLT model [101]. This chapter will end with Sect. 7.7 which contains a simplified form of the SLT model, developed by Pagonis et al. [122].

7.2 Overview of the LT Model

There have been several types of localized transition models in the literature. In the original work by Halperin and Braner [55], electrons are raised into the excited state of the electron trap during the thermal excitation stage, and from there they can either recombine at the recombination center, or they can be de-excited into the ground state. In this model, the conduction and valence bands participate

V. Pagonis, *Luminescence*, Use R!, https://doi.org/10.1007/978-3-030-67311-6_7

in the kinetics only during the irradiation stage, and direct transitions from the conduction band into the ground or excited state of the trap are not allowed. Most importantly, the rate of recombination in this model was assumed to depend on *both* the concentrations of electrons in the excited state of the trap and on the concentration of holes in the center. Chen and Kirsh [33] suggested that each electron can only recombine with its closest neighbor in the crystal, and modified the differential equation of Halperin and Braner [55], so that the rate of recombination is proportional *only* to the concentration of electrons in the excited state. This assumption when combined with the assumption of a quasi-equilibrium state leads to an analytical first order kinetics equation. Bull [25] used the same assumptions as Chen and Kirsh [33] and the same system of differential equations, to simulate characteristic TL glow curves in this localized model. The model parameters used by Bull [25] resulted in TL peaks obeying first order kinetics only.

During the past decade there has been extensive experimental and simulation work on luminescence from nanodosimetric materials. Recent Monte Carlo simulations have shown that in materials with very high density of defects, the nearest neighbor approximation may not be valid, and therefore each electron may interact with several of its neighbors (Pagonis and Truong [147]). In such cases of localized high densities of electrons and holes, one might expect that the original model by Halperin and Braner [55] would be a better approximation than the later model by Chen and Kirsh [33].

These previous studies were based on three different assumptions as follows. Firstly, it was assumed that the recombination rate is independent of the concentration of recombination centers and is proportional only to the concentration of electrons in the excited state of the trap. Secondly, it was assumed that the principle of detailed balance holds for these localized transitions (Chen and Pagonis [38]). A third common assumption was that the system is in quasi-equilibrium condition. When these three conditions are applied to the system of differential equations describing the localized transitions, it was shown that the resultant TL signals follow exclusively first order kinetics.

Kitis and Pagonis [75] examined critically these assumptions used in these previous studies. These authors simulated the localized transition model using the original version of the model by Halperin and Braner [55], without the simplifications leading to first order kinetics. They also derived new analytical expressions using the Lambert W function, by solving the system of differential equations in the localized transition model, by using the method of Kitis and Vlachos [82] for delocalized models. The analytical master equation was shown to describe a variety of optically and thermally stimulated phenomena.

These authors studied in detail the peak shape characteristics of the TL peaks, for a wide range of values of the parameters in the model, and showed how the analytical equations can described LM-OSL, CW-OSL, and isothermal TL (ITL) phenomena. The results of the simulations show that the TL peaks in the localized model have very similar characteristics with TL peaks derived from delocalized models, including non-first order kinetic characteristics.

7.3 The LT Model: Numerical and Analytical Solution

Kitis and Pagonis [75] examined critically the original model of Halperin and Braner [55], by including the differential equations for the irradiation, relaxation, and heating stages. Their study used the original model without the later modifications by Chen and Kirsh [33] and Bull [25], which lead to first order kinetics.

The energy band diagram for the localized transition model is shown in Fig. 7.1. During the heating stage electrons are raised into the excited state of electron trap, from which they can either recombine at the recombination center (transition A_m), or they can be de-excited into the ground state (transition A_D). As mentioned above, in this model the conduction and valence bands participate in the kinetics only during the irradiation stage.

The differential equations governing the traffic of free carriers during the heating stage are

$$\frac{dn}{dt} = -n\, p(t) + A_D\, n_e, \tag{7.1}$$

$$\frac{dm}{dt} = -A_m\, m\, n_e, \tag{7.2}$$

$$\frac{dn_e}{dt} = n\, p(t) - A_D\, n_e - A_m\, m\, n_e \tag{7.3}$$

The symbols in these equations are n (cm^{-3}) the concentration of trapped electrons, m (cm^{-3}) concentration of trapped holes. n_e is the concentration of electrons in the excited state of the trap, $p(t)$ is the rate of excitation from the ground state energy level of the trap into the excited state, from which the electron can either recombine with a recombination coefficient A_m ($cm^3\, s^{-1}$), or it can de-excite back into the ground state with a de-excitation coefficient A_D (s^{-1}). The rate of excitation $p(t)$ depends on the type of experiment (LM-OSL, CW-OSL, etc.), and is of course different for thermal and optical excitation. For example, in TL experiments the rate of thermal excitation is $p(t) = s\, \exp(-\frac{E}{kT})$, while in CW-OSL or CW-IRSL experiments the rate of optical excitation is constant λ (s^{-1}), and depends on the optical cross section of the trap and on the intensity of the stimulating light source. In the case of LM-OSL experiments, the rate of excitation is varied linearly with time according to $p(t) = \lambda t/P$, where λ (s^{-1}) is the rate of optical excitation and P is the total stimulation time during the LM-OSL experiment.

An important point to note in the above equations is that the recombination rate is written in the form $-A_m\, m\, n_e$, in agreement with the original model of Halperin and Braner [55]. In the later versions of the model by Chen and Kirsh [33] and Bull [25], this term is written as $-\gamma\, n_e$, where γ is a constant with dimensions of s^{-1}. In addition, the de-excitation term in the above equations is written here as $-A_D\, n_e$, while in these previous publications it is written as $-s\, n_e$, based on the assumption that the principle of detailed balance is valid for the model (i.e. $A_D = s$).

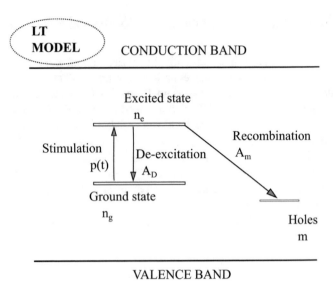

Fig. 7.1 Energy band diagram for the localized transition LT model

According to the principle of detailed balance (PDB), the de-excitation rate coefficient A_D must be equal to the frequency factor s. However, the PDB has been derived for a system in thermal equilibrium (Chen and Pagonis [38], their Chapter 2), and experimental studies have raised the question whether it is applicable for systems in non-thermal equilibrium (Klein [83], Thomsen [178]). Furthermore, theoretical work has shown that the PDB principle applies only to cyclic systems. As a consequence of these theoretical studies, the PDB will not be applicable in a multiple level model such as the localized model in Fig. 7.1.

Kitis and Pagonis [75] assumed that the PDB does not apply and studied the consequences of this assumption on the model. They used the original version of the model as written by Halperin and Braner [55], by writing the recombination rate in the form $-A_m\, m\, n_e$. By making the common assumption that only a few electrons are in the excited state at any given moment, i.e. $n_e << n$ and therefore from the conservation of charge $m \simeq n$, they obtained the differential equation describing the LT model:

The GOT equation for the LT model

$$I_{LOC} = -\frac{dn}{dt} = p(t)\,\frac{n^2}{r+n} \qquad (7.4)$$

---

where the *localized retrapping ratio r* is defined as

$$r = \frac{A_D}{A_m} \tag{7.5}$$

Equation (7.4) has the exact same mathematical form as the following GOT equation

$$I_{DEL} = p(t) \frac{n^2}{(N - n) R + n} \tag{7.6}$$

where the parameter R in Eq. (7.4) is the dimensionless quantity expressing the ratio of the retrapping and recombination coefficients A_n and A_m in the *delocalized* OTOR model:

$$R = \frac{A_n}{A_m} \tag{7.7}$$

If $r \gg n_0$, then $r \gg n$ (or $r/n \gg 1$) at all times t, the above differential Eq. (7.4) becomes

$$I_{LOC} = -\frac{dn}{dt} = p(t) \frac{n}{r/n + 1} \cong \frac{p(t)}{r} n^2 \tag{7.8}$$

and the LT model becomes a second order kinetics model.

If $r \ll n$ at all times, then the LT Eq. (7.4) becomes

$$I_{LOC} = -\frac{dn}{dt} = p(t) \frac{n}{r/n + 1} \cong p(t) n \tag{7.9}$$

which is the equation for first order kinetics.

Equation (7.4) was integrated by Kitis and Pagonis [75], who obtained the following analytical solution for the LT model in terms of the Lambert function:

Analytical solution of the LT model

$$I_{LOC}(t) = p(t) \frac{r}{W[k, e^z] + W[k, e^z]^2} \tag{7.10}$$

$$z_{LOC} = \frac{r}{n_0} - \ln\left[\frac{n_0}{r}\right] + \int_{t_0}^{t} p(t) dt \tag{7.11}$$

Equations (7.10) and (7.11) are master equations, similar to those obtained by Kitis and Vlachos [82] for the OTOR model, and by Kitis and Pagonis [73] for the localized tunneling recombination model of Jain et al. [62]. They are termed master equations because they provide a mathematical description for several types of stimulated luminescence phenomena, by selecting the appropriate expression for the excitation rate $p(t)$, as discussed previously in Chap. 6.

For TL and LM-OSL phenomena in the localized model, Eq. (7.11) for z becomes correspondingly:

$$z_{LOC-TL} = \frac{r}{n_0} - \ln\left[\frac{n_0}{r}\right] + s \int_{T_0}^{T} exp\left(-\frac{E}{k\,T'}\right) dT' \tag{7.12}$$

$$z_{LOC-LMOSL} = \frac{r}{n_0} - \ln\left[\frac{n_0}{r}\right] + \frac{\lambda\,t^2}{2\,P} \tag{7.13}$$

It is important to note that the r ratio in Eq. (7.5) is *not* a dimensionless quantity, but has units of cm^{-3}. By contrast, the parameter R in Eq. (7.7) is a dimensionless quantity expressing the ratio of the retrapping and recombination coefficients in the *delocalized* OTOR model (Kitis and Vlachos [82]). It is also noted that Halperin and Braner [55] assumed that the principle of detailed balance holds, so that $r = s/A_m$ in their model. In the present paper, r is used in the more general form of Eq. (7.5).

It is also necessary to note an important difference between the solutions of delocalized and localized models. In the case of *delocalized* models, the solution is based on *both* real branches of the Lambert W function. By contrast, in the case of *localized* models the solution in Eq. (7.10) is based *only* on the first real branch of the Lambert W function, for any value of r.

The following code plots the analytical solution of LT using the Lambert W function, for three values of the retrapping ratio $r = 10^4,\ 10^8,\ 10^9\ cm^{-3}$ and for an initial concentration $n_0 = 10^8\ cm^{-3}$. As r increases, the shape of the TL glow curve in Fig. 7.2 changes from the asymmetric shape for first order kinetics, to the symmetric shape for second order kinetics.

Code 7.1: Plot W(x) solution of LT model

```
rm(list=ls())
library(lamW)
## Plot the analytical solution of LT, using Lambert W-function
x1<-300:450
kB<-8.617E-5
no<-1E8
r<-1e4
En<-1
s<-1E13
beta<-1
k<-function(u) {integrate(function(p){exp(-En/(kB*p))},
300,u)[[1]]}
y1<-lapply(x1,k)
```

```
x<-unlist(x1)
y<-unlist(y1)
zTL<-(r/no)-log(no/r)+(s*y)
plot(x,r*s*exp(-En/(kB*x))/(lambertW0(exp(zTL))
        +lambertW0(exp(zTL))^2),type="l",col="red",
lwd=3,lty=1,xlab="Temperature, K",ylab="TL in LT model [a.u.]")
zTL<-(r/no)-log(no/r)+(s*y)
r<-1e8
zTL<-(r/no)-log(no/r)+(s*y)
lines(x,r*s*exp(-En/(kB*x))/(lambertW0(exp(zTL))
        +lambertW0(exp(zTL))^2),col="green",
        lwd=3,  lty=2,   xlab="Temperature, K",ylab="TL")
r<-1e9
zTL<-(r/no)-log(no/r)+(s*y)
lines(x,r*s*exp(-En/(kB*x))/(lambertW0(exp(zTL))
        +lambertW0(exp(zTL))^2),col="blue",
        lwd=3,  lty=3,   xlab="Temperature, K",ylab="TL")
legend("topright",bty="n",lwd=2, lty=c(NA,NA,1,2,3),
legend = c("Retrapping ratio r", expression("(cm"^"-3"*")"),
expression("10"^"4"*""),expression("10"^"8"*""),
expression("10"^"9"*"")),
        col=c(NA,NA,"red","green","blue"))
```

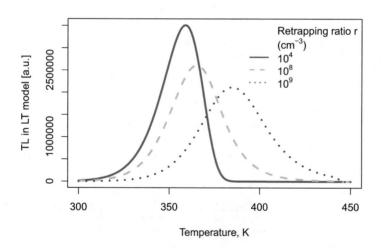

Fig. 7.2 Plots of the analytical solution of the LT model, using the Lambert W function, for three values of the retrapping ratio $r = 10^4$, 10^8, $10^9 \, cm^{-3}$ and $n_0 = 10^8 \, cm^{-3}$. As r increases, the shape of the TL glow curve changes from an asymmetric peak for first order kinetics, to a symmetric shape for second order kinetics. For details, see Kitis and Pagonis [75]

7.4   Simulations Using the LOC Functions in the R Package *RLumCarlo*

The previous section in this chapter was based on the use of differential equations to describe the luminescence signals associated with localized transitions. As discussed in the previous chapter, an alternative approach to the differential equations is the use of Monte Carlo methods. The MC approach for localized transition models is described in detail in Chap. 9.

In this section we give several examples of using the package *RLumCarlo* to simulate a variety of luminescence signals due to localized transitions within the LT model. Tables 7.1 and 7.2 summarize typical values of the parameters, and the available LOC functions in *RLumCarlo* (Kreutzer et al. [87]).

Table 7.1 Table of input parameters for LOC functions in *RLumCarlo*

Process	Symbol	Parameter in RLumCarlo function	Unit	Typical values
Localized TL	E	Thermal activation energy of the trap	eV	0.5–3
	s	Frequency factor of the trap	1/s	1E8–1E16
	Times	Sequence of time steps for simulation (heating rate 1 K/s)	s	0–700
	Clusters	Number of MC runs	1	1E1–1E4
	n_filled	Number of filled electron traps at the beginning of the simulation	1	1–1E5
	r	Localized retrapping ratio	1	0–1E5
Localized CW-IRSL	A	Optical excitation rate from ground state of the trap to the excited state	1/s	1E-3-1
	Times	Sequence of time steps for simulation	s	0–500
	Clusters	Number of MC runs	1	1E1–1E4
	n_filled	Number of filled electron traps at the beginning of the simulation	1	1–1E5
	r	Localized retrapping ratio	1	0–1E5
Localized ISO	E	Thermal activation energy of the trap	eV	0.5–3
	s	Frequency factor of the trap	1/s	1E8–1E16
	T	Temperature of the isothermal process	°C	20–300
	Times	Sequence of time steps for simulation	s	0–1000
	Clusters	Number of MC runs	1	1E1–1E4
	n_filled	Number of filled electron traps at the beginning of the simulation	1	1–1E5
	r	Localized retrapping ratio	1	0–1E5

(continued)

Table 7.1 (continued)

Process	Symbol	Parameter in RLumCarlo function	Unit	Typical values
Localized LM-OSL	A	Optical excitation rate from ground state of the trap to the excited state	1/s	1E-3-1
	Times	Sequence of time steps for simulation	s	0–3000
	Clusters	Number of MC runs	1	1E1–1E4
	n_filled	Number of filled electron traps at the beginning of the simulation	1	1–1E5
	r	Localized retrapping ratio	1	0–1E5

Table 7.2 Table of LOCalized transition functions available in the package *RLumCarlo*

Function name	Description
plot_RLumCarlo	Plots "RLumCarlo" modeling results (the averaged signal or the number of remaining electrons), with modeling uncertainties
run_MC_CW_IRSL_LOC	Simulation of CW-IRSL signals in the LT model, due to tunneling transitions from the excited state of the trap, into a recombination center (RC)
run_MC_ISO_LOC	Simulation of ITL signals in the LT model, due to tunneling from the excited state into the RC
run_MC_LM_OSL_LOC	Simulation of LM-IRSL signals in the LT model, due to tunneling from the excited state into the RC
run_MC_TL_LOC	Simulation of TL signals in the LT model, due to tunneling from the excited state into the RC

The following code simulates a CW-IRSL signal in Fig. 7.3 for a system of 10 clusters in the LT model, with each cluster consisting of $n_0 = 100$ electrons, with an IR excitation rate $A = 0.12\,\mathrm{s}^{-1}$, and a localized retrapping ratio $r = 10^{-7}$. The output is the CW-IRSL intensity.

Code 7.2: Single plot MC simulations for localized CW-IRSL (LT model)

```
##=================================================================##
## Example: MC simulations for localized CW_IRSL (LT model)
##=================================================================##
rm(list = ls(all=T))
library(RLumCarlo)
run_MC_CW_IRSL_LOC(
A = 0.12,
times = 0:100,
clusters = 10,
n_filled = 100,
r = 1e-7,
method = "seq",
output = "signal"
) %>%
#Plot results of the MC simulation
plot_RLumCarlo(legend = T)
 legend("top",bty="n",c("CW-IRSL signal","LOC function",
"LT model"))
```

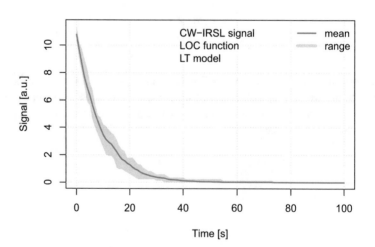

Fig. 7.3 Simulation of CW-IRSL signal in the LT model, using the function *run_MC_LM_IRSL_LOC* in the package *RLumCarlo*

The following code simulates an LM-IRSL signal for a system of 100 clusters, each cluster consisting of $n_0 = 50$ electrons, with an IR excitation rate $A = 0.1\,\mathrm{s}^{-1}$, and a localized retrapping ratio $r = 10^{-7}$. The output in Fig. 7.4 is the LM-IRSL intensity.

Code 7.3: Single plot MC simulations for localized LM-OSL

```
##===============================================================##
##===============================================================##
## MC simulations for localized LM-OSL
##===============================================================##
rm(list = ls(all=T))
library(RLumCarlo)
run_MC_LM_OSL_LOC(
A = 0.1,
times = 0:250,
clusters = 100,
n_filled = 50,
r = 1e-7,
method = "seq",
output = "signal"
) %>%
#Plot results of the MC simulation
plot_RLumCarlo(legend = T)
legend("top",bty="n",c("LM-IRSL signal","LOC function",
"LT model"))
```

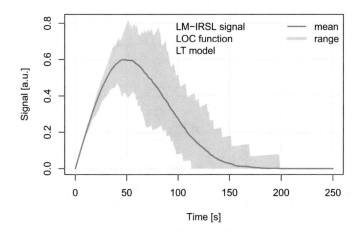

Fig. 7.4 Simulation of LM-IRSL signal in the LT model, using the function *run_MC_LM_IRSL_LOC* in the package *RLumCarlo*

The following code simulates the TL signal in the LT model, by using the function *run_MC_TL_LOC*, for a sample with thermal trap parameters $E = 1.0\,\text{eV}$, $s = 10^{12}\,\text{s}^{-1}$, and for three values of the localized retrapping ratio $r = 1, 10^3, 10^4$. The output in Fig. 7.5 is the TL intensity. As the value of r increases, the TL peak changes from the asymmetric shape for first order peaks, to the almost symmetric shape for second order kinetics.

Code 7.4: Localized TL with variable retrapping ratio r

```
##=====================================================##
## Localized TL with variable retrapping ratio r (LT model)
##=====================================================##
rm(list = ls(all=T))
library(RLumCarlo)
f<-function(rvar,vars,addTF){
  run_MC_TL_LOC(
    s = 1e12,
    E = 1,
    times = 0:300,
    r = rvar
  ) %>%
    #Plot results of the MC simulation
    plot_RLumCarlo(legend = F,plot_uncertainty=NULL,type="o",
  pch=vars,col=vars,add=addTF, xlim=c(50,275))}
f(1,1,FALSE)
f(1e3,2,TRUE)
f(1e4,3,TRUE)
legend("topright",bty="n",pch=c(NA,NA,NA,1:3),
col=c(NA,NA,NA,1:3),legend = c("Retrapping"  ,"Ratio r ",
  " ","1",  expression("10"^"3"*""),expression("10"^"4"*"")) )
```

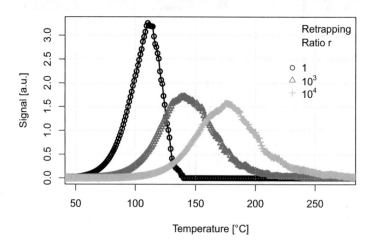

Fig. 7.5 Simulation of TL signal in the LT model, by using the function *run_MC_TL_LOC* in the package *RLumCarlo,* for three different values of the retrapping ratio *r*

7.5 The Mandowski SLT Model

Mandowski [101] developed a complex kinetic model for spatially correlated TL systems, which contains elements from both the localized transitions (LT) and delocalized transition models of TL. The complexity of this *semilocalized transition model* (*SLT*) is due partly to the presence of two distinct activation energies E and E_V shown in Fig. 7.6. There are two possible recombination transitions within the model, leading to two TL peaks denoted by L_B and L_C. The L_B peak corresponds to the intra-pair luminescence due to localized transitions, and the L_C peak corresponds to delocalized transitions involving the conduction band.

Mandowski [101] found that for certain parameters in the model, the L_C peak can exhibit non-typical double peak structure that resembles the TL displacement peaks previously found in spatially correlated systems by Monte Carlo simulations. This double peak structure of the L_C peaks was interpreted as follows: the first L_C peak corresponds to an increasing concentration of charges accumulating in the excited local level and undergoing intra-pair recombination. The second L_C peak was interpreted as relating to carriers wandering in the crystal, and undergoing recombination with holes in the recombination center via the conduction band.

When no retrapping occurs within the trap-recombination center pair (T-RC), the model produces a single L_C peak with no apparent double peak structure. When retrapping is present, it was found that the L_C peaks exhibit a double peak structure. Mandowski [101] also found that the amount of recombination in the T-RC pair also influences the shape and relative size of the delocalized peaks. He also studied the effect of the energy gap E_V on the double peak structure of the L_C peak.

The energy diagram for the SLT Model is shown in Fig. 7.6, together with the possible transitions between different states in Fig. 7.7. Since several of the transitions shown depend on the instantaneous number of carriers occupying traps (T) and recombination centers (RC), it is not possible to write kinetic equations

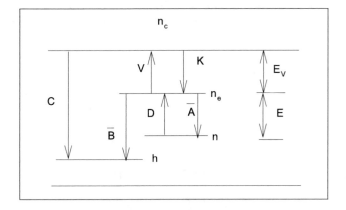

Fig. 7.6 The SLT model proposed by Mandowski [101]

using global concentrations of charge carriers in each level. In order to overcome this difficulty, Mandowski [101] introduced the following notation to denote the occupation of possible states in a single T-RC unit:

$$\left\{ \begin{array}{c} \overline{n_e} \\ \overline{n} \\ \overline{h} \end{array} \right\} \tag{7.14}$$

where $\overline{n_e}$ is the number of electrons in the local excited level, $\overline{n}$ is the number of electrons in the trap level T and $\overline{h}$ is the number of holes in the T-RC pair. By using this notation, the following possible states can be defined:

$$H_0^0 = \left\{ \begin{array}{c} 0 \\ 0 \\ 1 \end{array} \right\} \quad H_1^0 = \left\{ \begin{array}{c} 0 \\ 1 \\ 1 \end{array} \right\} \quad H_0^1 = \left\{ \begin{array}{c} 1 \\ 0 \\ 1 \end{array} \right\} \quad H_1^1 = \left\{ \begin{array}{c} 1 \\ 1 \\ 1 \end{array} \right\} \tag{7.15}$$

$$E_0^0 = \left\{ \begin{array}{c} 0 \\ 0 \\ 0 \end{array} \right\} \quad E_1^0 = \left\{ \begin{array}{c} 0 \\ 1 \\ 0 \end{array} \right\} \quad E_0^1 = \left\{ \begin{array}{c} 1 \\ 0 \\ 0 \end{array} \right\} \quad E_1^1 = \left\{ \begin{array}{c} 1 \\ 1 \\ 0 \end{array} \right\} \tag{7.16}$$

The symbols/variables $H_m^n(t)$ and $E_m^n(t)$ denote the concentrations of states (i.e. corresponding T–RC units) with full and empty RCs, respectively. They are time dependent variables denoting concentration (in cm^{-3}) of all T–RC units having n_e, n, and h charge carriers in the respective trap levels. For example, the first of these states H_0^0 is the concentration of all states {0 0 1}, meaning states with 0 electrons in the excited state, 0 trapped electrons in the ground state, and 1 hole in the RC. Similarly, E_0^0 means the concentration of all states of type {0 0 0}, in which a T-RC unit has no electrons in the excited or ground state, and an empty associated RC.

Mandowski [101] introduced the following simplifying assumptions in the SLT model: (a) The K-transitions shown in Fig. 7.6 are such that $K = 0$ (b) The creation of states with two active electrons (H_1^1, E_1^1) is unlikely. As a result of this assumption, the model does not contain differential equations for H_1^1, E_1^1. (c) Only H_1^0 states are initially filled within the model. By using these assumptions the following set of 7 equations were obtained:

$$\dot{H}_1^0 = -(D + Cn_c)\,H_1^0 + \bar{A}H_0^1, \tag{7.17}$$

$$\dot{H}_0^1 = DH_1^0 - (\bar{A} + \bar{B} + V + Cn_c)H_0^1, \tag{7.18}$$

$$\dot{H}_0^0 = VH_0^1 - Cn_c H_0^0, \tag{7.19}$$

$$\dot{E}_1^0 = Cn_c H_1^0 - DE_1^0 + \bar{A}E_0^1, \tag{7.20}$$

$$\dot{E}_0^1 = Cn_c H_0^1 + DE_1^0 - (\bar{A} + V)E_0^1, \tag{7.21}$$

$$\dot{E}_0^0 = \bar{B}H_0^1 + Cn_c H_0^0 + VE_0^1, \tag{7.22}$$

$$n_c = H_0^0 - E_0^1 - E_1^0 \tag{7.23}$$

The parameters in these equations and the corresponding electronic transitions are shown in Figs. 7.6 and 7.7. The activation energy within the T-RC pair is denoted by E, the energy barrier is E_V, and the corresponding frequency factors are v and v_V. A linear heating rate is assumed, and the retrapping coefficient r is defined as $r = \bar{A}/\bar{B}$. In addition to these parameters, the global concentration of traps in the crystal is denoted by N (cm^{-3}). The parameters $D(t) = vexp(-E/kT)$ and $V(t) = v_V exp(-E_V/kT)$ describe the thermal excitation probabilities in the SLT model. The variables H_m^n, E_m^n are related to the standard *global* variables n, n_e, n_c, h used in the localized transition (LT) model and in the OTOR mode of TL by the following equations:

$$n = H_1^0 + E_1^0, \tag{7.24}$$

$$n_e = H_0^1 + E_0^1, \tag{7.25}$$

$$n_c = H_0^0 - E_0^1 - E_1^0, \tag{7.26}$$

$$h = H_1^0 + H_0^1 + H_0^0 \tag{7.27}$$

Here n, n_e, n_c, h denote as usual the *global* concentrations of carriers in the trap level, excited level, conduction band, and recombination center correspondingly. The two recombination transitions within the model lead to two possible TL peaks L_B and L_C. The corresponding TL intensities for these two peaks are given by

$$L_C = Cn_c h = Cn_c(H_1^0 + H_0^1 + H_0^0), \tag{7.28}$$

$$L_B = \bar{B}H_1^0 \tag{7.29}$$

Figure 7.8 shows an example of the L_B and L_C peaks calculated with the model, for the following values of the parameters: $v = v_V = 10^{10}\,s^{-1}$, $\beta = 1\,K/s$, $C = 10^{-10}\,cm^3 s^{-1}$, $B = 10^5\,s^{-1}$, $E_V = 0.7\,eV$, and $\bar{A} = 10^4\,s^{-1}$. The initial concentration of trapped carriers is assumed to be $H_1^0 = n_o = 10^{17}\,cm^{-3}$ and all other initial concentrations are assumed to be zero. Furthermore, it is assumed that all states are initially full so that $n_o = N$, where N is the global concentration of traps in the crystal.

Fig. 7.7 Schematic representation of the possible transitions in the SLT model by Mandowski [101]

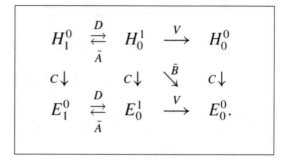

$$H_1^0 \underset{\bar{A}}{\overset{D}{\rightleftharpoons}} H_0^1 \overset{V}{\longrightarrow} H_0^0$$

$$c\downarrow \qquad c\downarrow \qquad \overset{\bar{B}}{\searrow} \quad c\downarrow$$

$$E_1^0 \underset{\bar{A}}{\overset{D}{\rightleftharpoons}} E_0^1 \overset{V}{\longrightarrow} E_0^0.$$

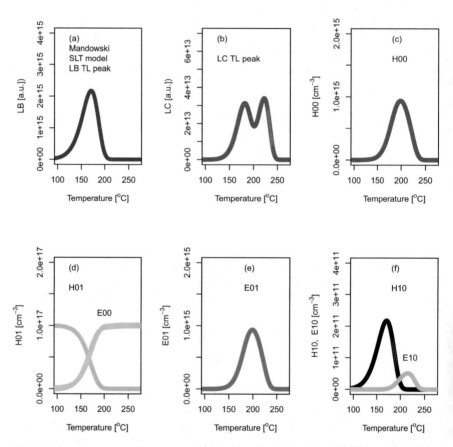

Fig. 7.8 Simulations of a TL process in the Mandowski SLT model, using the parameters given in the text, showing plots of the various concentrations H_m^n and E_m^n. (**a**) The L_B peak due to localized transitions in the model; (**b**) The double L_C peak from delocalized transitions; (**c–f**) The concentrations H_m^n and E_m^n as a function of temperature. *Caution:* Note that the numerical integration used in the *deSolve* package can often become numerically unstable for this type of model, see the discussion in the text

Caution The readers must be cautioned that the numerical integrations in the following code can become *numerically unstable* when using the original set of parameters in the model. The reason for these instabilities is located in the numerical integration solver *lsoda*, which is loaded with the package *deSolve* (Soetaert et al. [170]). See for example, the discussion of the corrupt Lotka-Volterra model in the vignettes of the package *deSolve,* available on the web at https://cran.r-project.org/web/packages/deSolve/vignettes/deSolve.pdf, page 47 (accessed on 9/22/2020).

The R codes shown in this book for the Mandowski and other models have been cross-checked for accuracy by the author, by running independent calculations using *Mathematica,* and comparing with the results of the R codes. While *deSolve* can become unstable in R with specific sets of parameters in the Mandowski model, the *Mathematica* codes do not show these types of instabilities.

Code 7.5:   Mandowski SLT model: simulation of TL experiment

```
# Mandowski SLT model: simulation of TL experiment
rm(list=ls())
library("deSolve")
TLMandowski <- function(t, x, parms) {
  with(as.list(c(parms, x)), {
  dH01<- - (s*exp(-E/(kb*(273+hr*t)))+C*(H00-E01-E10))*H01+A*H10
  dH10 <- s*exp(-E/(kb*(273+hr*t)))*H01-(A+B+sv*exp(-Ev/(kb*
     (273+hr*t)))+C*(H00-E01-E10))*H10
  dH00<-sv*exp(-Ev/(kb*(273+hr*t)))*H10-C*(H00-E01-E10)*H00
  dE01<-C*(H00-E01-E10)*H01-s*exp(-E/(kb*(273+hr*t)))*E01+A*E10
  dE10<-C*(H00-E01-E10)*H10+s*exp(-E/(kb*(273+hr*t)))*E01-
     (A+sv*exp(-Ev/(kb*(273+hr*t))))*E10
  dE00<-B*H10+C*(H00-E01-E10)*H00+sv*exp(-Ev/(kb*(273+hr*t)))*E10
     res <- c(dH01,dH10,dH00,dE01,dE10,dE00)
     list(res)   })}
enVar<-function(en){
parms   <- c(E =en, s=1e10, sv=1e10, kb=8.617*10^-5,
hr=1,  C=C,B=B,Ev=0.7,A=2*10^4)
y <- xstart <- c(H01 = 10^17,   H10=0,H00=0,E10=0,E01=0,E00=0)
out <-   ode(xstart, times, TLMandowski, parms, method = "lsoda",
 atol = 1,rtol=1)  }
C<-1e-10
B<-1e4
temps<-times  <- seq(0, 360,by=.5)
Lc<-Lb<-matrix(NA,nrow=length(times),ncol=4)
areaLb<-areaLc<-vector(length=4)
 en<-.95
```

```
a<-enVar(en)
  Lc<-(C*(a[,"H00"]-a[,"E01"]-a[,"E10"])*
          (a[,"H10"]+a[,"H01"]+a[,"H00"]))
  Lb<-B*a[,"H10"]
    areaLc<-sum(Lc)*0.5
    areaLb<-sum(Lb)*0.5
  par(mfrow=c(2,3))
  xlabs=expression("Temperature [""^""o""*""C]")
  plot(temps,Lb,typ="l",xlim=c(100,270),ylim=c(0,4e15),col="blue",
  pch=1,xlab=expression("Temperature [""^""o""*""C]"),lwd=5,
  ylab="LB [a.u.])")
  legend("topleft",bty="n",col="red",
  legend=c("(a)","Mandowski","SLT model","LB TL peak"))
  plot(temps,Lc,typ="o",xlim=c(100,270),ylim=c(0,.7e14),col="red",
  xlab=xlabs,ylab="LC [a.u.])")
  legend("topleft",bty="n",col="red",
  legend=c("(b)"," ","LC TL peak"))
  plot(temps,a[,"H00"],typ="o",xlim=c(100,270),ylim=c(0,2e15),
  xlab=xlabs,ylab=expression("H00 [cm""^-3*""]"),col="red",pch=1)
  legend("top",bty="n",c("(c)"," ","H00"))
  plot(temps,a[,"H01"],typ="o",xlim=c(100,270),ylim=c(0,2e17),
  xlab=xlabs,ylab=expression("H01 [cm""^-3*""]"),col="green")
  lines(temps,a[,"E00"],typ="o",col="gray",pch=3)
  legend("topleft",bty="n",c("(d)"," ","H01"))
  text(200,1.2e17,"E00")
  plot(temps,a[,"E01"],typ="o",xlim=c(100,270),ylim=c(0,2e15),
  xlab=xlabs,ylab=expression("E01 [cm""^-3*""]"),col="magenta")
  legend("top",bty="n",c("(e)"," ","E01"))
  plot(temps,a[,"H10"],typ="o",xlim=c(100,270),ylim=c(0,4e11),
  col="black",xlab=xlabs,ylab=expression("H10,   E10 [cm""^-3*""]"))
  lines(temps,a[,"E10"],typ="o",col="green",)
  legend("top",bty="n",c("(f)"," ","H10"))
  text(220,1e11, "E10")
```

Figure 7.9 shows a simulation of the SLT Mandowski model, using four different values of the activation energy parameter $E = 0.95$, 1.0, 1.05, 1.1 eV. Figure 7.9a shows the L_B peaks due to localized transitions in the model, while Fig. 7.9b show the corresponding L_C peaks from delocalized transitions.

Note that the values of the parameters $\bar{A}$, B, E used in the simulation of Fig. 7.9 are different from those used in the original Mandowski paper [101], which were $B = 10^3 \, s^{-1}$, $\bar{A} = 0$, $10^3 \, s^{-1}$, and $E = 0.9$ eV. As the E value in Fig. 7.9b increases, the double TL peak L_c shifts to higher temperatures, and the total area of the TL glow curve increases as well, as shown in Fig. 7.9c. This increase in the L_C areas with the E value is accompanied by a corresponding *decrease* in the L_B areas, while the *sum* of the areas $L_C + L_B$ remains constant, and equal to the total number of initially trapped electrons $H_1^0 = n_0 = 10^{17} \, cm^{-3}$ (+ symbols in Fig. 7.9c).

The physical picture that emerges from Fig. 7.9 is that there is strong competition between the localized pathway in the model (creating the L_B signals via the transition coefficient B) and the alternative delocalized pathway (creating the L_C

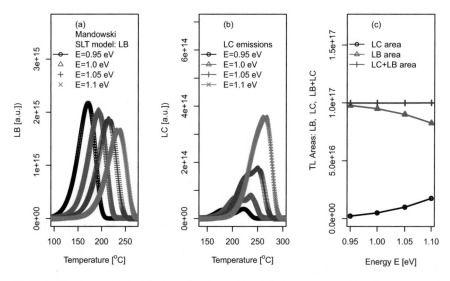

Fig. 7.9 Simulations of the Mandowski SLT model, using the parameters given in the text, and four different values of the parameter $E = 0.95, 1.0, 1.05, 1.1\,\text{eV}$. (**a**) The L_B peaks due to localized transitions in the model; (**b**) The double L_C peaks from delocalized transitions; (**c**) The areas under the L_C and L_B peaks, and their sum $L_C + L_B$ as a function of the activation energy E

signals via the transition coefficient C). Whatever charge is lost from one pathway, is gained by the alternative pathway in the model.

7.6 The Anomalous Heating Rate Effect

The SLT model has been used to explain the anomalous heating rate effect reported for YPO_4 doped with trivalent lanthanide ions and other materials (Mandowski and Bos [104]). For most dosimetric materials, the TL glow curve is shifted towards higher temperatures when higher heating rates β are used, while the area under the normalized TL signal (TL/β) vs temperature should remain constant and independent of the heating rate. The exact opposite effect was reported in the dosimetric material $YPO_4:Ce^{3+}$, Sm^{3+}, in which the area under the normalized curve TL/β vs temperature *increased* with increasing heating rate. It is also noted that a similar anomalous heating rate effect was reported by Kitis et al. [79] for Durango apatite.

The anomalous heating rate effect is a good example of the competition between the localized L_B pathway described in the previous section, and the alternative delocalized L_C pathway in the SLT model. The code below simulates a TL experiment with four heating rates $\beta = 1$–$4\,\text{K/s}$, with the results shown in Fig. 7.10a, b.

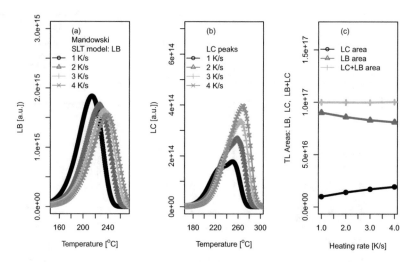

Fig. 7.10 Simulations of the anomalous heating rate effect using the Mandowski model, with the parameters given in the text, and for four different values of the heating rate $\beta = 1$–4 K/s. (**a**) The L_B peaks due to localized transitions in the model; (**b**) The L_C peaks from delocalized transitions. (**c**) The areas under the TL glow curves in (**a**, **b**) and their sum $L_C + L_B$, as a function of the heating rate

As the heating rate increases, the L_C pathways become the preferred route for charge towards the recombination center. In the delocalized OTOR model, we would expect that the area of the L_C peak would stay constant and equal to the initial number of trapped electrons, in this example $n_0 = 10^{17}$ cm^{-3}. What is observed instead is that the area of the L_C peak increases with the heating rate, while at the same time the area of the L_B peak decreases; the total area $L_B + L_C$ stays constant, as can be seen in Fig. 7.10c.

7.7 The Simplified SLT Model by Pagonis et al.

Pagonis et al. [122] presented a simplified SLT model, which also explains the anomalous heating rate effect as a competition phenomenon between two luminescence pathways, as shown in Fig. 7.11.

The system of differential equations in the model is

Code 7.6: Mandowski model: the anomalous heating rate effect

```
#Mandowski model: the anomalous heating rate effect
rm(list=ls())
```

```
library("deSolve")
TLMandowski <- function(t, x, parms) {
  with(as.list(c(parms, x)), {
  dH01<- -(s*exp(-E/(kb*(273+hr*t)))+C*(H00-E01-E10))*H01+A*H10
     dH10 <- s*exp(-E/(kb*(273+hr*t)))*H01-(A+B+sv*exp(-Ev/(kb*
     (273+hr*t)))+C*(H00-E01-E10))*H10
     dH00<-sv*exp(-Ev/(kb*(273+hr*t)))*H10-C*(H00-E01-E10)*H00
  dE01<-C*(H00-E01-E10)*H01-s*exp(-E/(kb*(273+hr*t)))*E01+A*E10
  dE10<-C*(H00-E01-E10)*H10+s*exp(-E/(kb*(273+hr*t)))*E01-
     (A+sv*exp(-Ev/(kb*(273+hr*t))))*E10
dE00<-B*H10+C*(H00-E01-E10)*H00+sv*exp(-Ev/(kb*(273+hr*t)))*E10
  res <- c(dH01,dH10,dH00,dE01,dE10,dE00)
  list(res)  })}
hrVar<-function(hr){
  parms  <- c(E =1.05, s=1e10, sv=1e10, kb=8.617*10^-5,
           hr=hr, C=C,B=B,Ev=0.7,A=2*10^4)
  y <- xstart <- c(H01 = 10^17,   H10=0,H00=0,E10=0,E01=0,E00=0)
  out <-ode(xstart,times, TLMandowski, parms, method = "lsoda",
           atol = 1,rtol=1) }
C<-1e-10
B<-1e4
times  <- seq(0, 300,by=.5)
Lc<-Lb<-temps<-matrix(NA,nrow=length(times),ncol=4)
areaLb<-areaLc<-vector(length=4)
for (i in 1:4){
  hr<-i
  a<-hrVar(hr)
  Lc[,i]<-(C*(a[,"H00"]-a[,"E01"]-a[,"E10"])*
           (a[,"H10"]+a[,"H01"]+a[,"H00"]))/hr
  Lb[,i]<-B*a[,"H10"]/hr
  temps[,i]<-hr*times
  areaLc[i]<-sum(Lc[,i])*hr*0.5
  areaLb[i]<-sum(Lb[,i])*hr*0.5
}
par(mfrow=c(1,3))
var<-c(NA,NA,NA,1:4)
matplot(cbind(temps),cbind(Lb),lty="solid",typ="o",pch=1:4,
xlim=c(150,270),ylim=c(0,3e15),lwd=2, col=1:4 ,
xlab=expression("Temperature ["^"o"*"C]"), ylab="LB [a.u.])")
legend("topleft",bty="n",pch=var,lty="solid",col=var,
legend=c("(a)",
"Mandowski","SLT model: LB","1 K/s","2 K/s","3 K/s","4 K/s"))
matplot(cbind(temps),cbind(Lc),lty="solid",typ="o",pch=1:4,
xlim=c(170,300),ylim=c(0,7e14),lwd=2,col=1:4,
xlab=expression("Temperature ["^"o"*"C]"), ylab="LC [a.u.])")
legend("topleft",bty="n",pch=var,lty="solid",col=var,
legend=c("(b)"," ","LC peaks","1 K/s","2 K/s","3 K/s","4 K/s"))
matplot(1:4,cbind(areaLc,areaLb,areaLb+areaLc),lty="solid",
typ="o",lwd=3, pch=1:3,col=1:3,ylim=c(0,1.7e17),
xlab="Heating rate [K/s]",ylab="TL Areas: LB,  LC,  LB+LC")
legend("topleft",bty="n",pch=c(NA,NA,1:3),lty="solid",col=c(NA,
NA,1:3),legend= c("(c)"," ","LC area","LB area","LC+LB area"))
```

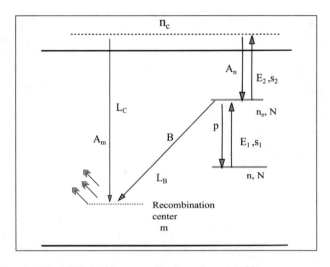

Fig. 7.11 The simplified SLT model proposed by Pagonis et al. [122]

$$\frac{dn}{dt} = -ns_1 \exp\left(-E_1/kT\right) + pn_e \tag{7.30}$$

$$\frac{dn_e}{dt} = A_n\left(N - n - n_e\right)n_c + ns_1 \exp\left(-E_1/kT\right) - pn_e - n_e s_2 \exp\left(-E_2/kT\right) - Bn_e \tag{7.31}$$

$$L_C = A_m m n_c \tag{7.32}$$

$$L_B = Bn_e \tag{7.33}$$

$$n_e + n + n_c = m \tag{7.34}$$

$$\frac{dm}{dt} = \frac{dn_e}{dt} + \frac{dn}{dt} + \frac{dn_c}{dt} \tag{7.35}$$

By assuming that the excited states relax quite rapidly compared to the time scales of TL experiments, we can set $dn_e/dt = 0$ and to obtain from Eq. (7.31):

$$n_e = \frac{A_n\left(N - n - n_e\right)n_c + ns_1 \exp\left(-E_1/kT\right)}{p + s_2 \exp\left(-E_2/kT\right) + B} \tag{7.36}$$

With the additional assumption $n_e \ll n$, so that $n_c = m - n$, the system of equations simplifies to a simple system of only two differential equations for n, m:

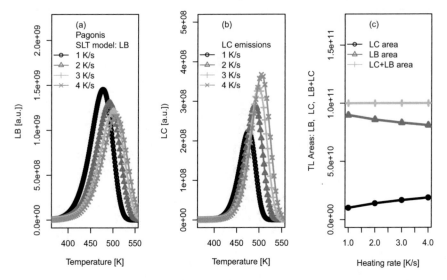

Fig. 7.12 Simulations of the anomalous heating rate effect using the simplified SLT model by Pagonis et al. [122], with the parameters given in the text, and for four different values of the heating rate $\beta = 1$–4 K/s. (a) The L_B peaks due to localized transitions in the model; (b) The L_C peaks from delocalized transitions; (c) The areas under the TL glow curves in (a, b) and their sum $L_C + L_B$, as a function of the heating rate

$$\frac{dn}{dt} = -ns_1 \exp(-E_1/kT) + p\frac{A_n(N-n)(m-n) + ns_1 \exp(-E_1/kT)}{p + s_2 \exp(-E_2/kT) + B}$$

(7.37)

$$\frac{dm}{dt} = -A_m m(m-n) - B\frac{A_n(N-n)(m-n) + ns_1 \exp(-E_1/kT)}{p + s_2 \exp(-E_2/kT) + B}$$

(7.38)

The following code solves Eqs. (7.37) and (7.38), with the result shown in Fig. 7.12 for four different values of the heating rate $\beta = 1 - 4$ K/s. Figure 7.12a, b show the L_C and L_B peaks. Figure 7.12c shows the areas under the TL glow curves and the sum of the areas $L_C + L_B$, as a function of the heating rate. The parameters used in these simulations are $p = 10^9\,\text{s}^{-1}$, $s_1 = 10^9\,\text{s}^{-1}$, $s_2 = 10^{12}\,\text{s}^{-1}$, $A_n = 10^{-8}\,\text{cm}^3\text{s}^{-1}$, $A_m = 10^{-6}\,\text{cm}^3\text{s}^{-1}$, $B = 10^7\,\text{s}^{-1}$, $N = 10^{13}\,\text{cm}^{-3}$, $E_1 = 0.8\,\text{eV}$, $E_2 = 0.5\,\text{eV}$.

These results from the simplified SLT model are similar to the results obtained in the previous section using the more complex Mandowski model.

Code 7.7: Pagonis SLT model: the anomalous heating rate effect

```
#Pagonis SLT model: the anomalous heating rate effect
rm(list=ls())
library("deSolve")
PagonisHR <- function(t, x, parms) {
  with(as.list(c(parms, x)), {
    dn<- -s1*n*exp(-E1/(kb*(273+hr*t)))+p*
      (An*(N-n)*(m-n)+s1*n*exp(-E1/(kb*(273+hr*t))))/
( s2*exp(-E2/(kb*(273+hr*t)))+p+B)
      dm<- -Am*m*(m-n)-B*
      (An*(N-n)*(m-n)+s1*n*exp(-E1/(kb*(273+hr*t))))/(
        s2*exp(-E2/(kb*(273+hr*t)))+p+B)
      res <- c(dn,dm)
    list(res)  })}
hrVar<-function(hr){
parms   <- c(p=p,s1=s1,s2=s2,An=An,Am=Am,
          B=B,N=N,E1=E1,E2=E2,kb=kb)
y <- xstart <- c(n = 1e11, m=1e11)
out <-  ode(xstart, times, PagonisHR, parms) }
p<-1e9
s1<-1e9
s2<-1e12
An<-1e-8
Am<-1e-6
B<-1e7
N<-1e13
E1<-0.8
E2<-0.5
kb<-8.617e-5
times   <- seq(0, 350,by=1)
Lc<-Lb<-temps<-matrix(NA,nrow=length(times),ncol=4)
areaLb<-areaLc<-vector(length=4)
for (i in 1:4){
  hr<-i
  a<-hrVar(hr)
  Lc[,i]<- Am*a[,"m"]*(a[,"m"]-a[,"n"])/hr
  Lb[,i]<-B* ((An*(N-a[,"n"])*(a[,"m"]-a[,"n"])+s1*a[,"n"]*
 exp(-E1/(kb*(273+hr*times)))))/( s2*exp(-E2/(kb*
(273+hr*times)))+p+B))/hr
    temps[,i]<-273+hr*times
    areaLc[i]<-sum(Lc[,i],rm.NA=TRUE)*hr
    areaLb[i]<-sum(Lb[,i],rm.NA=TRUE)*hr
}
par(mfrow=c(1,3))
var<-c(NA,NA,NA,1:4)
matplot(cbind(temps),cbind(Lb),lty="solid",typ="o",pch=1:4,
xlim=c(370,550),ylim=c(0,22e8),lwd=2, col=1:4 ,
xlab=expression("Temperature [K]"), ylab="LB [a.u.])")
```

```
legend("topleft",bty="n",pch=var,lty="solid",col=var,
legend=c("(a)",
"Pagonis","SLT model: LB","1 K/s","2 K/s","3 K/s","4 K/s"))
matplot(cbind(temps),cbind(Lc),lty="solid",typ="o",pch=1:4,
xlim=c(370,550),ylim=c(0,5e8),lwd=2,col=1:4,
xlab=expression("Temperature [K]"), ylab="LC [a.u.])")
legend("topleft",bty="n",pch=var,lty="solid",col=var,
legend=c("(b)"," ","LC emissions","1 K/s","2 K/s","3 K/s",
"4 K/s"))
matplot(1:4,cbind(areaLc,areaLb,areaLb+areaLc),lty="solid",
typ="o",lwd=3, pch=1:3,col=1:3,ylim=c(0,1.7e11),
xlab="Heating rate [K/s]",ylab="TL Areas: LB,  LC,  LB+LC")
legend("topleft",bty="n",pch=c(NA,NA,1:3),lty="solid",
col=c(NA,NA,
1:3),legend= c("(c)"," ","LC area","LB area","LC+LB area"))
```

Part III
Monte Carlo Simulations of Luminescence Signals

Chapter 8
Monte Carlo Simulations of Delocalized Transitions

8.1 Introduction

In this chapter we introduce Monte Carlo simulations of models based on delocalized transitions. In Chap. 9 we will extend our presentation to models based on localized transitions.

Section 8.2 presents a general discussion of MC methods, and how they are different from the deterministic differential equation approach we saw in previous chapters. This is followed in Sect. 8.3 by a summary and overview of MC applications in TL and OSL, and of previous research work in this area.

Section 8.4 presents a standard textbook example of how the fixed time interval MC method can be applied to the radioactivity process, and by extension also to first order kinetics CW-OSL process. Section 8.5 considers how luminescence processes can be described within the general framework of birth and death processes. Stochastic birth -death processes have been studied extensively in various branches of science, but there have been very few previous studies of them within the context of luminescence phenomena.

Sections 8.6–8.10 describe how MC methods are used to estimate the stochastic uncertainties in a luminescence model, and present several examples of luminescence phenomena as birth and death processes. Section 8.11 extends this MC method to the GOT model, in order to describe TL and OSL signals from large clusters of traps and centers. Section 8.12 is an example of a system consisting of small clusters, as one may encounter in nanodosimetric materials.

Finally, in Sect. 8.13 we present a MC simulation of irradiation processes within the GOT model, by using a fixed time interval method.

V. Pagonis, *Luminescence*, Use R!, https://doi.org/10.1007/978-3-030-67311-6_8

8.2 Deterministic Versus Stochastic Processes

The differential equation description of luminescence phenomena is well established, and analytical equations have been derived which describe the loss of charge
and trap filling processes. For a recent review of modeling work in this area, the
reader is referred to Pagonis et al. [127]. A second modeling approach of stimulated
luminescence phenomena is based on Monte Carlo (MC) techniques. MC methods
are used widely in many areas of science.

Why use a MC method? Almost always during TL and OSL experiments we
are dealing with a large number of traps and centers which participate in the
luminescence process. Such large systems can be described accurately with the
differential equations approach we saw in the previous chapters.

However, when we deal with a small number of traps and centers, we cannot treat
their concentrations as a continuous variable, and must consider the discrete nature
of the luminescence process. This discreteness will lead us into the probabilistic
nature of the luminescence process, and will introduce the ideas of noise and
randomness. In the limit of large numbers, we would anticipate to find the usual
results from the differential equations approach. One of the main ideas behind
the MC method is that many independent simulations of the same luminescence
process can proceed in different ways, and therefore will produce a distribution of
final states. These processes are characterized in general as *stochastic processes*,
as opposed to the differential equations method which describes *deterministic
processes*.

The MC method can be applied to TL and OSL models under certain physical
assumptions, and is based on the idea that differential equations for the continuous
variable become *difference* equations for the discrete variable. In this book we will
look at two different types of MC methods. In this chapter and in Chap. 9 we will
look at *MC methods with fixed time interval* between events. A second type of MC
methods is introduced in Chap. 10, and belongs to the general category of *Kinetic
MC Methods*, in which the time interval between events is variable.

A unique feature of the MC method, as opposed to the differential equation
approach, is that MC allows an estimation of the intrinsic theoretical uncertainty
of the intensity of the luminescence signal in the various models. By running and
averaging several MC variants, these uncertainties can be readily estimated. These
stochastic uncertainties are of importance in studies of nanodosimetric materials
and single grain luminescence experiments, in which one deals with small clusters
of spatially correlated traps and centers. Luminescence signals in these materials
can originate in interactions taking place within these small clusters, and therefore
may be very different from the corresponding bulk dosimetric materials which can
be described by the macroscopic differential equations approach.

Some of the advantages of the MC methods presented in this book are:

- The MC methods are easily generalized to many types of signals.
- They are fast and efficient, and avoid the numerical integrations necessary in
  the differential equations approach.

- They can produce accurate results in cases of low thermal or optical stimulation probability, where it is known that some of the analytical equations for luminescence phenomena might fail.

8.3 Overview of previous MC Research in TL and OSL

Some luminescent materials consist of nanoclusters containing only a few atoms. The synthesis and characterization of such nanodosimetric materials are an active research area, and it has been shown that their physical properties can be different from those of similar conventional microcrystalline phosphors (see, for example, Salah [164]; Sun and Sakka [172]; and the references therein). It has been suggested that traditional energy band models may not be applicable for some of these nanodosimetric materials, because of the existence of strong spatial correlations between traps and recombination centers. Such spatially correlated systems are also likely to be found in low-dimensional structures, as well as in materials which underwent high energy/high dose irradiations which create groups of large defects. The luminescence properties of such spatially correlated materials can be simulated by using Monte Carlo techniques.

Monte Carlo methods for the study of TL were presented in the papers by Mandowski [100–103], and by Mandowski and Światek [105–109]. These authors suggested that one must consider clusters of traps as separate systems, since the continuous differential equations are not valid. Kulkarni [89] described how the MC technique can be used to simulate TL and thermally stimulated conductivity (TSC) phenomena, with emphasis on the required calculation time, statistical errors and comparison with other methods.

These earlier Monte Carlo simulations of luminescence phenomena can be classified as Kinetic Monte Carlo (KMC) simulations. Typically these Monte Carlo calculations are performed with the total population of carriers simultaneously, and in each step of the Monte Carlo simulation one finds the lowest transition time for all possible transitions, and this is the only transition which is executed. Mandowski and Światek [107] simulated TL spectra for different trap parameters and different correlations between traps and recombination centers, and compared the resulting TL glow curves with the empirical general order kinetics model. For relatively small trap clusters, large differences were found between the modeled TL/TSC curves and the GOK expressions. These authors showed conclusively that spatially correlated effects become prominent for low concentrations of thermally disconnected traps and for situations with high recombination probability.

A different type of Monte Carlo simulation was carried out in the papers by Pagonis et al. [125, 130]. These authors solved the differential equations for several luminescence models, by using the MC method with a fixed time interval.

Later work by Pagonis and Kulp [140], and Pagonis et al. [141] presented kinetic Monte Carlo simulations for various luminescence models. Larsen et al. [92] presented a numerical Monte Carlo model that simulated the processes of charge loss, charge creation and charge recombination in feldspars. These authors also used the assumption of nearest neighbor tunneling. Their model assumed that the concentration of electrons and positive ions are equal at all times. Pagonis and Kulp [140], presented a different version of the MC model by Larsen et al. [92], in which the concentration of positive ions far exceeds that of electrons, and the results from their model compared well with the analytical equation originally derived by Tachiya and Mozumder [175, 176].

The analytical expressions developed by Tachiya and Mozumder [175, 176] were generalized recently by Pagonis et al. [141], in order to describe systems of arbitrary relative concentrations of donors and acceptors. They presented theoretical studies for arbitrary relative concentrations of electrons and acceptors in the solid. Two new differential equations were derived which describe the loss of charge in the solid by tunneling, and they were solved analytically. The analytical solution compared well with the results of Monte Carlo simulations carried out in a random distribution of electrons and positive ions.

Pagonis et al. [131] presented MC simulations of tunneling recombination in random distributions of defects, for CW-IRSL, TL, ITL, LM-IRSL tunneling phenomena. Specifically they presented a simple and fast MC approach which can be applied to these four phenomena, without the need for numerical integrations. They also showed that the method is also applicable to cases of truncated distributions of nearest neighbor distances, which characterize samples that underwent multiple optical or thermal pretreatments. The accuracy and precision of the MC methods were tested by comparing with experimental data from several feldspar samples.

Pagonis et al. [139] used a MC method to evaluate the percent coefficients of variation (CV[%]) for TL, CW-OSL, and LM-OSL signals from delocalized transitions. The results of the simulations showed that CW-OSL signals have the smallest CV values among the three stimulation modes, and therefore these signals are least likely to exhibit stochastic variations. The stochastic uncertainties in these phenomenological models were discussed in the context of single grain luminescence experiments and nanodosimetric materials, in which one deals with small numbers of charge carriers (Autzen et al. [7]).

Recently Lawless et al. [95] studied the inherent noise or uncertainty in the TL process, within the OTOR model. They developed the master equation for this model, and evaluated the probability distributions for trap populations during both irradiation and heating stages of a TL experiment.

8.4 The Simplest Luminescence MC Application: First Order CW-OSL Process

The basic Monte Carlo technique of solving differential equations is found in many standard textbooks of simulations in Statistical Physics (see, for example, Gould and Tobochnik [53]; Landau and Binder [90]; and the references therein). The prototype application of these so-called "brute force" Monte Carlo methods is radioactive decay, which is described by the well-known differential equation:

$$\frac{dn}{dt} = -\mu n \qquad (8.1)$$

where $n(t)$ is the concentration of remaining nuclei in the material at time t, and μ (in s^{-1}) is the decay probability per unit time. In this differential equation $n(t)$ represents a continuous variable. This equation is mathematically identical to the differential equation describing a first order CW-OSL process, which we discussed in Chap. 3. The variable $n(t)$ for the CW-OSL process refers to the concentration of electrons remaining in the material at time t.

The solution of Eq. (8.1) for both the radioactivity and the CW-OSL processes is an exponential decay function:

$$n(t) = n_0 e^{-\mu t} \qquad (8.2)$$

where n_0 is the initial number of electrons in the system at time $t = 0$. The corresponding CW-OSL intensity is found from:

$$I(t) = -dn/dt = \mu n(t) = \mu n_0 e^{-\mu t} \qquad (8.3)$$

In the case of the Monte Carlo simulations, the differential equation (8.1) becomes a *difference equation* for the *discrete* variable n:

$$\Delta n = -\mu n \Delta t \qquad (8.4)$$

The dimensionless probability P for the electron to recombine within a time interval Δt, is $P = \mu \, \Delta t$. In the MC technique, one chooses a suitable value of Δt so that $P < 1$. In the simplest form of the MC algorithm, one chooses an electron and the code generates a random number r uniformly distributed in the unit interval $0 < r < 1$. If $r \leq P$ the electron recombines, otherwise it does not; all non-recombined remaining electrons are tested during each time interval Δt, and several recombination events can take place during each time interval Δt. The value of the remaining electrons n is updated at the end of each time interval Δt, and the process is continued until there are no electrons left.

The following R code shows how to implement the MC method for Eq. (8.1). The simulated system starts with $n_0 = 500$ electrons present at time $t = 0$. The probability of decay per electron and per second is $\mu = 0.03 \, s^{-1}$, and the system is simulated for a total time of $t = 100$ s, and by using a fixed time interval $\Delta t = 1$ s. The Monte Carlo simulation is repeated $M = 100$ times, as described by the assignment of the variable $mcruns \leftarrow 100$. A matrix $nMatrix$ is defined to store the results of the MC simulations, with the number of columns equal to $mcruns$, and the number of rows equal to the time instants $t = 0, 1, 2, \ldots, 100$. Each row of $nMatrix$ corresponds to a different MC run. In this example, the matrix $nMatrix$

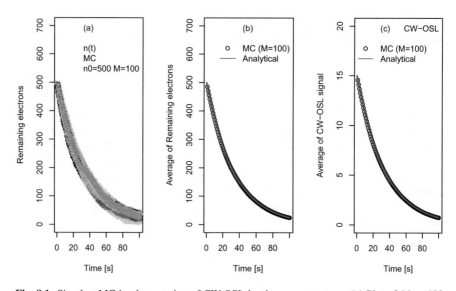

Fig. 8.1 Simplest MC implementation of CW-OSL luminescence process. (**a**) Plot of $M = 100$ MC runs with the same initial number of electrons $n_0 = 500$, simulating a total of 50,000 electrons. (**b**) Plot of the average of the $M = 100$ MC iterations in (**a**). The solid lines in (**b**) and (**c**) represents the analytical solution of the differential equation

will be a 100×500 matrix, and will contain all the results of the MC simulation. We also obtain the values of the CW-OSL signal $I(t) = \mu \, n(t)$, by evaluating the matrix $\mu \cdot nMatrix$.

The algorithm contains three nested loops as follows: the innermost j-loop examines each of the remaining n electrons in the system, the middle i-loop cycles through the time instants $t = 0, 1, 2, \ldots, 100$, and the outermost k-loop cycles over each of the $M = 100$ iterations.

The R-command *rowMeans* from the R library *MatrixStats* is used to evaluate the average of the rows in the matrix, and the result is the average of the $M = 100$ MC runs, which is stored in the variable *avgn*. The *plot* command is called three times, to plot all $M = 100$ results $n(t)$ of the remaining electrons, to plot the average *avgn* of the MC runs, and also to plot the average CW-OSL signal $\mu \cdot nMatrix$. The corresponding analytical solution of the differential equation (8.1) is plotted as a solid line in the graphs, for comparison purposes.

Since this system contains a large number of $n_0 = 500$ electrons, we expect the solution of the differential equation to coincide with the result of the MC simulation. The graphic output of the R code in Fig. 8.1 shows that this is indeed the case.

The R code in this section is rather slow and requires about 3 s for executing the main *for* loops. This code does not take advantage of the speed of vectorized R code, and a vectorized version will be presented in Sect. 8.8.

Code 8.1: Simple MC implementation of CW-OSL process

```r
# Simulate CW-OSL process using the simplest MC code
# Original Mathematica program by Vasilis Pagonis
# R version written by Johannes Friedrich, 2018
rm(list = ls(all=T))
options(warn=-1)
library(matrixStats)
# Define Parameters
mu <- 0.03 # probability of optical excitation per second
deltat <- 1
times <- seq(1, 100, deltat) # time sequence
n0   <- 500 # initial number of electrons at t=0
# Number of iterations of the Monte carlo process
mcruns <- 100
nMatrix   <- matrix(NA, nrow = length(times), ncol = mcruns)
# The 3 main Monte Carlo loops follow
system.time(invisible(
  for (k in 1:mcruns)
  { n <- n0
  for (t in 1:length(times)){
    for (j in 1:n){
      r <- runif(1)  # random number in (0,1)
      P <- mu*deltat
      if (r < P) n <- n - 1 # the electron has recombined
    }
    nMatrix[t,k] <- n  } } ))
# Take average of iterations
avgn <- rowMeans(nMatrix)
## plot MC and analytical solution n(t)
par(mfrow=c(1,3))
pch<-c(NA,NA,1,NA)
lty<-c(NA,NA,NA,"solid")
col<-c(NA,NA,"black","red")
matplot(x=times,nMatrix,xlab = "Time [s]",
        ylab = "Remaining electrons",ylim=c(0,700))
legend("topright",bty="n",legend=c("(a)"," ",
 "n(t) ","MC","n0=500 M=100"))
plot(x = times,       y = avgn,type = "p",pch = 1,,ylim=c(0,700),
     xlab = "Time [s]", ylab = "Average of Remaining electrons")
curve(n0*exp(-mu*x),0,max(t),add=TRUE,col="red",lwd=2)
legend("topright",bty="n",c("(b)      "," ","MC (M=100)",
 "Analytical"),pch=pch,lty=lty,col=col)
plot(x = times,   y = mu*avgn,type = "p",pch = 1,ylim=c(0,20),
     xlab = "Time [s]",   ylab = "Average of CW-OSL signal")
legend("topright",bty="n",c("(c)        CW-OSL"," ","MC (M=100)",
 "Analytical"),pch=pch,lty=lty,col=col)
curve(n0*mu*exp(-mu*x),from=0,max(t),add=TRUE,col="red",lwd=2)

##     user  system elapsed
##     2.95    0.00    2.96
```

Before we discuss how to vectorize the MC R code, we show how luminescence phenomena can be described within the more general framework of stochastic birth and death processes.

8.5   Luminescence Phenomena as Stochastic Birth and Death Processes

In the next few sections we take a different look at thermally and optically stimulated luminescence phenomena, by considering them as a stochastic pure birth and death process. Stochastic birth-death problems have been studied extensively in various branches of science. However, there have been very few previous studies of such problems within the context of thermally or optically stimulated luminescence phenomena (Pagonis et al. [139], Lawless et al. [95]). Recently Pagonis et al. [139] studied birth-death processes within the context of thermally or optically stimulated luminescence phenomena.

A stochastic birth-death process is such that only two types of state transitions are present, births and deaths. In a birth transition, the state variable of the system increases by one, while in a death process the state decreases by one. Birth and death processes have been studied within the context of continuous-time Markov processes, and they have been applied to the study of epidemics, biological/physiology systems, queuing theory, etc.

We will use the same notation as in the review article by Novozhilov et al. [118], and will denote the general birth rate by λ_n and the death rate by μ_n; these rates in general will depend on the population n, on the elapsed time t, and other physical parameters. We will use the symbols λ, μ to denote constant parameters. In very general terms, in describing a birth-death stochastic process, one considers a small time Δt in which only three types of transitions can take place: one death, or one birth, or no birth nor death. The probabilities of these three transitions are $\lambda_n \Delta t$, $\mu_n \Delta t$, and $1 - (\lambda_n + \mu_n) \Delta t$ and these correspond to a possible transition $n \rightarrow n+1$, $n \rightarrow n-1$, and $n \rightarrow n$, respectively.

Many birth-death phenomena can be described by equations of the general form:

$$\frac{dn}{dt} = \lambda_n - \mu_n \tag{8.5}$$

In the case of luminescence phenomena, $n(t)$ is the population of trapped charges at time t, and the initial condition for these processes is $n(0) = n_0$, where n_0 is the concentration of trapped charges at time $t = 0$. The difference equation of the corresponding *stochastic* processes is

$$\Delta n = \{\lambda_n - \mu_n\} \, \Delta t \tag{8.6}$$

Table 8.1 Examples of various luminescence processes and their corresponding stochastic birth-death processes

Stochastic process	Birth rate	Death rate	Luminescence process
Simple linear pure death	$\lambda_n = 0$	$\mu_n = \mu n$	CW-OSL ITL
Generalized simple linear pure death	$\lambda_n = 0$	$\mu_n = \mu(t) n$	TL LM-OSL
Simple nonlinear pure birth	$\lambda_n = \lambda_n(n)$	$\mu_n = 0$	Dose response GOT equation (8.7)
Generalized simple nonlinear pure death	$\lambda_n = 0$	$\mu_n = \mu(n)$	TL (MOK model) OSL (MOK model)

Processes in which $\lambda_n = 0$ and $\mu_n = \mu n$ where $\mu > 0$ is a constant, are referred to as *simple linear pure death* processes. Similarly, when $\lambda_n = \lambda n > 0$ and $\mu_n = 0$ with $\lambda > 0$ a constant, we refer to a *simple linear pure birth* process. When $\lambda_n = \lambda n$ and $\mu_n = \mu n$ with λ, μ constants, we refer to a *simple linear birth and death* process. When the birth and death rates depend on time t, i.e. when $\lambda_n = \lambda(t) n$ and $\mu_n = \mu(t) n$, we refer to these processes as *generalized simple linear birth and death processes*.

Table 8.1 summarizes the various luminescence processes and their corresponding stochastic birth-death processes.

Some examples of luminescence signals described by Eq. (8.5) were discussed previously in Chaps. 1–3. For example, both CW-OSL and CW-IRSL processes are described by $dn/dt = -\sigma I n$, where σI (in s^{-1}) is the rate of optical excitation with a source of constant light intensity, σ (cm^2) represents the optical cross section for the optical process, and I $(photons \, cm^{-2} s^{-1})$ is the constant photon flux. Clearly then, CW-OSL and CW-IRSL are simple linear pure death processes, in which $\lambda_n = 0$ and a death rate $\mu_n = \mu n = \sigma I > 0$. Isothermal luminescence signals (ITL) measured at a constant temperature T_{ISO} are also linear pure death processes; in this case $\lambda_n = 0$ and $\mu_n = n s \, \exp[-E/(kT_{ISO})] = n \mu > 0$.

However, the situation is different for TL and LM-OSL signals. TL signals are described by the Randall-Wilkins differential equation for first order TL kinetics $dn/dt = -n s \, \exp\{-E/[kT(t)]\}$. In this case $\lambda_n = 0$ and the death rate $\mu_n = s \, \exp\{-E/[kT(t)]\} n = \mu(t)n > 0$ varies with time (and the corresponding temperature of the sample). The Randall-Wilkins equation represents a *generalized simple linear death process*, with time dependent death rate $\mu(t) > 0$. Similarly, first order LM-OSL signals also represent a *generalized simple linear death process*, with time dependent death rate $\mu_n = At/P \, n = \mu(t) n > 0$, where $A = \sigma I$ (in s^{-1}) is the rate of optical excitation, t is the elapsed time and P (s) represents the total excitation time.

Similarly, irradiation processes are also examples of birth and death processes. For example, in the previous chapter we saw that the irradiation process in the OTOR model leads to the differential equation:

$$\frac{dn}{dt} = \lambda_n = \frac{R(N-n)X}{(N-n)R+n} \tag{8.7}$$

This equation represents a *pure nonlinear birth process*, in which the death rate $\mu_n = 0$, and the birth rate λ_n depends on both n and t.

In some of the above luminescence processes, analytical solutions exist for the corresponding deterministic differential equation, and they can be compared with the results of the MC simulations. However, in many cases there is no analytical equation available for the corresponding stochastic uncertainties of the luminescence process. In such situations, MC simulations are the only available means of estimating these stochastic uncertainties.

One of the main goals of the stochastic description of luminescence signals, is to obtain an estimate of the *stochastic uncertainties* associated with various experimental techniques. By running and averaging several MC simulations, the MC method allows an estimation of the *intrinsic stochastic uncertainty* of the intensity of the luminescence signal in the various models. These stochastic uncertainties can be described quantitatively by the *variances* σ_n^2 and the *percent coefficients of variation CV[%]* of the populations of trapped electrons $n(t)$ and of the corresponding luminescence signal. In a few cases, we present the *master equations* and their *analytical solutions* (when available), for the populations of trapped electrons during luminescence experiments. We will also provide examples of cases in which analytical expressions for the CV[%] is not available, and where MC is the only available method to evaluate stochastic uncertainties. An example is given of how to evaluate these stochastic uncertainties for the OTOR model, for which analytical equations are not available.

Next, we present examples of evaluating these stochastic uncertainties CV[%], and how these uncertainties may depend on the various parameters in the luminescence models. We first will summarize some of the basic concepts behind birth and death stochastic processes, and show how they can be applied to several luminescence phenomena.

8.6   A Linear Simple Death Process with Death Rate $\mu_n = n\mu$

Recently Pagonis et al. [139] and Lawless et al. [95] studied various luminescence phenomena within the context of stochastic birth-death processes. In this section we start the discussion with the *linear simple death process*, in which the difference equation is

$$\Delta n = -\mu \, n \, \Delta t \tag{8.8}$$

As mentioned in the last section and shown in Table 8.1, CW-OSL and ITL signals are examples of this type of process. The stochastic nature of a linear simple death process is described by the well-known master equation (Allen [5]):

Master equation for pure death processes with $\mu_n = n\mu$

$$\frac{dP_n}{dt} = \mu(n+1)P_{n+1} - \mu n P_n \qquad (8.9)$$

---

where $P_n(t)$ is the probability for the sample to contain exactly n trapped charges at time t. A common assumption for the initial conditions of this system of differential equations is that all trapped charges are initially in the same state, such that $P_{n_0}(0) = 1$ for $n = n_0$, and $P_n(0) = 0$ for $n \neq n_0$.

Equation (8.9) represents a system of coupled differential equations, which in general has to be solved numerically. For example, in a system of $n_0 = 1000$ initially trapped electrons at time $t = 0$, Eq. (8.9) is a system of 1000 coupled difference equations.

When $\mu = $ constant, the solution of Eq. (8.9) is found by using mathematical induction to be

$$P_j(t) = \binom{n_0}{j} \exp(-\mu j t) \left[1 - \exp(-\mu j t)\right]^{n_0 - j} \qquad (8.10)$$

where $j = 1 \ldots n_0$. This expression represents the standard form of the binomial distribution, with the probability coefficient $p(t) = \exp(-\mu j t)$:

Population $P_j(t)$ for a linear death problem when $\mu_n = \mu n$

$$P_j(t) = \binom{n_0}{j} p(t) \left[1 - p(t)\right]^{n_0 - j} \qquad (8.11)$$

$$p(t) = \exp(-\mu j t)$$

The *mean* of the population of trapped electrons in this binomial distribution is found from the general properties of the binomial distribution:

$$<n> = n_0 \exp(-\mu t) \qquad (8.12)$$

and the *variance* σ_n^2 of the population of trapped electrons is given by (Allen [5]):

Variance σ_n^2 of population $n(t)$ when $\mu_n = \mu n$

$$\sigma_n^2 = <n^2> - <n>^2 = n_0 \left[\exp(-\mu t) - \exp(-2\mu t) \right] \qquad (8.13)$$

Note that at time $t = 0$ the variance of this population $P_j(t)$ is $\sigma_n^2(0) = 0$. As time t increases during the CW-OSL experiment, the stochastic variance σ_n^2 will also increase.

By combining Eqs. (8.12) and (8.13), we find the dimensionless *coefficient of variation* (CV[%]) for the simple death process, expressed in %:

$$CV_n[\%] = 100 \frac{\sigma_n}{<n>} = 100 \frac{\sqrt{n_0 e^{-\mu t} \left(1 - e^{-\mu t} \right)}}{n_0 e^{-\mu t}} = 100 \sqrt{\frac{e^{\mu t} - 1}{n_0}} \qquad (8.14)$$

$CV_n[\%]$ of population $n(t)$ when $\mu_n = \mu n$

$$CV_n[\%] = 100 \sqrt{\frac{e^{\mu t} - 1}{n_0}} \qquad (8.15)$$

The equations discussed above, describe the mean $<n>$ and the variance σ_n^2 of the population of trapped electrons.

The following code plots the populations $P_j(t)$ from Eq. (8.11) for various times t, and the results are shown in Fig. 8.2. The number of electrons in this example in $n_0 = 100$ and the probability of recombination is $\mu = 0.03 \, \text{s}^{-1}$. Note that as the time t increases, the average of the distributions moves from right to left in Fig. 8.2. Initially at time $t = 0.01$ s the distribution is very narrow, then gets wider and finally gets narrow again at long times t. These results are consistent with the analytical equations for the stochastic mean number of particles $<n(t)>$, and for the stochastic standard deviation σ_n.

For additional examples of these types of simulations, the reader is referred to the recent paper by Lawless et al. [95].

From an experimental point of view, we are also interested in the variance σ_L^2 of the luminescence intensity L. Since $I(t) = L = -dn/dt$, the quantity σ_L^2 can be evaluated from:

$$\sigma_L^2 = <\left(\frac{dn}{dt} \right)^2> - <\frac{dn}{dt}>^2 \qquad (8.16)$$

Code 8.2: Populations P(j) of stochastic simple death process

```
# Populations P(j) of stochastic simple death process
rm(list = ls(all=T))
n0<-100
mu<-.03
x<-seq(1,100)
f<-function(u) {choose(n0,u)*exp(-mu*u*t)*
    ((1-exp(-mu*u*t))**(n0-u))}
times<-c(.01,1,20,40,110)
TF=c(FALSE,rep(TRUE,4))
for (i in 1:5){
  t<-times[i]
  area<-sum(unlist(lapply(x,f)))
  curve(choose(n0,round(x))*exp(-mu*x*t)*
((1-exp(-mu*x*t))**(n0-x))/area,
  1,100,ylim=c(0,.4),lwd=3,add=TF[i],col=i,lty=i,
  xlab="# of particles j",ylab="probability Pn(j,t)")}
legend("topleft",bty="n",c("t=.01 s","1 s ","20 s ","40 s ",
 "110 s"), col=1:5,lty=1:5,lwd=3)
legend("top",bty="n",c("Populations P(j,t)"," ",
"Stochastic death process"))
```

By using $dn/dt = -\mu n$ we obtain:

$$\sigma_L^2 = <(\mu n)^2> - <\mu n>^2 = \mu^2 \left(<n^2> - <n>^2\right) = \mu^2 \sigma_n^2 \qquad (8.17)$$

$$\sigma_L = \mu \sigma_n \qquad (8.18)$$

This tells us that in CW-OSL and ITL experiments, the variance of the luminescence signal σ_L^2 is directly proportional to the variance σ_n^2 of the population of filled traps. The corresponding coefficient of variation $CV_{CW}[\%]$ of the CW-OSL signal is

$$CV_{CW}[\%] = 100\frac{\sigma_{CW}}{I(t)} = 100\frac{\mu \sigma_n}{\mu n} = CV_n[\%] \qquad (8.19)$$

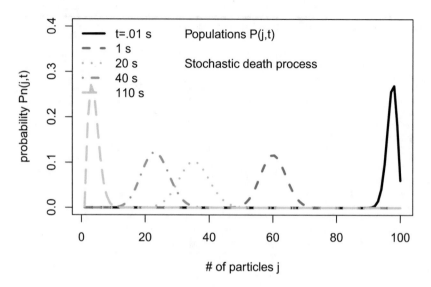

Fig. 8.2 Plots of $P_j(t)$ from Eq. (8.11), for a simple death stochastic process. As the time t increases from right to left in this plot, the width initially increases and then decreases with t. For more details see Lawless et al. [95]

$$CV_{CW}[\%] \text{ of CW-OSL signal with } \mu_n = n\mu$$

$$CV_{CW}[\%] = CV_n[\%] = \sqrt{\frac{e^{\mu t} - 1}{n_0}} \qquad (8.20)$$

The following code plots the stochastic average $< n(t) >$ for the number of particles, the stochastic standard deviation σ_n and the stochastic $CV[\%]$ from Eqs. (8.12), (8.13) and (8.15), respectively. The results are shown in Fig. 8.3.

As time t increases during a CW-OSL experiment, the value of $CV_{CW}[\%]$ in Fig. 8.3 also increases. This means that the most accurate part of a CW-OSL signal is the initial part near $t = 0$. Equation (8.15) also shows that for a given time instant t, the value of $CV_{CW}[\%]$ will be inversely proportional to $1/\sqrt{n_0}$, and as the initial number of electrons n_0 in the material increases, the stochastic uncertainty will become negligible and therefore the experimental uncertainty will become much larger than the stochastic uncertainty.

However, in the case of TL and LM-OSL experiments we will see that the two variances σ_L^2, σ_n^2 will not be proportional to each other, but will have a different dependence on time (or temperature).

Code 8.3: Plots of stochastic simple death process

```
rm(list = ls(all=T))
n0<-100
mu<-.03
par(mfrow=c(1,3))
curve(n0*exp(-mu*x),0,100,lwd=3,xlab="Time t, s",ylab="<n(t)>",
    ylim=c(0,140))
legend("top",bty="n",c("(a)"," ","Stochastic", "<n>"))
curve(sqrt(n0)*sqrt(exp(-mu*x)-exp(-2*mu*x)),0,100,lwd=3,
    xlab="Time t, s",ylab=expression(sigma),ylim=c(0,8))
legend("top",bty="n",c(expression("(b)"," ","Stochastic",sigma)))
 curve(100*sqrt(exp(mu*x)-1)/n0,1,100,xlab="Time t, s",lwd=3,
    ylab="CV[%]",ylim=c(0,7))
 legend("top",bty="n",c("(c)"," ","Stochastic","CV[%]"))
```

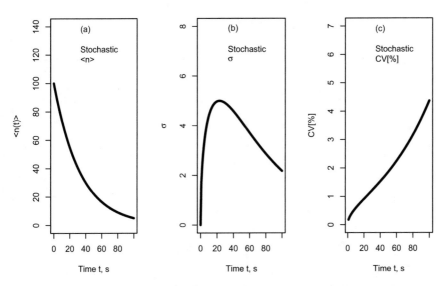

Fig. 8.3 (a) Plot of stochastic mean number of particles $< n(t) >$. (b) Plot of stochastic standard deviation σ_n . (c) Plot of $CV[\%]$. From Eq. (8.12), (8.13) and (8.15)

8.7　Vectorized MC Simulation of a First Order CW-OSL Process

The R code in Sect. 8.4 was rather slow, and required about 3 s to simulate $M = 100$ MC runs with $n_0 = 500$ electrons each, for a CW-OSL process. In this section we present a vectorized version of the code from Sect. 8.4.

The non-vectorized code does not take advantage of the speed of R, due to the presence of the three nested *for* loops, and also because of the *if* command contained in the innermost loop. We can improve the speed of the MC code rather dramatically, by vectorizing the innermost loop, i.e. by replacing the innermost *i*-loop and the *if* command it contains with a vectorized code. For a good discussion of how to vectorize this type of code, see the R book by Grolemund [54].

In the code that follows, the innermost *j*-loop and the *if* command are replaced by the R code *vec[vec>P]*. This code selects the elements of the array *vec,* by using the logical statement *vec>P.*

The following vectorized R code produces Fig. 8.3 is approximately 30 times faster than the code in Sect. 8.4. The results are shown in Fig. 8.4.

In a MC simulation, it is important to check the effect of using different values of Δt. The simulation results should stay unaffected when changing the value of Δt, as long as the probability $P \Delta t$ is less than 1.

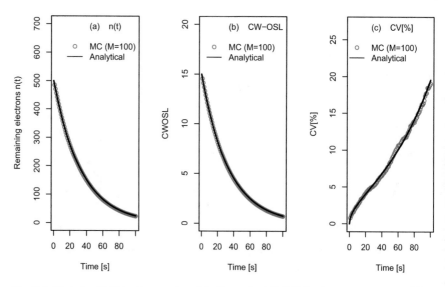

Fig. 8.4 Vectorized MC implementation of the first order CW-OSL luminescence process. (**a**) Plot of $M = 100$ MC runs with the same initial number of electrons $n_0 = 500$, simulating a total of 50,000 electrons. (**b**) Average of the $M = 100$ MC iterations in (**a**). (**c**) The corresponding $CV[\%]$. The solid lines represent the analytical solution of the differential equation. For more details and examples, see Pagonis et al. [139]

Code 8.4: Vectorized MC implementation of CW-OSL

```
# Vectorized MC code for first-order CW-OSL process
rm(list = ls(all=T))
options(warn=-1)
library(matrixStats)
mcruns<-300
n0<-500
mu<-.03
deltat<-1
tmax<-100
times<-seq(1,tmax,deltat)
nMatrix <-  matrix(NA, nrow = length(times), ncol = mcruns)
nMC<-rep(NA,length(times))
system.time(
    for (k in 1:mcruns){
        n<-n0                #initialize each of the M=100 MC runs
        for (t in 1:length(times)){
        vec<-rep(runif(n))    #create a vector vec,
        #containing n random numbers between 0 and 1
        P<-mu*deltat
        n<-length(vec[vec>P]) #if the random number in vec is >P,
        #then the corresponding electron survives
    nMC[t]<-n } # store number of electrons n in the vector nMC
    nMatrix[,k]<-nMC # store single run in column k of nMatrix
    })
#Find average of n(t), CW-OSL signal, and CV[%]
avgn<-rowMeans(nMatrix)
avgCWOSL<-mu*rowMeans(nMatrix)
sd<-rowSds(nMatrix)
cv<-100*sd/avgn
par(mfrow=c(1,3))
pch<-c(NA,NA,1,NA)
lty<-c(NA,NA,NA,"solid")
col<-c(NA,NA,2,1)
plot(times,avgn,ylab="Remaining electrons n(t)",
    xlab="Time [s]",ylim=c(0,700),col=2)
curve(n0*exp(-mu*x),0,tmax,add=TRUE,col=1,lwd=2)
legend("topright",bty="n",c("(a)      n(t)"," ","MC (M=100)",
 "Analytical"),pch=pch,lty=lty,col=col)
plot(times,avgCWOSL,ylab="CWOSL",xlab="Time [s]",ylim=c(0,20),
col=2)
legend("topright",bty="n",c("(b)      CW-OSL"," ","MC (M=100)",
 "Analytical"),pch=pch,lty=lty,col=col)
curve(n0*mu*exp(-mu*x),0,tmax,add=TRUE,col=1,lwd=2)
plot(times,cv,ylab="CV[%]",xlab="Time [s]",ylim=c(0,27),col=2)
curve(100*sqrt((exp(mu*x)-1)/n0),0,max(times),add=TRUE,
    col=1,lwd=2)
legend("topleft",bty="n",c("(c)      CV[%]"," ","MC (M=100)",
    "Analytical"),pch=pch,lty=lty,col=col)

    ##    user  system elapsed
    ##    0.41    0.00    0.41
```

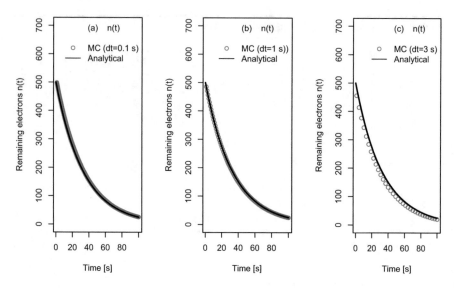

Fig. 8.5 The effect of the parameter *deltat* in the previous MC implementation of the first order CW-OSL luminescence process, with (**a**) *deltat* = 0.1 s, (**b**) *deltat* = 1 s, (**c**) *deltat* = 3 s. All three runs are carried out with the same parameters $M = 100$ MC runs, initial number of electrons $n_0 = 500$, $\mu = 0.03\,\text{s}^{-1}$

Figure 8.5 shows an example of the effect of the parameter *deltat* in the above MC simulation. The result of the simulation becomes more accurate for small *deltat* = 0.1 and 1 s, and becomes less accurate for the large value of the parameter *deltat* = 3 s.

In general, the results of the MC simulation should be independent of the number of initially trapped electrons n_0. Figure 8.6 shows the results of MC implementation of the first order CW-OSL luminescence process, with the same parameters $\mu = 0.03\,\text{s}^{-1}$, and with three different combinations of n_0 and M. The combinations of n_0 and M are such that the three runs contain the same total number of electrons $n_0 \times M = 10^4$. The results of Fig. 8.6 show that indeed the results do not depend on n_0, and therefore they can be considered *unbiased results*.

8.8   Linear Pure Death Process with $\mu_n = n\mu(t)$

As discussed above, the Randall-Wilkins equation for TL processes is a pure linear death process, in which $\mu(t) = s\,\exp\{-E/[kT(t)]\}$. In this case the parameter $\mu(t)$ is an explicit function of time, and the results of the MC method must account for the time dependence of $\mu(t)$.

In this more general case, the solution of the master equation is found to be (Allen [5], Pagonis et al. [139]):

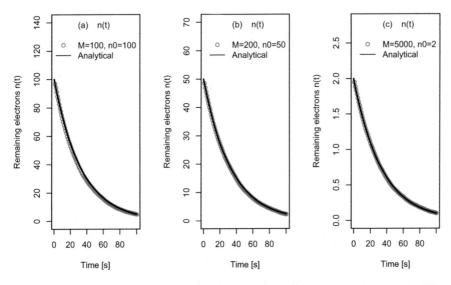

Fig. 8.6 The effect of the parameter n_0 in the previous MC implementation of the first order CW-OSL luminescence process, with (**a**) $n_0 = 100$ and $M = 100$ MC runs, (**b**) $n_0 = 50$ and $M = 200$, (**c**) $n_0 = 2$ and $M = 5000$. All three runs are carried out with the same parameters $\mu = 0.03\,\mathrm{s}^{-1}$ and the same total number of electrons $n_0 \times M = 10^4$

$$P_j(t) = \binom{n_0}{j} q(t)\,[1 - q(t)]^{n_0 - j} \tag{8.21}$$

where the function $q(t)$ satisfies the differential equation:

$$\frac{dq}{dt} = -\mu(t)q(t) \tag{8.22}$$

The solution of this differential equation is

$$q(t) = \exp\left[-\int_0^t \mu(t')dt'\right] \tag{8.23}$$

where t' is a dummy integration variable. By combining Eqs. (8.21) and (8.23), we find the general solution for the case of time dependent death rate $\mu(t)$:

Pure death process: the population $P_j(t)$ with $\mu_n = n\mu(t)$

$$P_j(t) = \binom{n_0}{j} \left(\exp\left[-\int_0^t \mu(t')dt' \right] \right) \left[1 - \exp\left[-\int_0^t \mu(t')dt' \right] \right]^{n_0-j}$$

(8.24)

The *mean* of this binomial distribution is found to be

$$<n> = n_0 \exp(-\int_0^t \mu(t')dt')$$

(8.25)

and the *variance* of the population $\sigma^2 = <n^2> - <n>^2$ is given by

Variance of population $n(t)$ with $\mu_n = n\mu(t)$

$$\sigma^2 = n_0 \left[\exp(-\int_0^t \mu(t')dt') - \exp(-2\int_0^t \mu(t')dt') \right]$$

(8.26)

The $CV[\%]$ in this more general case of $\mu(t) \neq$ constant, is given by

$CV[\%$ of population $n(t)$ with $\mu_n = n\mu(t)$

$$CV[\%] = 100\frac{\sigma_n}{<n>} = 100\sqrt{\frac{e^{\int_0^t \mu(t')dt'} - 1}{n_0}}$$

(8.27)

When $\mu(t) = \mu =$ constant, the last three equations revert to the previous three simpler equations (8.12), (8.13), and (8.14).

In the case of TL, $\mu(t) = s\, e^{-\frac{E}{kT(t)}}$ and the integral $\int_0^t \mu(t')dt'$ must be evaluated numerically. In the case of LM-OSL, the death rate $\mu(t) = \sigma\, I\, t/P$, and the integral $\int_0^t \mu(t')dt' = \sigma\, I\, t^2/(2P)$.

The variance of the luminescence signal σ_L^2 can be evaluated from:

$$\sigma_L^2 = < \left(\frac{dn}{dt}\right)^2 > - < \frac{dn}{dt} >^2 \tag{8.28}$$

By using Eq. (8.5):

$$\sigma_L^2 = < (\mu(t)\,n)^2 > - < \mu(t)\,n >^2 = \mu(t)^2 \left(< n^2 > - < n >^2 \right) = \mu(t)^2 \,\sigma_n^2 \tag{8.29}$$

$$\sigma_L = \mu(t)\,\sigma_n \tag{8.30}$$

In the case of TL,

$$\sigma_{TL} = \left(s\, e^{-\frac{E}{kT(t)}} \right) \sigma_n \tag{8.31}$$

In the case of LM-OSL,

$$\sigma_{LMOSL} = \left(\frac{\sigma I}{P} t \right) \sigma_n \tag{8.32}$$

8.9 Vectorized Code for First Order Kinetics TL Process

The details of implementing the Monte Carlo algorithm have been given already in Sect. 8.4, and recently in Pagonis et al. [131]. The following code shows an example of a first order kinetics TL glow curve obtained with this method, representing $M = 100$ MC runs for a system of $n_0 = 100$ initially trapped electrons. The results are shown in Fig. 8.7. In the following code the average and standard deviation σ of several MC runs is evaluated, in order to estimate the stochastic uncertainties in the MC method.

Figure 8.7a shows the results of the average $< n(t) >$ of $M = 100$ MC runs, while in Fig. 8.7b shows the corresponding average TL signal.

Figure 8.7c is a plot of the stochastic coefficient of variation $CV[\%]$ of the TL intensity, as a function of temperature. As the temperature increases, $CV[\%]$ also increases monotonically. The solid lines in Fig. 8.7 are the analytical solutions of the corresponding differential equation for the deterministic process, in terms of the Lambert W function (see the extensive discussion in Chap. 2).

Code 8.5: Vectorized MC implementation of TL

```
rm(list = ls(all=T))
options(warn=-1)
library(matrixStats)
mcruns<-100
n0<-500
s<-1e12
E<-1
kb<-8.617e-5
tmax<-150
deltat<-1
times<-seq(0,tmax,deltat)
nMatrix<-TLMatrix<-matrix(NA,nrow=length(times),ncol=mcruns)
nMC<-TL<-rep(NA,length(times))
system.time(
for (j in 1:mcruns){
  n<-n0
  for (t in 1:length(times)){
    vec<-rep(runif(n))
    P<-s*exp(-E/(kb*(t+273)))*deltat
    n<-length(vec[vec>P])
    nMC[t]<-n
    TL[t]<-n*P}
nMatrix[,j]<-nMC
TLMatrix[,j]<-TL
 })
#Find average n(t),average CW-OSL signal and CV[%]
avgn<-rowMeans(nMatrix)
avgTL<-rowMeans(TLMatrix)
sd<-rowSds(TLMatrix)
cv<-100*sd/avgTL
## Calculate the analytical error of TL in first order peak
x1<-times+273
k<-function(u) {integrate(function(p){s*exp(-E/(kb*p))},
 273,u)[[1]]}
y1<-lapply(x1,k)
x<-unlist(x1)
y<-unlist(y1)
errn<-sqrt(n0*(exp(-y)-exp(-2*y)))
nanalyt<-n0*exp(-y)
TLanalyt<-n0*s*exp(-E/(kb*x))*exp(-y)
# plots
par(mfrow=c(1,3))
pch<-c(NA,NA,1,NA)
lty<-c(NA,NA,NA,"solid")
col<-c(NA,NA,2,1)
plot(times,avgn,ylab="Remaining electrons n(t)",
col=2,xlab=expression("Temperature ["^"o"*"C]"),ylim=c(0,700))
lines(x-273,nanalyt,col=1)
legend("topleft",bty="n",c("(a)      n(t)"," ","MC",
 "Lambert Eq."),pch=pch,lty=lty,col=col)
```

```
plot(times,avgTL,ylab="TL",
 col=2,xlab=expression("Temperature ["^"o"*"C]"),ylim=c(0,23))
lines(x-273,TLanalyt,col=1)
legend("topleft",bty="n",c("(b)     TL"," ","MC",
 "Lambert Eq."),pch=pch,lty=lty,col=col)
plot(times,cv,ylab="CV[%]",ylim=c(0,150),
 col=2,xlab=expression("Temperature ["^"o"*"C]"))
lines(x-273,100*errn*s*exp(-E/(kb*x))/TLanalyt,col=1)
legend("topleft",bty="n",c("(c)     CV[%]"," ","MC",
 "Analytical"), pch=pch,lty=lty,col=col)
##      user  system elapsed
##      0.29    0.00    0.30
```

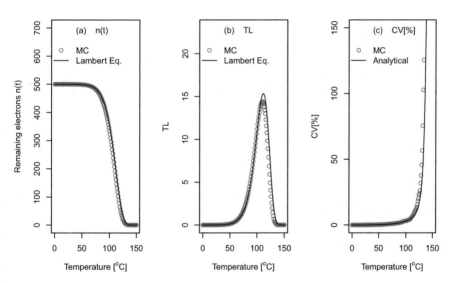

Fig. 8.7 Vectorized MC implementation of the first order TL luminescence process. (a) Plot of $< n(t) >$ for $M = 100$ MC runs with the same initial number of electrons $n_0 = 500$, simulating a total of 50,000 electrons; (b) Average of the corresponding TL signal. (c) The corresponding CV[%]. The solid lines represent the analytical equations. For details, see Pagonis et al. [139]

8.10 Vectorized MC Simulation of First Order LM-OSL Process

The following vectorized code shows an example of a first order kinetics LM-OSL curve, obtained with the same method as in the previous section, with the results shown in Fig. 8.8.

The LM-OSL simulation contains $M = 100$ MC runs for a system of $n_0 = 500$ initially trapped electrons, with an excitation rate $p(t) = \sigma I t/P = 0.2\,\mathrm{s}^{-1}$ and a

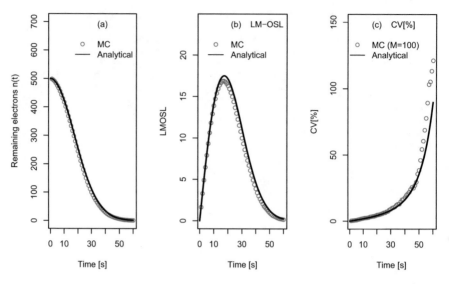

Fig. 8.8 Vectorized MC implementation of the first order LM-OSL luminescence process. (**a**) Plot of the mean $< n(t) >$ for $M = 100$ MC runs with the same initial number of electrons $n_0 = 500$, simulating a total of 50,000 electrons. (**b**) Average of the corresponding LM-OSL signal. (**c**) The corresponding $CV[\%]$. The solid lines represent the analytical equations from Chap. 3

total excitation period of $P = 60$ s. Figure 8.8a shows the results of the average $< n(t) >$ of the MC runs, while Fig. 8.8b shows the corresponding LM-OSL signal.

Figure 8.8c is a plot of the stochastic coefficient of variation CV[%] of the LM-OSL intensity, as a function of excitation time t. As the time increases, CV[%] also *increases* monotonically. The solid lines are the analytical solutions of the corresponding differential equation for the deterministic process (see the extensive discussion in Chap. 3).

8.11 Vectorized MC Simulation of TL in the GOT Model

Pagonis et al. [130] used the simple MC method described in the previous section, to solve the differential equation for the GOT model. An example of their method is presented in this section; this example also uses a fixed time interval.

As we saw in Chap. 2, the GOT differential equation for TL in the OTOR model is

$$I(t) = -\frac{dn}{dt} = s\, e^{-\frac{E}{kT(t)}} \frac{n^2}{(N-n)R + n} \tag{8.33}$$

where $R = A_n/A_m$ is the retrapping ratio in the GOT model, $E\ (eV)$ is the thermal activation energy of the trap, and $s\ (s^{-1})$ is the associated frequency factor. This equation becomes a difference equation:

Code 8.6: Vectorized MC implementation of LM-OSL

```
rm(list = ls(all=T))
options(warn=-1)
library(matrixStats)
mcruns<-100
n0<-500
tmax<-60
A<-0.2
deltat<-1
times<-seq(1,tmax,deltat)
nMatrix<-LMMatrix<-matrix(NA,nrow=length(times),ncol=mcruns)
nMC<-LM<-rep(NA,length(times))
system.time(
for (j in 1:mcruns){
  n<-n0
  for (t in 1:length(times)){
    vec<-rep(runif(n))
    P<-deltat*t*A/tmax
    n<-length(vec[vec>P])
    nMC[t]<-n
    LM[t]<-n*P}
nMatrix[,j]<-nMC
LMMatrix[,j]<-LM
 })
#Find average of n(t),LM-OSL signal and CV[%]
avgn<-rowMeans(nMatrix)
avgLM<-rowMeans(LMMatrix)
sd<-rowSds(LMMatrix)
cv<-100*sd/avgLM
par(mfrow=c(1,3))
pch<-c(NA,NA,1,NA)
lty<-c(NA,NA,NA,"solid")
col<-c(NA,NA,2,1)
plot(times,avgn,ylab="Remaining electrons n(t)",
xlab="Time [s]",ylim=c(0,700),col=2)
curve(n0*exp(-A*x^2/(2*tmax)),0,tmax,add=TRUE,
col=1,lwd=2)
legend("topright",bty="n",c("(a)      ",
" ","MC","Analytical"),
pch=pch,lty=lty,col=col)
plot(times,avgLM,ylab="LMOSL",xlab="Time [s]",
col=2,ylim=c(0,24))
legend("topright",bty="n",c("(b)      LM-OSL"," ","MC",
"Analytical"), pch=pch,lty=lty,col=col)
curve(A*n0*exp(-A*x^2/(2*tmax))*x/tmax,0,tmax,add=TRUE,
col=1,lwd=2)
```

```
plot(times,cv,ylab="CV[%]",xlab="Time [s]",ylim=c(0,150),
col=2)
curve(100*sqrt((exp(A*x^2/(2*tmax))-1)/n0),0,tmax,add=TRUE,
col=1,lwd=2)
legend("topleft",bty="n",c(" (c)       CV[%]"," ","MC (M=100)",
 "Analytical"), pch=pch,lty=lty,col=col)

 ##     user  system elapsed
 ##     0.07    0.00    0.08
```

$$\Delta n = -s\,e^{-\frac{E}{kT(t)}}\,\frac{n^2}{(N-n)R+n}\,\Delta t \qquad (8.34)$$

The code solves Eq. (8.34), by using the same MC method as in the previous sections. A fixed time interval $\Delta t = 1$ s is used in the simulation. The results are shown in Fig. 8.9.

The corresponding TL intensity $I_{TL}(T)$ is calculated from the values of Δn in Eq. (8.34) using the expression:

$$I_{TL}(T) = -\frac{1}{\beta}\frac{\Delta n}{\Delta t} = -\frac{1}{\beta}n\,s\,\exp\left[-\frac{E}{k_B T}\right] \qquad (8.35)$$

Figure 8.9a shows the results of the average $< n(t) >$ for $M = 100$ MC runs, while Fig. 8.9b shows the corresponding mean TL signal.

Figure 8.9c is a plot of the stochastic coefficient of variation CV[%] of the TL intensity, as a function of temperature. As the temperature increases, CV[%] also increases monotonically. The solid line in Fig. 8.9a, b is the analytical solution of the corresponding differential equation for the deterministic process, in terms of the Lambert W function (see the extensive discussion in Chap. 2). There are *no* analytical solutions for the stochastic coefficient of variation CV[%] shown in Fig. 8.9c. This is because of the nonlinear nature of the deterministic differential equation in the GOT model.

8.12   Monte Carlo Simulation of TL/OSL from a System of Small Trap Clusters

The MC method becomes most useful when one is dealing with a small number of electrons and traps, as in the case of nanodosimetric materials. The example below simulates such a system, with the details given in the paper by Pagonis et al. [130].

For a small system of trap clusters, one can make a clear distinction between *local variables* describing the internal structure of each cluster, and *global variables*

Code 8.7: Vectorized MC code for TL in GOT model

```
# GOT MODEL- Monte Carlo code for TL
rm(list = ls(all=T))
options(warn=-1)
library(matrixStats)
library(lamW)
mcruns<-100
n0<-500
N<-1000
s<-1e12
E<-1
R<-0.6
kb<-8.617e-5
tmax<-200
deltat<-1
times<-seq(1,tmax,deltat)
nMatrix<-TLMatrix<-matrix(NA,nrow=length(times),ncol=mcruns)
nMC<-TL<-rep(NA,length(times))
system.time(
for (j in 1:mcruns){
  n<-n0
  for (t in 1:length(times)){
    vec<-rep(runif(n))
    P<-s*exp(-E/(kb*(t+273)))*n/((N-n)*R+n)
    n<-length(vec[vec>P])
    nMC[t]<-n
    TL[t]<-n*P}
nMatrix[,j]<-nMC
TLMatrix[,j]<-TL
  })
#Find average of n(t), average TL signal and CV[%]
avgn<-rowMeans(nMatrix)
avgTL<-rowMeans(TLMatrix)
sd<-rowSds(TLMatrix)
cv<-100*sd/avgTL
# plots
par(mfrow=c(1,3))
pch<-c(NA,NA,1,NA)
lty<-c(NA,NA,NA,"solid")
col<-c(NA,NA,2,1)
k<-function(u) {integrate(function(p){exp(-E/(kb*p))},
300,u)[[1]]}
x1<-300:450
y1<-lapply(x1,k)
x<-unlist(x1)
y<-unlist(y1)
c<-(n0/N)*(1-R)/R
zTL<-(1/c)-log(c)+(s*n0/(c*N*R))*y
```

```
plot(times,avgn,ylab="Remaining electrons n(t)",
col=2,xlab=expression("Temperature ["^"o"*"C]"),ylim=c(0,700))
lines(x-273,(N*R/(1-R))/(lambertW0(exp(zTL))),col=1)
legend("topright",bty="n",c("(a)        n(t)"," ","MC",
 "Lambert Eq."),pch=pch,lty=lty,col=col)
plot(times,avgTL,ylim=c(0,14),ylab="TL",
col=2,xlab=expression("Temperature ["^"o"*"C]"))
# plots
lines(x-273,(N*R/((1-R)^2))*s*exp(-E/(kb*x))/
(lambertW0(exp(zTL))+lambertW0(exp(zTL))^2),col=1)
legend("topleft",bty="n",c("(b)        TL"," ","MC",
 "Lambert Eq."), pch=pch,lty=lty,col=col)
plot(times,cv,ylab="CV[%]",ylim=c(0,120),
col=2,xlab=expression("Temperature ["^"o"*"C]"))
legend("topleft",bty="n",c("(c)       CV[%]"," ","MC"))

    ##     user   system elapsed
    ##     0.34    0.00    0.34
```

which describe the whole group of clusters. The system simulated by these authors has some similarities to the one used by Mandowski and Światek [107], and consists of a large number of small clusters of traps and recombination centers. The local physical parameters characterizing each cluster are: the total number of traps per cluster N_{traps}, the number of initially filled traps per cluster n_{filled}, (in general $n_{filled} \leq N_{traps}$), and the instantaneous number of remaining filled traps in the cluster denoted by n_{local}. As the system develops in time during the optical or thermal stimulation process, the value of the local variable n_{local} will decrease from its initial value down to zero, as more recombinations take place within the cluster.

In terms of global variables, the system is described by the number of trap clusters in the system $N_{clusters}$. The total number of available traps in the system is given by the product $N = N_{clusters}N_{traps}$, while the total number of initially filled traps n_0 is given by $n_0 = N_{clusters}n_{filled}$ (with $n_0 \leq N$). The equalities $n_{filled} = N_{traps}$ and $n_0 = N$, denote a system with all traps initially filled. The total instantaneous number of filled traps is denoted by the global variable n, which is calculated as the sum of remaining filled traps n_{filled} over all clusters in the system, i.e.:

$$n = \sum_{all\ clusters} n_{local} \tag{8.36}$$

Clearly the variables n_{local}, n_{filled} and N_{traps} represent *local* variables characterizing each cluster in the system, while the *global* variables $N_{clusters}$, n_0, n and N characterize the whole system of trap clusters.

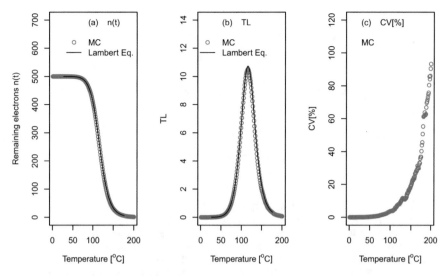

Fig. 8.9 Vectorized MC simulation of TL in a system of $M = 100$ large clusters of defects, based on the GOT equation. The total number of traps in each cluster is $N = 1000$, and initially $n_0 = 500$ of these traps are filled. (a) The average $n(t)$ (b) The average TL signal; (c) The corresponding CV[%]. The solid lines in (a) and (b) represent the analytical equations which are based on the Lambert function. For details, see Pagonis et al. [139]

Figure 8.10 shows schematically an example of such a system of small trap clusters, in which there are 4 traps in each cluster (shown as both open and solid circles), with only 3 of them being initially filled (shown as solid circles). One ensures the charge balance in the system, by assuming the existence of an equal number of 4 luminescence centers (shown as both open and solid stars), 3 of which have been activated (shown as solid stars).

In this example one could simulate, for instance, a large number of clusters in the system, resulting in a total number of initially filled traps. From a physical point of view, the activated luminescence centers may exist in physical proximity to the filled traps, since they both could have been created simultaneously during the irradiation process. As the system of trap clusters in Fig. 8.10 develops in time, the local variable n_{local} will vary from an initial value of n_{filled} to zero at the end of the thermal/optical excitation process. Similarly the global variable n will vary from an initial value of n_0 to its final value of zero at the end of the process.

The luminescence intensity from the overall system of trap clusters will consist of the sum of signals from all clusters in the system (Table 8.2).

The implicit physical assumption in this description is that each cluster acts as an independent entity as far as the luminescence process is concerned, since each electron participates only in local processes within the cluster. The result of the following R code shows that the TL intensity in this system of small clusters contains two peaks, while the solution of the corresponding stochastic differential equation for the corresponding system of large clusters contains a single TL peak. It

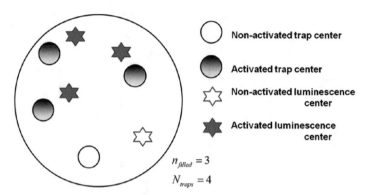

Fig. 8.10 Schematic representation of a small trap cluster, consisting of a total of four traps in each cluster ($N_{traps} = 4$ shown as both open and solid circles). Only three of these traps are initially filled ($n_{filled} = 3$ shown as solid circles). Charge balance in the system is ensured by assuming the existence of an equal number of four luminescence centers (shown as both open and solid stars), three of which have been activated (shown as solid stars). The solid is assumed to consist of a large number of clusters (e.g. $N_{clusters} = 10^5$). For a detailed description, see Pagonis et al. [130]

Table 8.2 Listing of local and global variables used in the simulations of luminescence from small clusters, and their typical values (see also Fig. 8.10 for a pictorial presentation of the local variables). From Pagonis et al. [130]

Variable type	Description	Typical value
Local		
$n_{local}(t)$	The number of remaining filled traps in the cluster	3
n_{filled}	The number of initially filled traps per cluster. ($n_{filled} \leq N_{traps}$)	3
N_{traps}	The total number of traps per cluster	4
Global		
$n(t)$	The total number of remaining filled traps in the system.	3×10^5
	This is calculated by summing over all clusters	
n_0	The total number of initially filled traps in the system	3×10^5
	n_0 is found from $n_0 = N_{clusters} \times n_{filled}$ (with $n_0 \leq N$)	
$N_{clusters}$	The number of trap clusters in the system	10^5

is noted that there are no analytical equations to describe these graphs in Fig. 8.11, and one must use MC to simulate such a system.

The parameters in this simulation are: $n_0 = 2$, $N = 3$, $s = 10^{12}$ s^{-1}, $E = 0.8$ eV, and a large retrapping ratio $R = 100$. The large value of R is the cause of the double peak structure shown in Fig. 8.11.

This result in Fig. 8.11 has certain similarity to the double peak structure obtained in the SLT model by Mandowski [101], although the physical descriptions of the two models are very different.

We now consider a MC simulation of the irradiation process within the OTOR/GOT model.

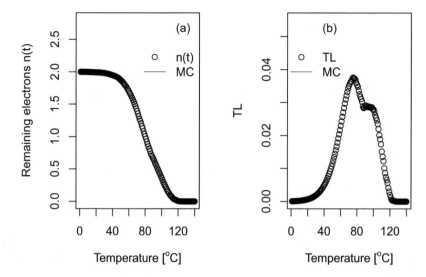

Fig. 8.11 Simulation of TL signal from a system of $M = 3000$ small clusters of defects, based on the GOT equation. The total number of traps in each cluster is $N = 3$, and initially $n_0 = 2$ of these traps are filled. The double peak structure is caused by the large retrapping ratio $R = 100$. For a more detailed description, see Pagonis et al. [130]

8.13 Irradiation Process as a Nonlinear Pure Birth Problem

In this section we provide an example of MC simulation for the irradiation of a dosimetric material, within the OTOR model.

The irradiation process can be described by the following differential equation, which is derived by applying the QE conditions to the irradiation stage of the OTOR model (Lawless et al. [94], their Eq. 7):

$$\frac{dn}{dt} = \frac{(N-n)R}{(N-n)R+n}X \tag{8.37}$$

The symbols in this equation are the same as in the OTOR system discussed so far in this book, with the additional symbol of X $(cm^{-3}s^{-1})$ representing the rate of production of electron-hole pairs in the system, per unit volume and per unit of time. As usual, R is the dimensionless retrapping ratio in the OTOR model, such that $R = A_n/A_m$, and $n(t)$, N are the instantaneous occupancy and total concentration of traps in the sample. The quantity $\alpha = \frac{(N-n)R}{(N-n)R+n}$ on the right hand side of Eq. (8.37) is dimensionless, and is also < 1. This quantity represents the ratio of the concentration of trapped electrons $(N-n)R$, over the total concentration $(N-n)R+n$ that they will either be retrapped or lost in the recombination centers during the irradiation process.

As we saw in Chap. 4, the analytical solution of this equation is found in terms of the Lambert function W:

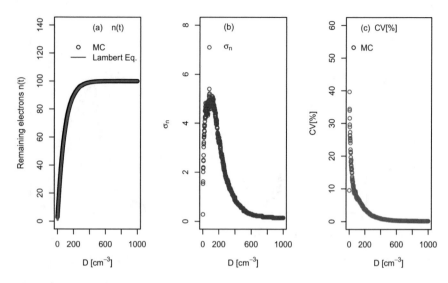

Fig. 8.12 MC simulation of irradiation process in the GOT model, as a nonlinear pure birth problem. The parameters are $n_0 = 1$, $M = 100$ MC runs, $R = 1.2$, $N = 100$. (a) Plot of the trapped electrons $n(t)$. The solid line is the Lambert W solution discussed in Chap. 4. (b) Plot of the uncertainty σ_n for the population $n(t)$. (c) Plot of the corresponding $CV[\%]$. There are no analytical expressions for (b) and (c)

$$n/N = 1 + W[(R - 1) \exp(R - 1 - RXt/N)]/(1 - R) \qquad (8.38)$$

The MC simulations simplify by dividing both sides of Eq. (8.37) by X, to obtain:

$$\frac{dn}{d(Xt)} = \frac{(N - n)R}{(N - n)R + n} \qquad (8.39)$$

We now change our time parameter from t to the new time parameter $D = Xt$, which has units of concentration (cm^{-3}) and which is proportional to the dose received by the sample. The previous equation in stochastic form becomes:

$$\Delta(n) = \frac{(N - n)R}{(N - n)R + n} \Delta D \qquad (8.40)$$

The following R code solves this difference equation with the parameters $N = 10^{10}$, $X = 10^5 \, s^{-1}$, $R = A_n/A_m = 1.2$, $n_0 = 10^5$ and the results are shown in Fig. 8.12.

The plot of $n(t)$ in Fig. 8.12a shows good agreement between the MC code and the solution of the differential equation for the irradiation process (see the detailed discussion in Chap. 4). Figure 8.12b shows the uncertainty σ_n, and Fig. 8.12c shows the corresponding of $CV[\%]$ as a function of irradiation dose D. As the irradiation proceeds, the value of the stochastic coefficient $CV[\%]$ decreases continuously with irradiation time from about 30% to a value of about 1%.

There are no analytical solutions for the σ_n and the $CV[\%]$ results in this example, since the corresponding differential equation is nonlinear.

Code 8.8: Vectorized Irradiation MC code in GOT model

```
#Vectorized Irradiation MC code in GOT model
rm(list = ls(all=T))
options(warn=-1)
library(matrixStats)
library(lamW)
mcruns<-100
deltat<-1
tmax<-1000
Dvalues<-seq(1,tmax,deltat)
nMatrix <-  matrix(NA, nrow = length(Dvalues), ncol = mcruns)
nMC<-rep(NA,length(Dvalues))
N<-100
n0<-1
R<-1.2
system.time(
  for (k in 1:mcruns){
    n<-n0              #initialize each of the M=100 MC runs
    for (t in Dvalues){
      vec<-rep(runif(n))    #create a vector vec,
      #containing n random numbers between 0 and 1
      P<-R*(N-n)*(1/n)/(R*(N-n)+n)*deltat
      dn<-length(vec[vec<P]) # if the random # in vec is <P,
      n<-n+dn #then increase the number of filled traps
      nMC[t]<-n+dn } #store number of electrons n in vector nMC
  nMatrix[,k]<-nMC # store single MC run in column k of nMatrix
  })
#Find average of n(t), CW-OSL signal, and CV[%]
avgn<-rowMeans(nMatrix)
sd<-rowSds(nMatrix)
cv<-100*sd/avgn
par(mfrow=c(1,3))
pch<-c(NA,NA,1,NA)
lty<-c(NA,NA,NA,"solid")
xlabs=expression("D [cm"^"-3*"]")
plot(Dvalues,avgn,ylab="Remaining electrons n(t)",
     xlab=xlabs,ylim=c(0,140))
legend("topright",bty="n",c("(a)     n(t)"," ","MC",
                            "Lambert Eq."),pch=pch,lty=lty)
lines(Dvalues,N*(1+lambertW0((R-1)*exp(R-1-R*Dvalues/N))/(1-R)),
     col="red",lwd=3)
plot(Dvalues,sd,ylab=c(expression(sigma[n]*" ")," ","MC"),
xlab=xlabs, ylim=c(0,8),col="blue")
legend("topleft",bty="n",legend=c(expression("(b)"," ",
sigma[n]*" ")),pch=pch,lty=lty,col="blue")
plot(Dvalues,cv,ylab="CV[%]", xlab=xlabs,ylim=c(0,60),col="red")
legend("topleft",bty="n",c("(c)  CV[%]"," ","MC"),pch=pch)

  ##    user  system elapsed
  ##    0.74    0.00    0.74
```

Chapter 9
Monte Carlo Simulations of Localized Transitions

9.1 Introduction

In this chapter we discuss MC simulations of luminescence phenomena produced by *localized* transitions, as described by the TLT and LT models discussed in Chaps. 6 and 7. The MC simulations in this chapter are based on the use of a fixed time interval between events. In Chap. 10 we will study MC simulations with a variable time interval, also referred to as kinetic Monte Carlo (KMC) methods.

In Sect. 9.2 we provide an example of MC simulations for TL signals in the excited state tunneling (EST) model, which was previously studied in Sect. 6.7. Similar simulations for CW-IRSL and LM-IRSL signals in the EST model are found in Sects. 9.3 and 9.4.

This chapter concludes in Sect. 9.5, where we present a MC simulation for the LT model.

9.2 MC Simulations of TL Based on the Excited State Tunneling (EST) Model

In Chap. 6 we saw the differential equation in the excited state tunneling (EST) model, for thermally and/or optically stimulated luminescence processes is

$$\frac{\partial n\left(r', t\right)}{\partial t} = -p(t)\, n\left(r', t\right) \frac{s_{tun}}{B \exp\left[(\rho')^{-1/3}\, r'\right]} \tag{9.1}$$

© The Author(s), under exclusive license to Springer Nature Switzerland AG 2021
V. Pagonis, *Luminescence*, Use R!, https://doi.org/10.1007/978-3-030-67311-6_9

In a TL experiment, the rate of excitation is given by $p(t) = s_{th} \exp[-E/(k_B T)]$, where s_{th}, E are the thermal parameters of the trap and T is the temperature of the sample.

Pagonis et al. [131] presented detailed MC simulations based on the EST model. In the MC method presented by these authors, Eq. (9.1) becomes a difference equation:

$$\Delta n\left(r', t\right) = -s_{th} \exp\left[-E/(k_B T)\right] n\left(r', t\right) \frac{s_{tun}}{B \exp\left[(\rho')^{-1/3} r'\right]} \Delta t \tag{9.2}$$

where Δt is an appropriate time interval, e.g. $\Delta t = 1$ s. As discussed by Pagonis et al. [131], the total concentration of trapped electrons $n(t)$ and the intensity of the TL signal are found by summing over all possible distances r', according to

$$n(t) = \sum_{r'=0}^{r'=2.2} n\left(r', t\right) \Delta r' \tag{9.3}$$

$$I(t) = N \sum_{r'=0}^{r'=2.2} 3 \left(r'\right)^2 \exp\left[\left(r'\right)^3\right] s_{eff}\left(r'\right) e^{-E/kT} \exp\left[-s_{eff}\left(r'\right) \int_0^t e^{-E/kT} dt'\right] \tag{9.4}$$

where we defined an effective frequency factor $s_{eff}\left(r'\right)$ by:

$$s_{eff}\left(r'\right) = \frac{s_{th} s_{tun}}{B} \exp\left[-\left(\rho'\right)^{-1/3} r'\right] \tag{9.5}$$

and $\Delta r'$ is an appropriate small distance interval, e.g. $\Delta r' = 0.1$. From a practical point of view, the dimensionless distance parameter r' does not need to extend to infinity, since the peak-shaped nearest distribution function $g(r') = 3\left(r'\right)^2 \exp\left[-\left(r'\right)^3\right]$ becomes practically zero for values larger than $r' = 2.2$.

The previous two equations are the basis of the MC method, and the software implementation in the code is as follows: the overall evolution of the system is followed for both the time variable t and for each value of the dimensionless distance r', by using two iterative loops. The inner loop is executed using a time variable t, and for a constant value of the distance parameter r'. The outer loop repeats the inside loop for all possible discrete values of the parameter r', which are stored in the vector $r[k]$.

At time $t = 0$ there are n_0 filled traps, and the distribution of nearest neighbors is given by the peak-shaped function $g(r')$. A random number is generated as in several previous MC codes, uniformly distributed in the unit interval $0 \leq r < 1$. If $r \leq P\Delta t$ (where $P = s_{eff}(r')\exp[-E/(k_B T)]$), the electron recombines radiatively, otherwise it does not; all non-recombined remaining electrons in the system are tested in this manner during each time interval Δt. At the end of each time interval Δt, the program stores the values of $n(r', t)$ and $\Delta n(r', t)/\Delta t$. This process is now repeated for the next value of the distance r' in the outer software loop. Finally, the contributions from all distances are added, resulting in the simultaneous evaluation of the discrete-value functions $n(t)$ and $I(t)$. Both iterative loops are executed until there are no particles left in the system. An array $signal[t,k,c]$ is used here, which contains the TL intensity at time t, for each discrete value $r[k]$ of the parameter r', and for each MC run characterized by the index c in the outermost loop of the code.

Figure 9.1 shows the simulated TL signals in the EST model, for the parameters $\rho' = 5 \times 10^{-3}$, $M =20$ MC runs, $n_0 = 100$ initially trapped electrons, $E = 1.43$ eV, and $s = s_{th}\, s_{tun}/B = 3.5 \times 10^{13}\,\text{s}^{-1}$. Figure 9.1a shows the result from the first MC run, by using the command $matplot(signal[,,1])$. Each curve in Fig. 9.1a corresponds to a partial TL glow curve, corresponding to a different value of r'. These partial TL glow curves are discussed in some detail in Chap. 12. Figure 9.1b shows the sum of the individual curves in Fig. 9.1a, resulting in the overall TL signal. This sum is obtained with the command $rowMeans(sum\_signal)$, and it has been normalized to its maximum. The dashed line in Fig. 9.1b represents the following approximate analytical KP-TL equation from Chap. 6 (Kitis and Pagonis [73]), also normalized to its maximum value:

$$I_{\text{TL}}(t) = \frac{I_0\, F(t)^2\, e^{-\rho'(F(t))^3}\,\left(E^2 - 6k^2 T^2\right)}{Eks T^2 z - 2k^2 s T^3 z + \exp\left(E/kT\right)E\beta} \tag{9.6}$$

where

$$F_{\text{TL}}(t) = \ln\left(1 + \frac{z\, skT^2}{\beta E}\, e^{-\frac{E}{kT}}\left(1 - \frac{2kT}{E}\right)\right) \tag{9.7}$$

We previously used these equations in Sect. 6.10, in a least squares routine used to fit experimental TL data for a feldspar sample.

The code presented in this section is also the prototype for the development of the R package $RLumCarlo$ (Kreutzer et al. [87]). The main difference with the code used in $RLumCarlo$ is that the latter uses a compiled code in C++, which speeds up the calculations considerably.

Code 9.1: Vectorized MC code for tunneling TL transitions (TLT model)

```
# Vectorized MC code for tunneling transitions (TLT model)
# Original Mathematica program by Vasilis Pagonis
# R version written by Johannes Friedrich, 2018
rm(list = ls(all=T))
rho <- 5e-3
En<-1.43
s<-3.5e12
kB<-8.617e-5
deltat <- 1
times <- seq(0, 500, deltat)
# In this example time = temperature, i.e. a heating rate = 1 K/s
r <- seq(0, 2, 0.1)
clusters <- 20
n0<-100
signal<-array(0,dim=c(length(times),
  ncol = length(r), clusters))
# Run MC simulation
system.time(invisible(for(c in 1:clusters)
{
  for(k in 1:length(r)){
    n <- n0
    for (t in 1:length(times)){
      P <- s*exp(-En/(kB*(t+273)))*exp(-rho^(-1/3) * r[k])
      vec<-rep(runif(n))
      n<-length(vec[vec>P*deltat])
      signal[t,k,c] <- n * P * 3 * r[k]^2 * exp(-r[k]^3) }}})
)
par(mfrow=c(1,2))
# plot an example: the result from the first cluster
matplot(signal[,,1],type = "l",lty="solid",
ylab = "Partial TL glow curves for constant(r')",
ylim=c(0,3.2),xlab=expression("Temperature ["^"o"*"C]"),lwd = 2)
legend("topleft",bty="n",legend=c("(a)"," ","Partial TL",
"glow curves"))
# add the signals from all clusters
sum_signal <- sapply(1:clusters, function(y){
  vapply(1:length(times), function(x){
    sum(signal[x,,y])
  }, FUN.VALUE = 1)  })
# add the signals from all r values
TL <- rowMeans(sum_signal)
# plot and normalize the TL signal
plot( x = times, y = TL/max(TL),type = "l", lwd = 3,
ylim=c(0,1.6), xlab=expression("Temperature ["^"o"*"C]"),
ylab="Average TL signal")
legend("topleft",bty="n",legend=c("(b)"," ",
"Sum of partial TL","glow curves"))
## plot analytical solution Kitis--Pagonis
```

```
z<-1.8
T<-times+273
TLanalyt<-exp(-rho*( (log(1+z*s*kB*((T**2.0)/
  abs(En))*exp(-En/(kB*T)))*(1-2*kB*T/En)))**3.0))*
(En**2.0-6*(kB**2.0)*(T**2.0))*( (log(1+z*s*kB*((T**2.0)/
    abs(En))*exp(-En/(kB*T))*(1-2*kB*T/En)))**2.0)/
  (En*kB*s*(T**2)*z-2*(kB**2.0)*s*z*(T**3.0)+
exp(En/(kB*T))*En)
lines(times,TLanalyt/max(TLanalyt),lty="dashed",col="red",
  lwd=3)
```

```
##     user  system elapsed
##     1.47    0.00    1.46
```

9.3 Tunneling MC Simulations for CW-IRSL Signals

In this section we provide an example of a MC simulation of CW-IRSL signals in the EST model. By following the same method as in the previous section, the differential equation for a CW-IRSL process is

$$\frac{\partial n\left(r',t\right)}{\partial t} = -p(t)\,n\left(r',t\right)\frac{s_{tun}}{B\exp\left[(\rho')^{-1/3}\,r'\right]} \tag{9.8}$$

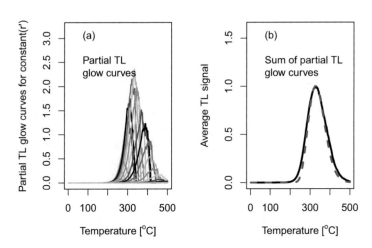

Fig. 9.1 MC simulation of TL signals in the TLT model, for the parameters $\rho' = 5 \times 10^{-3}$, $M = 20$ MC runs, $n_0 = 100$ initially trapped electrons, $E = 1.43$ eV, and $s = 3.5 \times 10^{13}\,\text{s}^{-1}$. (a) Example of partial TL glow curves evaluated for each distance r'. (b) The sum of the partial TL glow curves from (a), normalized to its maximum. The dashed line in (b) represents the approximate analytical KP-TL equation from Chap. 6, also normalized to its maximum value

In a CW-IRSL experiment, the rate of excitation is given by $p(t) = \sigma I$, where σ, I are the optical cross section and the intensity of the exciting IR source correspondingly. In the MC method presented by Pagonis et al. [131], Eq. (9.8) becomes a difference equation:

$$\Delta n\left(r', t\right) = -\sigma I n\left(r', t\right) \frac{s_{tun}}{B \exp\left[(\rho')^{-1/3} r'\right]} \Delta t \tag{9.9}$$

and the total concentration of trapped electrons $n(t)$ and the intensity of the CW-IRSL signal are found by summing over all possible distances r', according to

$$n(t) = \sum_{r'=0}^{r'=2.2} n\left(r', t\right) \Delta r' \tag{9.10}$$

$$I(t) = N \sum_{r'=0}^{r'=2.2} 3\left(r'\right)^2 \exp\left[(r')^3\right] s_{eff}\left(r'\right) \exp\left[-s_{eff}\left(r'\right) t\right] \tag{9.11}$$

where we defined an effective frequency factor $s_{eff}\left(r'\right)$ by

$$s_{eff}\left(r'\right) = \frac{\sigma I s_{tun}}{B} \exp\left[-(\rho')^{-1/3} r'\right] = A \exp\left[-(\rho')^{-1/3} r'\right]$$

where $A = \sigma I s_{tun}/B$ is a constant with units of s^{-1}. Figure 9.2 shows the simulated CW-IRSL signals in the TLT model, for the parameters $\rho' = 5 \times 10^{-3}$, $M = 10$ MC runs, $n_0 = 500$ initially trapped electrons, and $A = 2\,s^{-1}$. Figure 9.2a shows the result from the first MC run, by using the command *matplot(signal[„1])*. Each curve in Fig. 9.1a corresponds to a partial CW-IRSL curve, corresponding to a different value of r'. These partial CW-IRSL curves are discussed in some detail in Chap. 12. Figure 9.2b shows the *sum* of the partial CW-IRSL curves in (a), resulting in the overall CW-IRSL signal. This sum is obtained with the command *rowMeans(sum_signal)*, and it has been normalized to its maximum. The dashed line in Fig. 9.2b represents the following approximate analytical KP-CW equation discussed in Chap. 6 (Kitis and Pagonis [73], also normalized to its maximum value:

$$I_{CW-IRSL}(t) = \frac{I_0 F(t)^2 e^{-\rho'(F(t))^3}}{1 + zA t} \tag{9.12}$$

where

$$F_{CW-IRSL}(t) = \ln\left(1 + zA t\right) \tag{9.13}$$

Code 9.2: Vectorized MC code for tunneling CW-IRSL transitions (TLT model)

```r
# Vectorized MC code for tunneling transitions (TLT model)
# Original Mathematica program by Vasilis Pagonis
# R version written by Johannes Friedrich, 2018
rm(list = ls(all=T))
rho <- 5e-3
A<-2
deltat <- 1
times <- seq(1, 400, deltat)
# In this example time = temperature, i.e. a heating rate=1 K/s
r <- seq(0, 2.2, 0.1)
clusters <- 10
n0<-500
signal<-array(0,dim=c(length(times),
                      ncol = length(r), clusters))
# Run MC simulation
system.time(invisible(for(c in 1:clusters)
{
  for(k in 1:length(r)){
    n <- n0
    for (t in 1:length(times)){
      P <- A*exp(-rho^(-1/3) * r[k])
      vec<-rep(runif(n))
      n<-length(vec[vec>P*deltat])
      signal[t,k,c] <- n * P * 3 * r[k]^2 * exp(-r[k]^3) }}})
)
par(mfrow=c(1,2))
# plot an example: the result from the first cluster
matplot(signal[,,1],type = "l",lty="solid",
ylab = "Partial CW-IRSL curves",
ylim=c(0,14),xlab=expression("Time [s]"),lwd = 2)
legend("topleft",bty="n",legend=c("(a)"," ","Partial CW-IRSL",
" curves"))
# add the signals from all clusters
sum_signal <- sapply(1:clusters, function(y){
  vapply(1:length(times), function(x){
    sum(signal[x,,y])
  }, FUN.VALUE = 1)  })
# add the signals from all r values
TL <- rowMeans(sum_signal)
# plot and normalize the TL signal
plot( x = times, y = TL/max(TL),type = "l", lwd = 3,
ylim=c(0,1.4), xlab=expression("Time [s]"),
       ylab="Average signal")
legend("topleft",bty="n",legend=c("(b)"," ",
"Sum of partial","CW-IRSL curves"))
## plot analytical solution Kitis--Pagonis
z<-1.8
```

```
CWanalyt<-exp(-rho*( (log(1+z*A*times))**3.0))*
  ( (log(1+z*A*times))**2.0)/(1+z*A*times)
lines(times,CWanalyt/max(CWanalyt),lty="dashed",col="red",lwd=3)

   ##     user  system elapsed
   ##     1.39    0.00    1.39
```

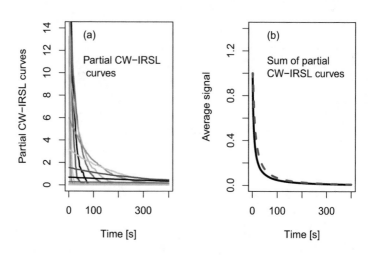

Fig. 9.2 MC simulation of CW-IRSL signals in the TLT model, for the parameters $\rho' = 5 \times 10^{-3}$, $M = 10$ MC runs, $n_0 = 500$ initially trapped electrons, and IR excitation rate $A = 2\,\text{s}^{-1}$. (**a**) Example of partial CW-IRSL curves evaluated for each distance r'. (**b**) The sum of the partial CW-IRSL curves from (**a**), normalized to its maximum. The dashed line in (**b**) represents the approximate analytical KP-CW equation from Chap. 6, also normalized to its maximum value (Kitis and Pagonis [73])

9.4 Tunneling MC Simulations for LM-IRSL Signals

In this section we provide an example of a simulation of LM-IRSL signals in the EST model. We follow the same method as in the previous two sections, and the results are shown in Fig. 9.3a, b. The following analytical equation by Kitis and Pagonis [73] is shown as a dashed line in Fig. 9.3b:

$$I_{\text{LM-IRSL}}(t) = \frac{I_0\,F(t)^2\,e^{-\rho'(F(t))^3}}{1 + zA\,t^2/(2P)} \tag{9.14}$$

where

$$F_{\text{LM-IRSL}}(t) = \ln\left(1 + zA\,t^2/(2P)\right) \tag{9.15}$$

Code 9.3: Vectorized MC code for tunneling LM-IRSL transitions (TLT model)

```r
# Vectorized MC code for tunneling transitions (TLT model)
# Original Mathematica program by Vasilis Pagonis
# R version written by Johannes Friedrich, 2018
rm(list = ls(all=T))
rho <- 5e-3
A<-2                    # IR excitation rate in s^-1
deltat <- 1
times <- seq(0, 400, deltat)
r <- seq(0, 2.2, 0.1)
clusters <- 10
n0<-500
signal<-array(0,dim=c(length(times),
ncol = length(r), clusters))
# Run MC simulation
system.time(invisible(for(c in 1:clusters)
{  for(k in 1:length(r)){
    n <- n0
    for (t in 1:length(times)){
      P <- A*t/max(times)*exp(-rho^(-1/3) * r[k])
      vec<-rep(runif(n))
      n<-length(vec[vec>P*deltat])
  signal[t,k,c] <- n * P * 3 * r[k]^2 * exp(-r[k]^3) }}}))
par(mfrow=c(1,2))
# plot an example: the result from the first cluster
matplot(signal[,,1],type = "l",lty="solid",
ylab = "Partial LM-IRSL curves",
ylim=c(0,7),xlab=expression("Time [s]"),lwd = 2)
legend("topleft",bty="n",legend=c("(a)"," ","Partial LM-IRSL",
" curves"))
# add the signals from all clusters
sum_signal <- sapply(1:clusters, function(y){
  vapply(1:length(times), function(x){
    sum(signal[x,,y])
  }, FUN.VALUE = 1)  })
# add the signals from all r values
TL <- rowMeans(sum_signal)
# plot and normalize the TL signal
plot( x = times, y = TL/max(TL),type = "l", lwd = 3,
ylim=c(0,1.6), xlab=expression("Time [s]"),
ylab="Average LM-IRSL signal")
legend("topleft",bty="n",legend=c("(b)"," ",
"Sum of partial","LM-IRSL curves"))
## plot analytical solution Kitis--Pagonis
z<-1.8
```

```
LManalyt<-exp(-rho*( (log(1+z*A*times^2/(2*max(times))))**3.0))*
times*((log(1+z*A*times^2/(2*max(times))))**2.0)/(1+z*A*times^2/
(2*max(times)))
lines(times,LManalyt/max(LManalyt),lty="dashed",col="red",lwd=3)
```

```
##     user  system elapsed
##     1.74    0.00    1.75
```

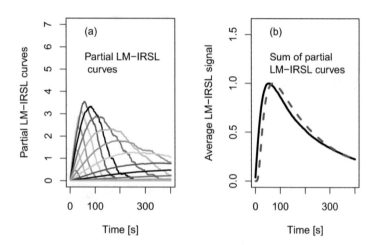

Fig. 9.3 MC simulation of LM-IRSL signals in the TLT model, for the parameters $\rho' = 5 \times 10^{-3}$, $M = 10$ MC runs, $n_0 = 100$ initially trapped electrons, and $A = 2\,s^{-1}$. (**a**) Example of partial LM-IRSL curves evaluated for each distance r'. (**b**) The sum of the partial LM-IRSL curves from (**a**), normalized to its maximum. The dashed line in (**b**) represents the approximate analytical equation by Kitis and Pagonis [73], also normalized to its maximum value

9.5   MC Simulation of Localized Transitions in the LT Model

As we saw in Chap. 7, the differential equation describing localized transitions in the LT model is

$$\frac{dn}{dt} = -p(t)\frac{n^2(t)}{n(t) + r} \tag{9.16}$$

where both $n(t)$ and the retrapping ratio r have units of concentrations (cm^{-3}), and $p(t)$ is the stimulation function, which is different for each type of experiment. Note the absence of the total number of traps parameter N in the differential equation for the LT model, due to the localized nature of the transitions.

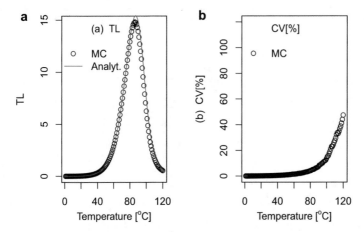

Fig. 9.4 (**a**) MC simulation of TL signals in the LT model, for the parameters $r = 10^2$, $M = 100$ MC runs, $n_0 = 500$, $E = 1\,eV$, and $s = 10^{13}\,s^{-1}$. The solid line is the analytical solution by Kitis and Pagonis [73]. (**b**) The corresponding coefficient of variation *CV[%]*

The stochastic equation for TL in this model is

$$\Delta n = -p(t)\frac{n^2}{n+r}\Delta t \qquad (9.17)$$

where both n and r are now dimensionless quantities. The structure of the following vectorized code is very similar to the code for *delocalized* transitions, which we used extensively in the previous chapter. We simulate $M = 100$ MC runs consisting of $n_0 = 500$ initially trapped trapped electrons, with a retrapping ratio of $r = 10^2$. The heating rate is taken here as $\beta = 1$ K/s.

The solid line in Fig. 9.4a represents the following analytical Lambert Eq. (7.10), developed by Kitis and Pagonis [73], which we saw in Chap. 7:

$$I_{LOC-TL}(t) = p(t)\,\frac{r}{W[k, e^z] + W[k, e^z]^2} \qquad (9.18)$$

where

$$z_{LOC-TL} = \frac{r}{n_0} - \ln\left[\frac{n_0}{r}\right] + s \int_{T_0}^{T} exp\left(-\frac{E}{kT}\right) dT \qquad (9.19)$$

Figure 9.4b shows the corresponding CV[%], which increases continuously with the sample temperature.

Code 9.4:   Vectorized MC code for TL in localized TL transitions (LT model)

```
# Vectorized MC code for localized TL transitions (LT model)
rm(list = ls(all=T))
options(warn=-1)
library(matrixStats)
library(lamW)
mcruns<-100
n0<-500
s<-1e13
E<-1
kb<-8.617e-5
r<-1e2
tmax<-120
deltat<-1
times<-seq(1,tmax,deltat)   # heating rate = 1 K/s
nMatrix<-TLMatrix<-matrix(NA,nrow=length(times),ncol=mcruns)
nMC<-TL<-rep(NA,length(times))
system.time(
   for (j in 1:mcruns){
     n<-n0
     for (t in 1:length(times)){
       vec<-rep(runif(n))
       P<-s*exp(-E/(kb*(t+273)))*n/(r+n)*deltat
       n<-length(vec[vec>P])
       nMC[t]<-n
       TL[t]<-nMC[t]*P}
     nMatrix[,j]<-nMC
     TLMatrix[,j]<-TL })
# Find average avgn, average TL signal,CV[%]
avgn<-rowMeans(nMatrix)
avgTL<-rowMeans(TLMatrix)
sd<-rowSds(TLMatrix)
cv<-100*sd/avgTL
# plots
par(mfrow=c(1,2))
pch<-c(NA,NA,1,NA)
lty<-c(NA,NA,NA,"solid")
col<-c(NA,NA,"black","red")
plot(times,avgTL,ylab="TL",
     xlab=expression("Temperature ["^"o"*"C]"))
# plot analytical solution
k<-function(u) {integrate(function(p){exp(-E/(kb*p))},
300,u)[[1]]}
yl<-lapply(times+273,k)
x<-unlist(273+times)
```

```
y<-unlist(y1)
zTL<-(r/n0)-log(n0/r)+(s*y)
lines(x-273,r*s*exp(-E/(kb*x)))/(lambertW0(exp(zTL))
+lambertW0(exp(zTL))^2),type="l",col="red")
legend("topleft",bty="n",c(" (a)   TL"," ","MC",
                      "Analyt."),pch=pch,lty=lty,col=col)
plot(times,cv,ylab=" (b)   CV[%]",ylim=c(0,120),
        xlab=expression("Temperature ["^"o"*"C]"))
legend("topleft",bty="n",c("CV[%]"," ","MC"),
pch=pch,lty=lty,col=col)

##    user  system elapsed
##    0.25    0.00    0.25
```

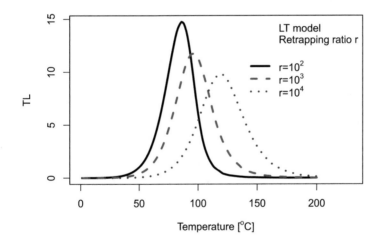

Fig. 9.5 MC simulation of TL signals in the LT model, for three different values of the retrapping ratio $r = 10^2$, 10^3, 10^4 cm^{-3}

Figure 9.5 shows an overlay of the TL glow curves evaluated using the above code, from three different values of the retrapping ratio $r = 10^2$, 10^3, 10^4.

Chapter 10
Kinetic Monte Carlo Simulations

10.1 Introduction to KMC Methods

In the previous two chapters we looked at examples of MC methods with a fixed time interval. In this chapter we present several examples of a different MC technique, the kinetic Monte Carlo (KMC) method.

The KMC method is a technique for simulating the time evolution of random processes, which occur with known transition rates between various states. The basic concept behind them is that the time intervals between random events follow an exponential distribution. Specifically, if a random process occurs with a constant rate p (times per second, or per some other unit of time), then the *waiting time between events* Δt is given by

$$\Delta t = -\frac{\ln(r)}{p} \tag{10.1}$$

where r is a random number between 0 and 1. For a detailed discussion of the waiting time concept, the reader is referred to the book by Allen [5].

In Sects. 10.2–10.4 we give examples of KMC methods for solving stochastic birth and death processes.

Section 10.5 presents a different example of the KMC method, which approaches quantum tunneling processes and the anomalous fading effect from a microscopic point of view.

The general steps in the KMC algorithms in this chapter can be described as follows:

- Initialization step: define the initial number of particles $n(0)$ in the system, the transition rates between the various states of the system, the starting time, the maximum time interval t_{max}, etc.
- Single Monte Carlo step: Generate random numbers to determine which reaction will take place next, as well as the time interval. The time interval for the next event to take place is exponentially distributed according to Eq. (10.1).

© The Author(s), under exclusive license to Springer Nature Switzerland AG 2021
V. Pagonis, *Luminescence*, Use R!, https://doi.org/10.1007/978-3-030-67311-6_10

- Update the time variable $t = t + \Delta t$ by using the randomly generated time interval Δt from the previous step, and update the number of particles, based on the reaction that occurred.
- Iterate the above steps to obtain the time evolution $n(t)$ for the number of particles, and repeat until the number of particles is zero, or the simulation time has been exceeded.
- Obtain several MC variants of $n(t)$, and take their average and their standard deviation σ. Since the times for the events are randomly produced, evaluation of the mean $< n(t) >$ and of σ must be done carefully, by binning the time intervals within, e.g. 1 s bins.
- Compare the stochastic $< n(t) >$ with the solution of the corresponding deterministic differential equation. For a few linear problems, it may also be possible to evaluate the standard deviation σ analytically and to compare it with the value from the KMC method.

10.2   Stochastic Simple Linear Death Process Using KMC

As a first simple example, let us consider the CW-OSL process we studied in the previous chapter. As discussed previously in detail, this is a simple linear death process with constant μ. In the KMC simulation, we can consider all the particles at once, and the recombination rate for the system of n particles is $p = n\mu$ (see, for example, the discussion by Kulkarni [89] and also by Mandowski and Świątek [105]).

The following R code simulates this stochastic death process, by using the exponential distribution of waiting times given by Eq. (10.1). The deterministic analytical solutions for $n(t)$ and for the stochastic coefficient of variation $CV[\%]$ were discussed previously in Chap. 8.

The code contains a *for* loop that is executed as long as there are particles left in the system, i.e. as long as $n > 0$. The time interval for the next event to take place is determined by $\Delta t = -\ln(r)/p$, and the time parameter t is updated in each step of the loop according to

$$t' = t + \Delta t = t - \frac{\ln(r)}{p} = t - \frac{\ln(r)}{n\,\mu} \tag{10.2}$$

In each step where a recombination has occurred, the number of particles is reduced from n to $n - 1$, and the variable times t are stored in the parameter *allt*. The command *cut* is used to collect all times in intervals of 1 s, i.e. the code collects together all times between 0 and 1 s, between 1 and 2 s, etc. This process is referred to as the *binning* of the times t. The command *tapply* is used to find the mean value of the counts $< n(t) >$ within each binning time interval.

The parameters *singlerun* and *allt* represent the mean value of $n(t)$ in each 1-s interval and the time values for each MC run, respectively.

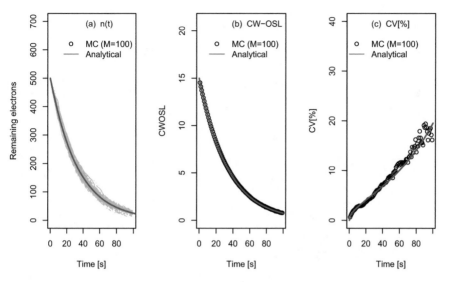

Fig. 10.1 Simulation of CW-OSL as a stochastic death process, using KMC

The *for* loop is evaluated for many MC runs, as determined by the parameter *mcruns*. Each single MC run represents a row of the matrix variable *nMatrix*, and the commands *rowMeans* and *rowSds* are used to evaluate the $< n(t) >$ and the standard deviation σ, which are stored in *avgn* and *sd*, respectively. The corresponding CW-OSL signal $L(t)$ is calculated from the product $\mu \cdot avgn$, since $L(t) = -dn/dt = \mu\,n$.

Figure 10.1a shows the plot of *nMatrix* containing all the MC runs (as green circles), while the command *lines* is used to plot the average *avgn* as the black solid line. The analytical solution is $n(t) = n_0 \exp(-\mu t)$ and is plotted as a red line in Fig. 10.1b, coinciding with the MC average.

As we saw in the previous chapter, the following analytical solution exists for the *CV[%]* in the simple death problem:

$$CV[\%] = 100\sqrt{\frac{e^{\mu t} - 1}{n_0}} \qquad (10.3)$$

This analytical equation is shown as a red line in Fig. 10.1c and is compared with the value of *CV[%]* obtained using the KMC runs (black circles). Good agreement is seen between the KMC and the analytical *CV[%]* equation.

Code 10.1: Stochastic CW-OSL process using KMC

```
#  Stochastic CW-OSL process using KMC
rm(list = ls(all=T))
library(matrixStats)
mcruns<-100
n0<-500
mu<-.03
tmax<-100
times<-(1:tmax)
nMatrix<-matrix(NA,nrow=tmax-1,ncol=mcruns)
system.time(
for (k in 1:mcruns)
{n<-n0
 allt<-t<-0.5
    for(i in 1:n){
    P<-mu
    t<-t+rexp(1)/(P*n)
    n<-n-1
    allt<- rbind(allt,t)
    }
 depth.class <- cut(allt,times, include.lowest = TRUE)
 singlerun <- tapply(seq(from=n0,to=0,by=-1), depth.class,
mean,na.rm = FALSE)
 singlerun<-as.vector(singlerun)
 nMatrix[,k]<-singlerun
})
par(mfrow=c(1,3))
matplot(nMatrix,typ="l",col="green",ylim=c(0,700),
xlab="Time [s]",ylab="Remaining electrons")
timesMC<-seq(from=1,to=tmax-1,by=1)
avgn<-rowMeans(nMatrix,na.rm=TRUE)
avgCWOSL<-mu*avgn
sd<-rowSds(nMatrix,na.rm=TRUE)
cv<-100*sd/avgn
pch<-c(NA,NA,1,NA)
lty<-c(NA,NA,NA,"solid")
col<-c(NA,NA,"black","red")
lines(timesMC,avgn,ylab="Remaining electrons n(t)",
     xlab="Time [s]")
curve(n0*exp(-P*x),from=0,to=150,add=TRUE,col="red",lwd=2)
legend("topright",bty="n",c("(a)  n(t)"," ","MC (M=100)",
  "Analytical"),pch=pch,lty=lty,col=col)
```

```
plot(timesMC,avgCWOSL,ylab="CWOSL",xlab="Time[s]",ylim=c(0,21))
legend("topright",bty="n",c("(b)    CW-OSL"," ","MC (M=100)",
    "Analytical"),pch=pch,lty=lty,col=col)
curve(n0*mu*exp(-mu*x),from=0,max(times),add=TRUE,
col="red",lwd=2)
plot(timesMC,cv,ylab="CV[%]",xlab="Time [s]",ylim=c(0,40))
curve(100*sqrt((exp(mu*x)-1)/n0),0,max(times),add=TRUE,
    col="red",lwd=2)
legend("topleft",bty="n",c("(c)    CV[%]"," ","MC (M=100)",
    "Analytical"),pch=pch,lty=lty,col=col)

##    user  system elapsed
##    0.48    0.00    0.48
```

10.3   Stochastic First Order LM-OSL Process Using KMC

In this section, we compare the results of the MC simulation with the analytical solutions of the deterministic differential Eq. (3.9), which was discussed in Chap. 3:

$$\frac{dn(t)}{dt} = -\frac{\sigma I}{P}t\, n(t) = -\frac{A}{P}t\, n(t) \tag{10.4}$$

The analytical solution is

$$L(t) = n_0\frac{A}{P}t\,\exp\left(-\frac{A}{2P}t^2\right) \tag{10.5}$$

The R code in the previous section can easily be modified to simulate a stochastic LM-OSL process, by replacing the transition probability with the appropriate mathematical expression $p = nAt/P$ and the time t' for the next event is now given by

$$t' = t + \Delta t = t - \frac{\ln(r)}{p} = t - \frac{\ln(r)}{nAt/P} \tag{10.6}$$

The results of the MC simulation are shown in Fig. 10.2, together with the analytical solutions for the stochastic LM-OSL process, which were discussed previously in Sect. 8.10. The numerical values of the parameters in this example are $n_0 = 500$ trapped electrons at time $t = 0$, the maximum stimulation time $P = 100$ s, the stimulation rate for the LM-OSL process is $A = 0.1\,\mathrm{s}^{-1}$, and $M = 100$ MC runs.

Code 10.2: Stochastic LM-OSL process using KMC

```
# Stochastic LM-OSL process using KMC
rm(list = ls(all=T))
library(matrixStats)
mcruns<-100
n0<-500
tmax<-100
A<-.1
nMatrix<-matrix(NA,nrow=tmax-1,ncol=mcruns)

times<-seq(1,tmax-1,1)
system.time(
for (k in 1:mcruns)
{n<-n0
 allt<-t<-1
    for(i in 1:n-1){
    P<-A*t/tmax
    t<-t+rexp(1)/(P*n)
    n<-n-1
    allt<- rbind(allt,t)
    }
 depth.class <- cut(allt,seq(1:tmax), include.lowest = TRUE)
 singlerun <- tapply(seq(from=n0,to=0,by=-1), depth.class,
mean,na.rm = FALSE)
 singlerun<-as.vector(singlerun)
 nMatrix[,k]<-singlerun
})
par(mfrow=c(1,3))
matplot(nMatrix,typ="l",col="green",ylim=c(0,700),
xlab="Time [s]",ylab="Remaining electrons")
avgn<-rowMeans(nMatrix,na.rm = TRUE)
cv<-100*rowSds(nMatrix,na.rm=TRUE)/avgn
# plots
pch<-c(NA,NA,1,NA)
lty<-c(NA,NA,NA,"solid")
col<-c(NA,NA,"black","red")
lines(seq(from=1,to=tmax-1,by=1),avgn,lwd=2,col="black")
curve(n0*exp(-A*x^2/(2*(tmax-1))),1,tmax,lwd=3,
col="red",add=TRUE)
legend("topright",bty="n",c("(a)   n(t) for LM-OSL",
" ","MC","Analytical"),pch=pch,lty=lty,col=col)
avgLM<-A*(times/(tmax-1))*as.numeric(avgn)
plot(times,avgLM,ylab="LMOSL",xlab="Time [s]",ylim=c(0,15))
legend("topright",bty="n",c("(b)   LM-OSL"," ","MC",
"Analytical"), pch=pch,lty=lty,col=col)
curve(A*n0*exp(-A*x^2/(2*tmax))*x/tmax,0,tmax,add=TRUE,
col="red",lwd=2)
```

```
plot(times,cv,ylab="CV[%]",xlab="Time [s]",ylim=c(0,120))
curve(100*sqrt((exp(A*x^2/(2*tmax))-1)/n0),0,tmax,add=TRUE,
col="red",lwd=2)
legend("topleft",bty="n",c("(c)   CV[%]"," ","MC (M=100)",
 "Analytical"),  pch=pch,lty=lty,col=col)

##     user  system elapsed
##     0.39    0.00    0.39
```

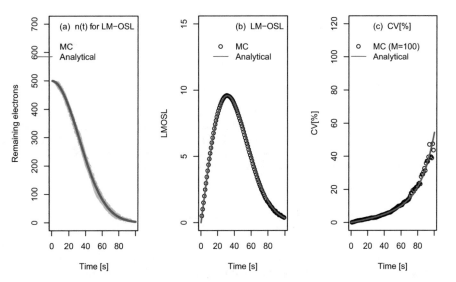

Fig. 10.2 Simulation of stochastic LM-OSL process using KMC method. The parameters are $n_0 = 500$ trapped electrons at time $t = 0$, maximum stimulation time $P = 100$ s, LM-OSL stimulation rate $A = 0.1\,\mathrm{s}^{-1}$, and $M = 100$ MC runs. (**a**) Plot of the $M = 100$ MC runs, (**b**) plot of the average MC $< n(t) >$, and (**c**) plot of the coefficient of variation $CV[\%]$. The solid lines in these graphs represent the analytical solutions for the LM-OSL deterministic process

10.4   Stochastic First Order Kinetics TL Process Using KMC

The R code in the previous section can easily be modified to simulate a first order stochastic TL process, by replacing the transition probability with the appropriate mathematical expression $p = n\,s\,\exp(-E/kT)$, and the time t' for the next event is now given by

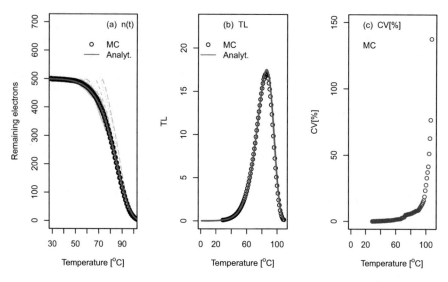

Fig. 10.3 Simulation of stochastic first order TL process using a KMC method. The parameters are $n_0 = 500$ trapped electrons at time $t = 0$, maximum temperature $T = 110\,°C$, frequency factor $s = 10^{13}\,s^{-1}$, energy $E = 1\,eV$, and heating rate $\beta = 1\,K/s$. (a) Plot of the $M = 100$ MC runs, (b) plot of the average MC $< n(t) >$, and (c) plot of the coefficient of variation CV[%]. The solid lines represent the analytical solutions for the TL deterministic process

$$t' = t + \Delta t = t - \frac{\ln(r)}{p} = t - \frac{\ln(r)}{n\,s\,\exp(-E/kT)} \tag{10.7}$$

The numerical values of the parameters in this example are $n_0 = 500$ trapped electrons at time $t = 0$, maximum temperature $T = 110\,°C$, starting temperature $T = 30\,°C$, frequency factor $s = 10^{13}\,s^{-1}$, energy $E = 1\,eV$, heating rate $\beta = 1\,K/s$, and $M = 100$ MC runs. The results of the MC simulation are shown in Fig. 10.3, together with the analytical solutions for the stochastic TL process, which were discussed previously in Sect. 8.9.

10.5   The Microscopic Description of Quantum Tunneling in Luminescent Materials

Larsen et al. [92] presented a microscopic description of quantum tunneling in luminescent materials, using a MC model. One of their main assumptions was that the number densities of donors and acceptors are equal at any time, and these authors were able to get good agreement with experiment only when they assumed that the crystal consisted of small sub-volumes, and charge carriers were only allowed to recombine within these nanocrystals.

Code 10.3: Stochastic first order TL process using KMC

```r
# Stochastic first order TL process using KMC
rm(list = ls(all=T))
library(matrixStats)
mcruns<-100
n0<-500
tmax<-110
s<-1e13
E<-1
kb<-8.617e-5

times<-seq(1,tmax-1,1)
nMatrix<-matrix(NA,nrow=tmax-1,ncol=mcruns)

system.time(
for (k in 1:mcruns)
{n<-n0
 allt<-t<--30
    for(i in 1:n-1){
    P<-s*exp(-E/(8.617e-5*(t+273)))
    t<-t+rexp(1)/(P*n)
    n<-n-1
    allt<- rbind(allt,t)
    }
 depth.class <- cut(allt,seq(1:tmax), include.lowest = TRUE)
 singlerun <- tapply(seq(from=n0,to=0,by=-1), depth.class,
mean,na.rm = FALSE)
 singlerun<-as.vector(singlerun)
 nMatrix[,k]<-singlerun
})
# Calculate analytical solution
x1<-times+273
k<-function(u) {integrate(function(p){s*exp(-E/(kb*p))},
                          273,u)[[1]]}
y1<-lapply(x1,k)
x<-unlist(x1)
y<-unlist(y1)
errn<-sqrt(n0*(exp(-y)-exp(-2*y)))
nanalyt<-n0*exp(-y)
TLanalyt<-n0*s*exp(-E/(kb*x))*exp(-y)
# plot MC and analytical
par(mfrow=c(1,3))
matplot(nMatrix,typ="l",col="green",xlim=c(30,100),
ylab="Remaining electrons",
ylim=c(0,700),xlab=expression("Temperature ["^"o"*"C]"))
avgn<-rowMeans(nMatrix,na.rm = TRUE)
cv<-100*rowSds(nMatrix,na.rm=TRUE)/avgn
pch<-c(NA,NA,1,NA)
lty<-c(NA,NA,NA,"solid")
```

```
col<-c(NA,NA,"black","red")
lines(seq(from=1,to=tmax-1,by=1),avgn,lwd=2,col="black",
typ="p",pch=1)
legend("topright",bty="n",c("(a)   n(t)"," ","MC",
  "Analyt."),pch=pch,lty=lty,col=col)
lines(x-273,nanalyt,col="red",lwd=2)
k<-function(u){s*exp(-E/(kb*(u+273)))}
y<-lapply(times,k)
TLMC<-as.numeric(y)*as.numeric(avgn)
plot(times,TLMC,col="black",ylab="TL",
ylim=c(0,23),xlab=expression("Temperature  ["^"o"*"C]"))
lines(x-273,TLanalyt,col="red",lwd=2)
legend("topleft",bty="n",c("(b)   TL"," ","MC",
  "Analyt."), pch=pch,lty=lty,col=col)
plot(times,cv,ylab="CV[%]",ylim=c(0,150),col="blue",
     xlab=expression("Temperature ["^"o"*"C]"))
legend("topleft",bty="n",c("(c)   CV[%]"," ","MC"))

##     user  system elapsed
##     0.41    0.00    0.41
```

Pagonis and Kulp [140] provided a different version of this previous model, in which the number density of acceptors far exceeds that of donors. The new version of the model describes the loss of charge due to ground state tunneling, as well as the charge creation by natural irradiation of samples, and the modeling results compared well with the approximate analytical equations developed by Kitis and Pagonis [73], and Pagonis and Kitis [134] for feldspars. The results from the model were also compared with experimental data from time-resolved infrared stimulated luminescence (TR-IRSL) in these materials.

As discussed in detail in Pagonis and Kulp [140], during the simulation each of the electrons in the crystal is examined, and the distances of this electron from all holes are calculated. The minimum distance to the nearest neighbor is found, and the Monte Carlo algorithm generates all possible random fading times for these minimum distances.

Only the event corresponding to the *shortest* of all the possible times takes place, i.e. the donor–acceptor pair corresponding to this shortest time is allowed to recombine. The distances between each donor and each acceptor are re-evaluated after the recombined pair is removed from the system in the simulation, and the minimum fading time is used to update the total time elapsed from the beginning of the simulation. This process is repeated until there are no more donors left in the

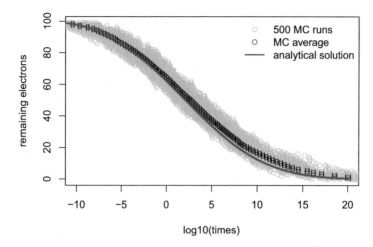

Fig. 10.4 Simulation of microscopic description of ground state quantum tunneling in luminescent materials with the parameters given in the text. The gray area indicates the results from $M = 500$ simulations of the same system, and the solid circles indicate the average of the 500 runs. The standard error of the 500 runs is about equal to the drawing size of individual circles. The solid line represents the analytical Eq. (6.13). For more details, see Pagonis and Kulp [140]

system. The physical picture in this model is that close-by pairs are more likely to recombine first, and farther-away pairs are likely to recombine later.

The following code uses the *FNN* package to evaluate the nearest neighbor distances, and the parameter *recomb_time* represents all possible times. The final graph in this code reproduces Fig. 2 in the paper by Pagonis and Kulp [140]. The parameters in the simulation are $\rho' = 10^{-5}$, $s = 3 \times 10^{15}\,\mathrm{s}^{-1}$, number of donors $n_{DONORS} = 100$, number of acceptors $n_{ACCEPTORS} = 1222$, $\alpha = 4 \times 10^9\,\mathrm{m}^{-1}$ and cube size $d = 200\,\mathrm{nm}$.

Code 10.4: Microscopic description of quantum tunneling

```
# Original Mathematica program by Vasilis Pagonis
# R version written by Johannes Friedrich, 2018
# The code reproduces Fig. 2 of Pagonis and Kulp (2010)
rm(list = ls(all = TRUE)) # empties the environment
library("plot3D")
library("FNN")
## Define Parameters ----
sideX <- 200e-9 # length of quader in m
```

```r
sideX_nm <- sideX*1e9 # length of quader in nm
s_tun <- 3e15
alpha <- 4e9
N_pts <- 100
clusters <- 50
rho_prime <- 1e-5

N_centers <- as.integer(rho_prime*sideX^3 *3 *alpha^3/(4*pi))
rho <- N_centers/ sideX^3
all_times_matrix <- matrix(NA,nrow = N_pts,ncol = clusters)
## Run MC -----
for(j in 1:clusters) {
  if(j %% 100 == 0) print(j)
  n_pts <- N_pts
  n_centers <- N_centers
  xyz_traps <- data.frame(
    x = sample(1:sideX_nm, N_pts, replace = TRUE),
    y = sample(1:sideX_nm, N_pts, replace = TRUE),
    z = sample(1:sideX_nm, N_pts, replace = TRUE)
  )
  xyz_centers <- data.frame(
    x = sample(1:sideX_nm, n_centers, replace = TRUE),
    y = sample(1:sideX_nm, n_centers, replace = TRUE),
    z = sample(1:sideX_nm, n_centers, replace = TRUE)
  )
  ### r calc distances ------
  for(i in 1:(n_pts)){
    ## find next neighbors with package FNN
    dist <- FNN::get.knnx(data = as.matrix(xyz_centers),
                          query = as.matrix(xyz_traps),
                          k = 1)
    all_dist <- as.data.frame(dist)
    P <- runif(n = length(all_dist$nn.dist), min = 0, max = 1)
    # P <- runif(n = 1, min = 0, max = 1)
    recomb_time <- - s_tun^(-1) * exp(alpha*all_dist$nn.dist *
    1e-9) * log(1-P)
    e_remove <- which.min(recomb_time)
    h_remove <- all_dist$nn.index[e_remove]
    ##remove index from data.frame
    xyz_centers <- xyz_centers[-h_remove,]
    xyz_traps <- xyz_traps[-e_remove,]
    all_times_matrix[i,j] <- recomb_time[e_remove]
  } # end n_pts loop
  all_times_matrix[,j] <- cumsum(all_times_matrix[,j])
} ## end cluster-loop
all_times_matrix <- log10(all_times_matrix)
### plot results ------
```

```
times_avg <- rowMeans(all_times_matrix)
matplot(x = all_times_matrix,
        y = (N_pts-1):0,xlim = c(-10,20), col = "grey",
        ylab = "remaining electrons",
        xlab = "log10(times)",pch = 1)
points(
  x = times_avg,  y = (N_pts-1):0, col = "blue")
sd <- apply(all_times_matrix, 1, sd)
sd_error <- sd/sqrt(N_pts)
## plot error bars
arrows(times_avg-sd_error,
       (N_pts-1):0,
       times_avg+sd_error,
       length=0.05,angle=90, code=3)
t <- 10^seq(-15,20,1)
lines(
  x = log10(t),
  y = N_pts * exp(-rho_prime * log(1.8 * s_tun * t)^3),
col = "red",lwd=3)
legend("topright",bty="n",
       legend = c("500 MC runs", "MC average",
       "analytical solution"),
       col = c("grey", "blue", "red"),
       pch = c(1,1,NA),lwd = c(NA,NA,2))
```

Part IV
Comprehensive Luminescence Models

Chapter 11
Comprehensive Luminescence Models for Quartz

11.1 Introduction

Quartz is one of the most important and ubiquitous geological materials, and its luminescence process is very complex (see, for example, the review paper by Preusser et al. [160]). Several comprehensive models have been developed in order to explain complex luminescence mechanisms and phenomena in quartz. These rather complex models have been developed in order to help us elucidate the luminescence mechanisms, to improve and understand experimental protocols, and to understand the behavior of different types of quartz samples that undergo thermal and optical treatments in the laboratory.

In this chapter we will describe several quartz models, and how to implement them, by using two approaches: we use the R program *KMS* developed by Peng and Pagonis [151] and also provide examples of using the R package *RlumModel* by Friedrich et al. [48].

In Sects. 11.2 and 11.3 we provide a general description of the *Bailey2001* model (Bailey [10]) and the *Pagonis2008* model (Pagonis et al. [149]), implemented in the suite of *KMS* program. These two models and their variants have been used extensively during the past 20 years, to model a variety of quartz phenomena. In many cases, it has been possible to provide quantitative agreement between the models and experimental data in quartz. The specific functions, which can be called to simulate sequences of experimental events in these two models, are summarized and explained in Sect. 11.4.

Section 11.5 provides a simulation of the history of natural quartz samples, as pertaining to luminescence phenomena. Examples are provided of how to modify the models by changing one or more of the original model parameters. TL signals in quartz are strongly affected by the phenomenon of thermal quenching, which is discussed and modeled in Sect. 11.6. This is followed in Sects. 11.7 and 11.8 by demonstrations of the dose response of TL and OSL signals from quartz samples, and how annealed quartz samples can exhibit superlinear dose response in certain dose ranges.

V. Pagonis, *Luminescence*, Use R!, https://doi.org/10.1007/978-3-030-67311-6_11

Section 11.9 covers the important phenomenon of phototransfer, which has been documented in many dosimetric materials including quartz. Another important phenomenon in quartz studies is the predose effect, which has been explained on the basis of a thermal transfer of holes between hole reservoirs and the luminescence center in quartz; this important class of phenomena are simulated and explained in Sect. 11.10. Next, is a simulation of the important experimental technique of pulse annealing, in Sect. 11.11. An example of the SAR protocol is given in Sect. 11.12.

This chapter is concluded in Sect. 11.13, with several examples of the extensive R package *RLumModel*, developed by Friedrich et al. [48].

11.2   General Description of the Bailey [10] Model

This section describes the comprehensive quartz model developed by Bailey [10]. Figure 11.1 shows the energy level diagram for this model. The parameters used in this model were arrived at on the basis of empirical data from several different types of quartz samples.

This model has been successful in simulating several TL and OSL phenomena in quartz [9, 10]. Level 1 in the model represents the 110 °C TL shallow electron trap, which gives rise to a TL peak at 110 °C when measured with a heating rate of 5 K/s. This TL peak has been the subject of numerous studies, because of its importance in predose dating and retrospective dosimetry (Bailiff [12]). Within the model, the 110 °C TL level is assigned a photostimulation probability, since it has been shown to be light sensitive (Wintle and Murray [191]).

Level 2 represents a generic 230 °C TL level, typical of such TL peaks found in many sedimentary quartz samples. It is assumed that this TL trap is not light sensitive, and thus it is not assigned a photostimulation probability.

Levels 3 and 4 are usually termed the fast and medium OSL components (Bailey et al. [11]), and they yield TL peaks at 330 °C as well as give rise to OSL signals. The photostimulation rates for these levels are discussed in some detail in the original paper by Bailey [10]. The model does not contain any of the slow OSL components that are known to be present in quartz (Singarayer and Bailey [167, 168]) and that were incorporated in the later versions of the model [9].

Level 5 is a deep, thermally disconnected electron center. Such a level is known to be necessary in order to explain several TL and OSL phenomena, based on competition between energy levels. The model contains also four hole trapping centers that act as recombination centers for optically or thermally released electrons. Levels 6 and 7 are thermally unstable, non-radiative recombination centers, similar to the *hole reservoirs* first introduced by Zimmerman [197], in order to explain the predose sensitization phenomenon in quartz. Level 8 is a thermally stable, radiative recombination center termed the *luminescence center* (L in the Zimmerman model). Holes can be thermally transferred from the two hole reservoirs (levels 6 and 7) into the luminescence center via the valence band.

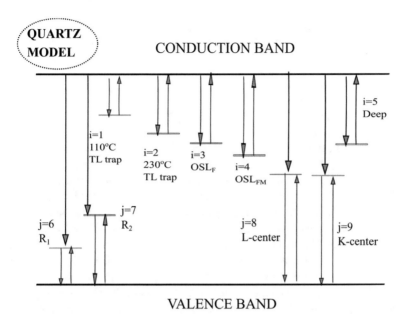

Fig. 11.1 Schematic diagram of the Bailey model, consisting of a total of nine energy levels. The arrows indicate possible transitions (after Bailey [10])

Level 9 is a thermally stable, non-radiative recombination center termed a killer center (labeled K in the Zimmerman model). The parameters are defined as follows: N_i are the concentrations of electron traps or hole centers (cm^{-3}), n_i are the concentrations of trapped electrons or holes (cm^{-3}), s_i are the frequency factors (s^{-1}), E_i are the electron trap depths below the conduction band or hole center energy levels above the valence band (eV), A_i $(i = 1, \ldots, 5)$ are the conduction band to electron trap transition coefficients $(cm^3 s^{-1})$, A_j $(j = 6, \ldots, 9)$ are the valence band to hole trap transition coefficients $(cm^3 s^{-1})$, and B_j $(j = 6, \ldots, 9)$ are the conduction band to hole center transition coefficients $(cm^3 s^{-1})$. The photo-eviction constants are θ_{0i} at $T = \infty$, and the thermal assistance energies are E_i^{th}.

The equations to be solved in the *Bailey2001* model [10] are as follows:

$$\frac{dn_i}{dt} = n_c(N_i - n_i)A_i - n_i P\theta_{0i}e^{(-\frac{E_i^{th}}{k_B T})} - n_i s_i e^{(-\frac{E_i}{k_B T})} \quad (i = 1, \ldots 5), \qquad (11.1)$$

$$\frac{dn_j}{dt} = n_v(N_j - n_j)A_j - n_j s_j e^{(-\frac{E_j}{k_B T})} - n_c n_j B_j \quad (j = 6, \ldots 9), \qquad (11.2)$$

$$\frac{dn_c}{dt} = R - \sum_{i=1}^{5}\left(\frac{dn_i}{dt}\right) - \sum_{j=6}^{9} n_c n_j B_j, \qquad (11.3)$$

$$\frac{dn_v}{dt} = \frac{dn_c}{dt} + \sum_{i=1}^{5}\left(\frac{dn_i}{dt}\right) - \sum_{j=6}^{9}\left(\frac{dn_j}{dt}\right), \tag{11.4}$$

and the luminescence intensity is defined as

$$L = n_c n_8 B_8 \eta(T), \tag{11.5}$$

with $\eta(T)$ representing the luminescence efficiency and R denoting the pair production rate [10]. A value of $R = 5 \times 10^7 \, \mathrm{cm}^{-3} \, \mathrm{s}^{-1}$ in this model corresponds to an approximate dose rate of 1 Gy/s. The underlying assumption of Eq. (11.5) is that the TL/OSL emission uses one and the same radiative recombination center type.

The luminescence efficiency is affected by the phenomenon of thermal quenching in thermoluminescence, which is discussed in some detail by Bailey [10] and by Nanjundaswamy et al. [115]. These authors suggested a Mott–Seitz mechanism, based on the existence of an activation energy W. This energy W represents the energy barrier that must be overcome, for an excited state electron to transition non-radiatively to the ground state, with the emission of phonons. The thermal quenching phenomenon was first described by Wintle [190], and further discussed by Petrov and Bailiff [152, 153] and several other authors. In this book, we discussed the effect of thermal quenching in Chap. 5.

The luminescence efficiency is given by the expression:

$$\eta(T) = \frac{1}{1 + K \exp(-W/kT)} \tag{11.6}$$

where $K = 0.8 \times 10^7$ is a dimensionless constant and W is the thermal quenching activation energy for quartz with a value of 0.64 eV. It should be noted that when the luminescence efficiency is given by Eq. (11.6), the initial rise method provides an evaluation not of E but of $E - W$ [152]. Further work along the same lines, on the thermal quenching of luminescence in Al_2O_3:C, was given by Akselrod et al. [4]. Examples of using the *Bailey2001* model are given in Sects. 11.5–11.12 of this chapter.

11.3   General Description of the Pagonis et al. [149] Model

Several studies have documented a different type of *thermally transferred OSL* (TT-OSL) signal, which has been used as the basis of a new OSL dating procedure introduced by Wang et al. [188]. By using this TT-OSL signal, it was found possible to extend the dating range for fine-grained quartz extracted from Chinese loess by almost an order of magnitude. It is believed that the production of this type of TT-OSL signal involves a single charge transfer mechanism, involving a slowly

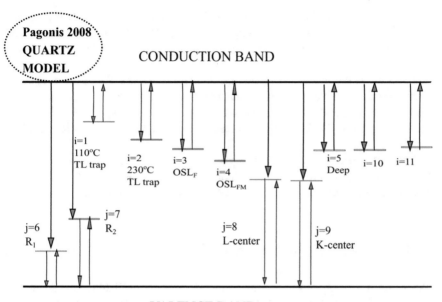

Fig. 11.2 The quartz model of Pagonis et al. [149]. Levels 10 and 11 are the additional levels introduced to the original model by Bailey [10]

bleachable OSL trap corresponding to a "source TL trap" around 315 °C. This single transfer mechanism was suggested by Adamiec et al. [1] and was modeled by Pagonis et al. [149].

The TT-OSL signal is usually obtained after a high temperature preheat (260 °C for 10 s), following an optical bleach at 125 °C for 270 s to deplete the fast and medium OSL components. The TT-OSL signal is typically measured for 90 s at 125 °C, in order to avoid the effect of retrapping of electrons in the 110 °C TL trap. The luminescence sensitivity related to this TT-OSL measurement is determined by the OSL response to a subsequent test dose. The TT-OSL signal is thought to consist of two components, a dose-dependent component termed the recuperated OSL (ReOSL), and a second component believed to be dose independent, which is termed the basic transferred OSL (BT-OSL). Wang et al. [188] proposed a method that used multiple aliquots to construct the dose response curve and used it for D_e determination.

Pagonis et al. [149] used a modified version of the model by Bailey [10], to simulate the dose response of the OSL, TT-OSL, and BT-OSL signals, and compared directly the experimental results with the model output.

The TT-OSL model of Pagonis et al. [149] is based on the previous model by Bailey [10], and Fig. 11.2 shows the corresponding energy level diagram. The original model consists of 5 electron traps and 4 hole centers. This model was expanded by Pagonis et al. [149] to include two additional levels. Levels 10 and 11 in Fig. 11.2 are the two new levels added to the original Bailey model and were introduced in order to simulate the experimentally observed TT-OSL and BT-OSL signals.

Level 10 in the model represents the source trap for the TT-OSL signal and is a less thermally stable trap, which saturates at very high doses. It is assumed that electrons are thermally transferred into the fast component trap (level 3) from level 10. This trap (level 10) is assumed to be emptied optically in nature by long sunlight exposure.

Level 11 is believed to contribute most of the BT-OSL signal in quartz; these are light-insensitive traps that are more thermally stable than either level 3 or level 10. Several examples of using the model are given in Sects. 11.5–11.12.

11.4   Using the R Program KMS

Table 11.1 is a listing of these various functions that are available in the R program *KMS*, described by Peng and Pagonis [151]. These functions are applicable to any of the three available *KMS* models *Bailey2001*, *Bailey2004*, and *Pagonis2008* [9, 10, 149]. In this chapter we will focus mostly in two of the models: the *Bailey2001* and *Pagonis2008* models.

Table 11.1 The various functions available in the R program *KMS*, described by Peng and Pagonis [151]

Functions called in KMS programs
setInis()
Set all center populations equal to zero for quartz crystallization
setPars()
Initialize the model using appropriate kinetic parameters
irradiate(*temp, tim, doseRate*)
Irradiate at *temp* °C for *tim* s with a dose rate of *doseRate* Gy/s
heatAt(*temp, tim*)
Heat at *temp* °C for *tim* s
heatTo(*temp1, temp2, hRate*)
Heat from *temp1* °C to *temp2* °C with a heating rate of *hRate* °C/s
stimOSL(*temp, tim, pValue, nChannel*)
OSL stimulation at temp °C for *tim* s with a photon stimulation flux of *pValue* s^{-1}cm^{-2}
The OSL signal is evaluated at the equally spaced number of channels *nChannel*
stimTL(*lowTemp, upTemp, hRate, nChannel*)
TL stimulation from *lowTemp* °C to *upTemp* °C with a heating rate of *hRate* °C/s,
the number of equally spaced channels is *nChannel*

It is important to note that these functions are *the same* for all three models in KMS; this gives large flexibility in interchanging the models, by using the same R code.

The R functions provided in program KMS save the center populations of electrons and holes internally, as global variables in the vector *inis*. Once a simulation event is completed, all center populations in vector *inis* are automatically updated, and the user does not need to pass the initial center populations as arguments between functions in simulating a sequence of events. In this way events can be simulated elegantly by using compact R commands, and the number of arguments to be typed is very small.

The model must be initialized using the R function *setPars( )*; this function creates a vector of kinetic parameters stored in a global vector *pars*. A direct way to modify kinetic parameters is by changing the elements of vector *pars* manually inside the function. A second way of modifying parameters is by using replacement functions. For example, the total concentration N_1 of electrons trap for level 1 can be modified to be 10^8 cm^{-3}, by using the R command *pars[N1]<-1e8*. Replacement functions of this kind are useful when we want to study the effect of individual parameters, or when we want to simulate several versions of kinetic parameters by using a Monte Carlo approach, as performed in previous studies.

The parameter *res* contains the results of functions like *stimTL* and *stimOSL*. We can refer to the values of concentrations during the TL experiment by examining, for example, the parameter *res["n1"]* for the concentration n_1 in level 1 of the model, *res["tlx"]* for the temperature values, *res["tly"]* for the TL intensities, etc.

11.5   Simulations of the History of Natural Quartz Samples

In this section we simulate natural quartz samples using the *Bailey2001* and *Pagonis2008* models. We also provide a method by which the codes in the rest of this chapter can be shortened significantly, by storing the concentrations of traps and centers for a natural quartz sample.

The exact steps in the simulation of a *natural* sedimentary quartz sample are given by Bailey [10] and are shown in Table 11.2. Both steps 5a and 5b in this table give similar results, even though they use very different values of the irradiation dose rates. In addition, one must keep in mind that there is not a unique set of parameters in these comprehensive models for all quartz samples, i.e. different quartz samples may require modifications to the parameters in the model.

As a first example, we simulate the history of a natural sedimentary quartz sample using the *Pagonis2008* model. In the first part of the code, the steps given in Table 11.2 are simulated for the natural sample, and the TL glow curve is measured in the laboratory. The result is shown in Fig. 11.3a. Note that the 110 °C TL peak

Table 11.2 The steps in simulating a *natural* sedimentary quartz sample, in the *Bailey2001* model in *KMS* [10]. Step 5a is used in the original *Bailey2001* model, and step 5b is used in the *Pagonis2008* model in KMS

	Steps in simulation of *natural* sedimentary quartz sample in the Bailey2001 model [10]
1	All electron and hole concentrations set to zero during the crystallization process
2	Geological dose of 1000 Gy with a dose rate of 1 Gy/s at 20 °C
3	Heat to 350 °C (simulation of geological time)
4	Illumination at 200 °C for 100s, repeated exposures to sunlight over a long period
5a	Burial dose of 20 Gy at 0.01 Gy/s at 220 °C
5b	Burial dose of 20 Gy at a very low natural dose rate of 3×10^{-11} Gy/s at 20 °C

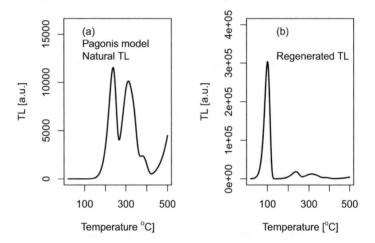

Fig. 11.3 Simulation of the natural history of a quartz sample, using the *Pagonis2008* model in *KMS* programs. (**a**) The TL glow curve, after the natural sample is heated in the laboratory with a heating rate of 5 K/s. Note the absence of the 110 °C TL peak. (**b**) The TL glow curve, after the natural sample is irradiated with 10 Gy and then heated in the laboratory with a heating rate of 5 K/s. Note the restored 110 °C TL peak in (**b**), which is missing in (**a**)

is missing from the TL glow curve of the natural sample, because it is thermally unstable over long periods of time in nature, and the natural dose rate is very low at 10^{-11} Gy/s (or about 1 Gy/Ka).

The concentrations n_i of traps and centers ($i = 1 \ldots 9$) and also n_c, n_v at the end of the natural sample simulation are stored in the variable *storeNat*. These concentrations are used in the second part of the code, by re-initializing the concentrations with the code line *inis<-storeNat*. This is done in order to not repeat the code for the natural sample.

In the second part of the code, the TL glow curve is measured in the laboratory, after the sample is irradiated with a dose of 10 Gy delivered with a dose rate of 1

Code 11.1: Natural history of quartz sample Pagonis2008 model

```
# Use Pagonis model to simulate quartz sample history and
# TL measured in lab after irradiating with 10 Gy at 1 Gy/s
rm(list=ls())
library("deSolve")
source("Pagonis_Model.R")
setPars()
setInis()
irradiate(temp=20, tim=1000, doseRate=1) #1000 Gy at 1 Gy/s
heatAt(temp=20, tim=60)                   #Relaxation
heatTo(temp1=20, temp2=350, hRate=5)     # Heat to 350 degC
heatTo(temp1=350, temp2=200, hRate=-5)    # Cool down to RT
stimOSL(temp=200, tim=100, pValue=2.0, nChannel=1000) #Bleach
irradiate(temp=20, tim=20/1e-11, doseRate=1e-11)# Burial dose
heatAt(20,60)
storeNat<-inis #Store concentrations for natural sample
TL<-stimTL(lowTemp=20,upTemp=500,hRate=5,nChannel=480) #TL
heatTo(temp1=500,temp2=20,hRate=-5) # Cool down
tlx<-TL[,"tlx"]
tly<-TL[,"tly"]
par(mfrow=c(1,2))
plot(tlx,tly,type="l",ylim=c(0,16000),lwd=2,
xlab=expression("Temperature "^"o"*"C]"), ylab = "TL [a.u.]")
legend("topleft",bty="n",legend=c("(a)","Pagonis model",
"Natural TL"))
# repeat irradiation and TL for lab irradiation
inis<-storeNat    # restore concentrations for natural sample
irradiate(temp=20, tim=10, doseRate=1)
heatAt(temp=20, tim=60)
TL<-stimTL(lowTemp=20,upTemp=500,hRate=5,nChannel=480)
heatTo(temp1=500,temp2=20,hRate=-5)
tlx<-TL[,"tlx"]
tly<-TL[,"tly"]
plot(tlx,tly,type="l",xlab=expression("Temperature["^"o"*"C]"),
ylab = "TL [a.u.]",ylim=c(0,400000),lwd=2)
legend("topright",bty="n",legend=c("(b)", "","Regenerated TL"))
```

Gy/s. The result is shown in Fig. 11.3b. Note that in this sample the 110 °C TL peak is now present, since the laboratory irradiation is carried out in a sort time, and with the much higher dose rate of 1 Gy/s.

By changing only one line in the previous code, we can simulate the same sequence of events using the *Bailey2001* model. Specifically, we can simply replace the line *source("Pagonis_Model.R")* with *source("Bailey01_Model.R")*, and the rest of the code will remain unchanged (Fig. 11.4). This shows the flexibility of the code in the *KMS* suite of R programs.

Peng and Pagonis [151] discussed how to run the *KMS* models by modifying the parameters in the model. In the models, the line *setPars()* initializes the original values of the parameters in the model. By using, for example, the sequential lines:

- *setPars()*
- *pars["N6"] «- 3e10*

in the code, we can modify the value of the global parameter N_6, and we can perform the simulation with this modified value. The symbol «- is the scoping assignment used in R, which is associated with setting global variables in a code. In this example, N_6 is the total concentration of holes in level 6 of the model, which represents the hole reservoir R_1. Any number of the parameters in the models can be modified at once, by using this method. We will use this method in some of the codes presented in this chapter.

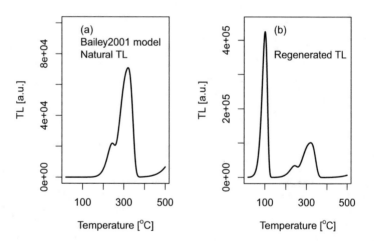

Fig. 11.4 Simulation of the natural history of a quartz sample, using the *Bailey2001* model in *KMS*. Compare these results with the similar results in Fig. 11.3, which were obtained using the *Pagonis2008* model

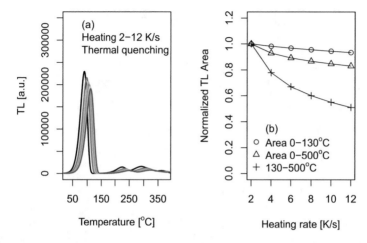

Fig. 11.5 Simulation of the thermal quenching of TL signal in quartz, demonstrated by using various heating rates in the *Pagonis2008* model of *KMS*. (**a**) The TL glow curves for heating rates 2–12 K/s and (**b**) the areas under different parts of the TL glow curves in (**a**), showing the thermal quenching effects

11.6 Thermal Quenching of TL Signals in Quartz

A model of thermal quenching based on the Mott-Seitz mechanism was already presented in Chap. 5. The following code simulates the effect of heating rate on the shape and area under the TL peaks in quartz, by simulating heating rates $\beta = $ 2–12 K/s during the TL measurement. The TL glow peaks are shown in Fig. 11.5a.

Figure 11.5b shows that as the heating rate is increased, the area of the TL signal at 110 °C drops by only about 5%, showing that thermal quenching has a small effect on the TL glow peaks in this temperature range.

By contrast, Fig. 11.5b shows that the integrated TL signal between 130 °C and 500 °C drops by about 50%, while the total integrated TL signal in the region 0–500 °C drops by almost 15% as the heating rate increases from 2 K/s to 12 K/s.

The effect of thermal quenching can be reduced significantly by using very low heating rates, less than 1 °C/s, and therefore the use of low heating rates is highly recommended in experimental work, when thermal quenching may be present in the samples.

11.7 Dose Response of Quartz Samples in the *KMS* Models

In the example that follows, we use the *Baileys2001* model of the *KMS* programs, to simulate laboratory irradiation with a series of doses and subsequent measurement of the TL glow curves.

Code 11.2:   Thermal quenching of TL signal in quartz (KMS)

```r
# Use Pagonis model to simulate thermal quenching in quartz
# TL measured in laboratory with different heating rates
rm(list=ls())
library("deSolve")
source("Pagonis_Model.R")
beta<-2*seq(1:6)
tlylow<-tlyhigh<-vector(length = length(beta))
tlx<-tly<-matrix(nrow=480,ncol=length(beta))
for (i in 1:6){
  setPars()
  setInis()
  irradiate(temp=20, tim=1000, doseRate=1) #1000 Gy at 1 Gy/s
  heatAt(temp=20, tim=60)                   #Relaxation
  heatTo(temp1=20, temp2=350, hRate=5)      # Heat to 350 degC
  heatTo(temp1=350, temp2=200, hRate=-5)    # Cool down to RT
  stimOSL(temp=200, tim=100, pValue=2.0, nChannel=1000) #Bleach
  irradiate(temp=20, tim=20/1e-11, doseRate=1e-11)# Burial dose
  heatAt(20,60)
  irradiate(temp=20, tim=100, doseRate=1)
heatAt(temp=20, tim=60)
TL<-stimTL(lowTemp=20,upTemp=500,hRate=beta[i],nChannel=480)
tlx[,i]<-TL[,"tlx"]
tly[,i]<-TL[,"tly"]/i
tlyhigh[i]<-sum(tly[,i][130:400])
tlylow[i]<-sum(tly[,i][1:130])}
par(mfrow=c(1,2))
matplot(tlx,tly,typ="l",lty="solid",lwd=2,
xlab=expression("Temperature ["^"o"*"C]"), ylab = "TL [a.u.]",
ylim=c(0,350000),xlim=c(30,380))
legend("topleft",bty="n",legend=c("(a)","Heating 2-12 K/s",
"Thermal quenching"))
plot(beta,tlylow/max(tlylow),pch=1,typ="o",ylim=c(0,1.2),
xlab="Heating rate [K/s]", ylab = "Normalized TL Area")
lines(beta,colSums(tly)/max(colSums(tly)),pch=2,typ="o")
lines(beta,tlyhigh/max(tlyhigh),pch=3,typ="o")
legend("bottomleft",bty="n",legend=expression("(b)",
"Area 0-130"^"o"*"C","Area 0-500"^"o"*"C","Area
130-500"^"o"*"C"),pch=c(NA,1,2,3))
```

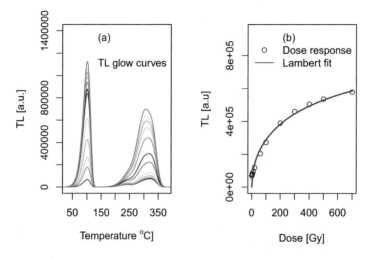

Fig. 11.6 Simulation of the dose response curve of the 330 °C TL peak in a sedimentary quartz sample. (**a**) The TL glow curves and (**b**) the dose response curve, fitted with the Lambert dose response equation we saw in Chap. 4

The TL dose response curve of the 330 °C peak is simulated in Fig. 11.6a for a sedimentary quartz sample. The dose response curve is fitted in Fig. 11.6b with the Lambert analytical equations, which were derived in Chap. 4.

The following code simulates the OSL dose response in quartz samples. Typically the OSL signal in quartz is measured at an elevated temperature of 125 °C, in order to avoid interference from the shallow traps in the material. In the example of Fig. 11.7 we simulate the dose response of the first 1 s of the OSL signal.

For a simulation of the multiple aliquot additive-dose technique for TL, the reader is referred to the paper by Peng and Pagonis [151], which contains the appropriate R code and discusses the simulation results in some detail.

11.8 Superlinear Dose Response of Annealed Quartz Samples

Figure 11.8a shows a simulation of the TL glow curves in quartz, measured after irradiating a sample with doses in the range 0–2 Gy. Figure 11.8b shows the nonlinear dose dependence of the TL signal at 330 °C in this dose range, simulated using the *Bailey2001* model. Finally, Fig. 11.8c shows the same data as in (b) on a log–log scale, showing a superlinearity index of $g(D) = 1.39$ as obtained from a linear least squares fit to the log–log data.

Code 11.3: TL dose response of quartz sample (KMS)

```
# Dose Response of 110degC TL peak in quartz
rm(list=ls())
library("deSolve")
library("minpack.lm")
library("lamW")
source("Bailey01_Model.R")
setPars()
setInis()
irradiate(temp=20, tim=1000, doseRate=1) #1000 Gy at 1 Gy/s
heatAt(temp=20, tim=60)                   #Relaxation
heatTo(temp1=20, temp2=350, hRate=5)      # Heat to 350 degC
heatTo(temp1=350, temp2=200, hRate=-5)    # Cool down to RT
stimOSL(temp=200, tim=100, pValue=2.0, nChannel=1000) #Bleach
irradiate(temp=20, tim=20/1e-11, doseRate=1e-11)# Burial dose
heatAt(20,60)       #Store concentrations for natural sample
storeNat<-inis  # in variable storeNat
reDose<-c(1,3,5,10,20,60,100,200,300,400,500,700)
tlm<-vector(length=length(reDose))
tlx<-tly<-matrix(nrow=480,ncol=length(reDose))
for (i in seq(length(reDose))) {
inis<-storeNat
irradiate(temp=20,tim=reDose[i],doseRate=1)
heatAt(temp=20,tim=60)
res<-stimTL(lowTemp=20,upTemp=500,hRate=5,nChannel=480)
heatTo(temp1=500,temp2=20,hRate=-5)
tlx[,i]<-res[,"tlx"]
tly[,i]<-res[,"tly"]
tlm[i]<-approx(x=res[,"tlx"],y=res[,"tly"],xout=330)$y }
par(mfrow=c(1,2))
matplot(tlx,tly,type="l",lty="solid",ylab = "TL [a.u.]",
xlab=expression("Temperature "^"o"*"C]"),ylim=c(0,1.4e6),
xlim=c(30,380))
legend("topright",bty="n",legend=c("(a)"," ","TL glow curves"))
plot(reDose,tlm,type="p",xlab="Dose [Gy]",ylab="TL [a.u]",
ylim=c(0,.95e6))
legend("topright",bty="n",legend=c("(b)","Dose response",
"Lambert fit"),pch=c(NA,1),lty=c(NA,NA,"solid"))
t <-reDose
y <-tlm
fit_data <-data.frame( t ,y)
#plot(fit_data,ylim=c(0,max(y)))
fit <- minpack.lm::nlsLM(
formula=y~N*(1+lambertW0((abs(R)-1)*exp(abs(R)-1-b*t))/
(1-abs(R))),
data=fit_data,start = list(N=5* max(y),R=0.1, b = .002))
N_fit <- coef(fit)[1]
```

```
R_fit <- abs(coef(fit)[2])
b_fit <- coef(fit)[3]
## plot analytical solution
curve(
N_fit*(1+lambertW0((R_fit-1)*exp(R_fit-1-b_fit*x))/(1-R_fit)),
0,700,col = "blue",add=TRUE,lwd=2)
cat("\nfitted N: ", N_fit)
cat("\nfitted R: ", R_fit)
cat("\nfitted D0: ", 1/b_fit, " Gy")

##
## fitted N:   768919.4
## fitted R:   4.819477e-08
## fitted D0:   1045.857  Gy
```

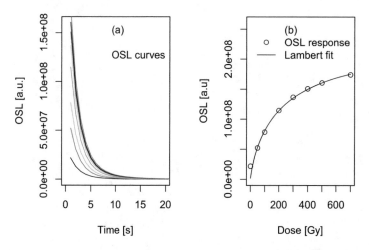

Fig. 11.7 Simulation of the dose response curve of the OSL signal measured at 125 °C, in the *Bailey2001* model. (**a**) The CW-OSL curves and (**b**) the dose response curve, fitted with the Lambert function we saw in Chap. 4

11.9   Simulation of the Phototransfer Effect Using the *Bailey2001* Model

In this section we present a simulation of the phototransfer phenomenon, using the *Bailey2001* model in the *KMS* suite of programs. The simulation consists of the following simple steps:

Code 11.4: Dose Response of quartz OSL signal, Bailey2001 model

```
# Dose Response of quartz OSL signal measured at 125degC
rm(list=ls())
library("deSolve")
library("minpack.lm")
library("lamW")
source("Bailey01_Model.R")
setPars()
setInis()
irradiate(temp=20, tim=1000, doseRate=1) #1000 Gy at 1 Gy/s
heatAt(temp=20, tim=60)                   #Relaxation
heatTo(temp1=20, temp2=350, hRate=5)     # Heat to 350 degC
heatTo(temp1=350, temp2=200, hRate=-5)   # Cool down to RT
stimOSL(temp=200, tim=100, pValue=2.0, nChannel=1000) #Bleach
irradiate(temp=20, tim=20/1e-11, doseRate=1e-11)# Burial dose
heatAt(20,60)     #Store concentrations for natural sample
storeNat<-inis   # in variable storeNat
reDose<-c(1,50,100,200,300,400,500,700)
oslm<-vector(length=length(reDose))
oslx<-osly<-matrix(nrow=100,ncol=length(reDose))
for (i in seq(length(reDose))) {
    inis<-storeNat #Restore natural concentrations
    irradiate(temp=20,tim=reDose[i],doseRate=1)
    heatAt(temp=20,tim=60)
    heatTo(20,125,hRate=5)
    res<-stimOSL(temp=125, tim=100, pValue=2.0, nChannel=100)
    oslx[,i]<-res[,"oslx"]
    osly[,i]<-res[,"osly"]
    oslm[i]<-res[,"osly"][1]}
par(mfrow=c(1,2))
matplot(oslx,osly,type="l",lty="solid",xlim=c(0,20),
    xlab="Time [s]",ylab = "OSL [a.u.]",ylim=c(0,1.6e8))
legend("topright",bty="n",legend=c("(a)"," ","OSL curves"))
plot(reDose,oslm,type="p",xlab="Dose [Gy]",ylab="OSL [a.u]",
    ylim=c(0,2.6e8))
legend("topright",bty="n",legend=c("(b)","OSL response",
    "Lambert fit"),pch=c(NA,1),lty=c(NA,NA,"solid"))
t <-reDose
y <-oslm
fit_data <-data.frame( t ,y)
fit <- minpack.lm::nlsLM(
    formula=y~N*(1+lambertW0((abs(R)-1)*exp(abs(R)-1-b*t))/
                (1-abs(R))),  data = fit_data,
    start = list(N= max(y),R=0.1, b = .002))
N_fit <- coef(fit)[1]
R_fit <- abs(coef(fit)[2])
b_fit <- coef(fit)[3]
```

```
curve(
  N_fit*(1+lambertW0((R_fit-1)*exp(R_fit-1-b_fit*x))/(1-R_fit)),
  1,700,add=TRUE )
cat("\nfitted N: ", N_fit)
cat("\nfitted R: ", R_fit)
cat("\nfitted Dc: ", 1/b_fit, " Gy")

  ##
  ## fitted N:   200028153
  ## fitted R:   0.2111977
  ## fitted Dc:   504.679  Gy
```

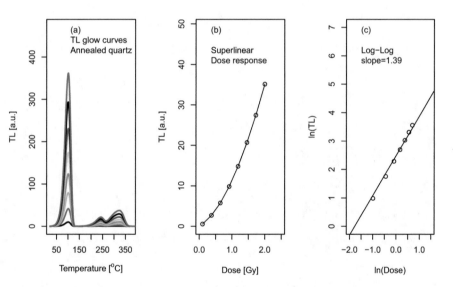

Fig. 11.8 Simulated superlinear dose response of 330 °C TL peak in annealed quartz using the *Bailey2001* model. The sample was annealed for 1 h at 700 °C, before measuring its TL dose response between 0.1 and 2 Gy. (**a**) The TL glow curves at different doses; (**b**) the superlinear dose dependence of the TL signal at 330 °C; (**c**) the data in (**b**) is plotted on a log–log scale and fitted with a linear function, yielding a superlinearity index of $g(D) = 1.39$

Code 11.5: Superlinearity in annealed quartz samples

```r
# Superlinear TL dose response in annealed quartz
rm(list = ls(all=T))
library("deSolve")
source("Bailey01_Model.R")
reDose<-seq(0.1,2,by=1.9/7)
setInis()
setPars()
irradiate(temp=20, tim=1000, doseRate=1) #1000 Gy at 1 Gy/s
heatAt(temp=20, tim=60)                    #Relaxation
heatTo(temp1=20, temp2=350, hRate=5)       # Heat to 350 degC
heatTo(temp1=350, temp2=200, hRate=-5)     # Cool down to RT
stimOSL(temp=200, tim=100, pValue=2.0, nChannel=1000) #Bleach
irradiate(temp=20, tim=20/1e-11, doseRate=1e-11)#Burial 100 Gy
heatTo(20,700,hRate=5)
heatAt(700,3600)
heatTo(700,20,hRate=-5)
storeNat<-inis
tlm<-vector(length=8)
tlx<-tly<-matrix(nrow=480,ncol=8)
# natural dose rate =0.1 Gy
for (i in seq(8)) {
  inis<-storeNat
  irradiate(temp=20,tim=reDose[i],doseRate=1)
  heatAt(temp=20,tim=60)
  res<-stimTL(lowTemp=20,upTemp=500,hRate=5,nChannel=480)
  tlx[,i]<-res[,"tlx"]
  tly[,i]<-res[,"tly"]
  tlm[i]<-approx(x=res[,"tlx"],y=res[,"tly"],xout=330)$y }
par(mfrow=c(1,3))
matplot(tlx,tly,type="l",ylim=c(0,470),xlim=c(20,380),
lty="solid",lwd=2,xlab=expression("Temperature ["^"o"*"C]"),
ylab = "TL [a.u.]")
legend("topright",bty="n",legend=c("(a)","TL glow curves",
"Annealed quartz"))
plot(reDose,tlm,type="o",xlab = "Dose [Gy]",ylab = "TL [a.u.]",
ylim=c(0,1.4*max(tlm)),xlim=c(0,2.5))
legend("topleft",bty="n",legend=c("(b)"," ","Superlinear",
"Dose response"))
x<- log(reDose)
y<- log(tlm)
rangeData<-cbind(x,y)
lm(y~x)$coefficients
plot(x, y, xlab = "ln(Dose)",ylab = "ln(TL)",ylim=c(0,7),
xlim=c(-2,1.5))
legend("topleft",bty="n",legend=c("(c)"," ","Log-Log",
"slope=1.39"))
abline(lm(y~x))

  ## (Intercept)           x
  ##    2.481778    1.397661
```

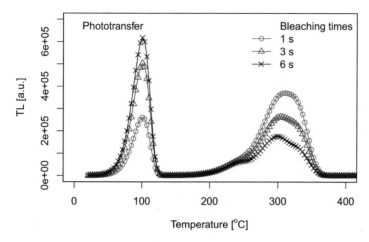

Fig. 11.9 Simulation of phototransfer process in the *Bailey2001* model. As the optical stimulation time increases, the height of the TL peak at 110 °C increases, while simultaneously the 330 °C TL peak decreases. For a more detailed study of this, and other quartz phenomena, see Pagonis et al. [126]

1. Simulation of the quartz natural sample as discussed in the text.
2. Optically bleach sample in laboratory, for 1 s at 20 °C.
3. Measure TL signal.
4. Repeat steps 1–3 for a different bleaching time.

Figure 11.9 shows the results of simulating the phototransfer effect using the *Bailey2001* model, for three optical bleaching times of 1, 3, and 6 s. The results in Fig. 11.9 show that as the bleaching time increases, the height of the 110 °C TL peak increases, while the corresponding height of the TL peak at 330 °C increases. There is clearly a charge transfer taking place; in this case, charge is transferred from the optically sensitive deeper traps responsible for the 330 °C TL peak, to the shallower traps responsible for the 110 °C TL peak.

For a more comprehensive simulation study of luminescence dating methods used for quartz, see Pagonis et al. [126]. These authors simulated the equivalent dose (ED) estimation for several methods that have been applied for ceramic materials containing quartz, namely the additive-dose TL, predose technique, phototransfer protocol, SAR-OSL protocol, and SAR-TL protocol. These extensive simulations estimated both the intrinsic precision and the intrinsic accuracy of these experimental techniques and were carried out using the quartz model by Pagonis et al. [149], which consists of 11 electron and hole traps and centers.

Code 11.6: Phototransfer phenomenon using Bailey2001 model

```r
#Phototransfer simulation using Bailey2001 model
rm(list = ls(all=T))
library("deSolve")
source("Bailey01_Model.R")
photo<-function(x,colr,ln,pchs){
  setInis()
  setPars()
  pars["N6"] <- 3e10 # change the parameter N6 in Bailey model
  # N6 = total concentration of holes in hole reservoir R1
  irradiate(temp=20, tim=1000, doseRate=1) #1000 Gy at 1 Gy/s
  heatAt(temp=20, tim=60)                     #Relaxation
  heatTo(temp1=20, temp2=350, hRate=5)      # Heat to 350 degC
  heatTo(temp1=350, temp2=200, hRate=-5)    # Cool down to RT
  stimOSL(temp=200, tim=100, pValue=2.0, nChannel=1000) #Bleach
  # Large Burial dose 100 Gy
  irradiate(temp=20, tim=200/1e-11, doseRate=1e-11)
  # end natural sample simulation, store concentrations
# Next is the optical Bleach
stimOSL(temp=20, tim=x, pValue=2.0, nChannel=1000)
heatAt(temp=20, tim=60)
TL<-stimTL(lowTemp=20,upTemp=500,hRate=5,nChannel=150) #TL
heatTo(temp1=500,temp2=20,hRate=-5) # Cool down
tlx<-TL[,"tlx"]
tly<-TL[,"tly"]
par(new=TRUE)
plot(tlx,tly,type="o",xlab=expression("Temperature  ["^"o"*"C]"),
 ylab = "TL [a.u.]",xlim=c(0,400),ylim=c(0,7e5),col=colr,lty=ln,
lwd=1,pch=pchs)
}    #end function photo
for (i in 1:3)
  {bleachTime<-c(1,3,6)
    colr<-c("red","blue","black")
    ln<-rep("solid",3)
    pchs<-c(1,2,4)
    photo(bleachTime[i],colr[i],ln[i],pchs[i])}
legend("topleft",bty="n","Phototransfer")
legend("topright",bty="n",lty=c(NA,ln),col=c(NA,colr),lwd=1,
 pch=c(NA,pchs),  c("Bleaching times","1 s","3 s","6 s"))
```

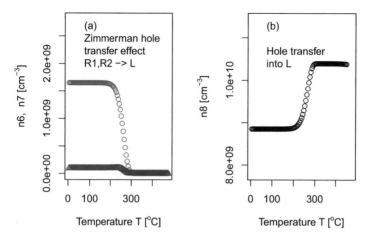

Fig. 11.10 The Zimmerman thermal transfer process for holes in quartz. (**a**) The concentrations of holes n_6 and n_7 in hole reservoirs R_1 and R_2 decrease as the temperature T is increased. (**b**) The concentration of hole n_8 in recombination center L also increases, indicating a thermal transfer of holes

11.10   The Predose Effect and the Zimmerman Hole Transfer

As discussed previously in this chapter, the comprehensive models *Bailey2001* and *Pagonis2008* for quartz include two hole reservoirs denoted by R_1 and R_2. These reservoirs were proposed early by Zimmerman [197] and are believed to be responsible for a thermal transfer of holes from the thermally shallow R_1 reservoir to the thermally deeper R_2 reservoir. Holes can also be thermally transferred from either R_1 or R_2 to the luminescence center L in these models.

Pagonis et al. [148] presented a simulation of this thermal transfer mechanism, and the following code provides a simple demonstration. The simulation consists of the geological history of a natural sample, which is then heated in the laboratory from room temperature to 500 °C. The parameter N_6 in the *Bailey2001* model is set here to a higher value of $N_6 < -3 \times 10^{10}$, as discussed in Pagonis et al. [148].

The simulation monitors the change in the concentrations of holes (n_6 and n_7) in the Zimmerman reservoirs R_1 and R_2, respectively, with the temperature. The results are shown in Fig. 11.10a, in which the concentrations n_6 and n_7 decrease as the temperature is increased, while at the same time in Fig. 11.10b the concentration of holes n_8 in the recombination center L increases.

For more details of the Zimmerman thermal hole transfer effect, see the detailed simulations by Pagonis et al. [126, 148]. The Zimmerman effect is closely related to the thermal activation curves (TACs), which are routinely measured for quartz and other dosimetric materials. For a simulation of TACs in quartz and a comparison with experimental data, see the study by Kitis et al. [77].

Code 11.7: The Zimmerman model: thermal transfer of holes in quartz

```r
# The Zimmerman thermal transfer of holes in quartz
rm(list=ls())
library("deSolve")
source("Bailey01_Model.R")
setInis()
setPars()
pars["N6"] <- 3e10    # change the concentration of holes in R1
irradiate(temp=20, tim=1000, doseRate=1) #1000 Gy at 1 Gy/s
heatAt(temp=20, tim=60)                   #Relaxation
heatTo(temp1=20, temp2=350, hRate=5)      # Heat to 350 degC
heatTo(temp1=350, temp2=200, hRate=-5)    # Cool down to RT
stimOSL(temp=200, tim=100, pValue=2.0, nChannel=1000) #Bleach
irradiate(temp=20, tim=100/1e-11, doseRate=1e-11)#Burial 100 Gy
storeNat<-inis
# Set up parameters and vectors
testDose<-.1
nTemps<-15
initTemp<-100
finalTemp<-500
actTemp<-seq(initTemp,finalTemp,by=(finalTemp-initTemp)/
            (nTemps-1))
# Loop for activation temperatures
inis<-storeNat    # Reset concentrations for natural sample
res2<-heatTo(25,500,hRate=1)    # heat to activation temperature
par(mfrow=c(1,2))
xlabel<-expression("Temperature T ["^"o"*"C]")
ylabel<-expression("n6,  n7 [cm"^"-3"*"]")
plot(res2[,"time"],res2[,"n6"],xlab = xlabel
,ylab =ylabel,ylim=c(0,1.7*max(res2[,"n6"])),pch=1,col="red")
legend("topleft",bty="n",legend=c("(a)","Zimmerman hole",
"transfer effect","R1,R2 -> L"))
par(new = TRUE)
plot(res2[,"time"],res2[,"n7"],xlab = xlabel
,ylab =ylabel,ylim=c(0,1.7*max(res2[,"n6"])),pch=2,col="blue")
plot(res2[,"time"],res2[,"n8"],xlab = xlabel
,ylab =expression("n8 [cm"^"-3"*"]"),
ylim=c(.75*max(res2[,"n8"]),1.1*max(res2[,"n8"])))
legend("topleft",bty="n",legend=c("(b)"," ","Hole transfer",
"into L"))
```

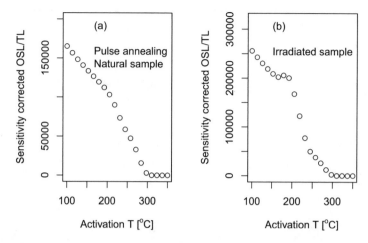

Fig. 11.11 Simulation of the experimental pulse annealing protocol of Wintle and Murray [192] for (**a**) a natural quartz sample and (**b**) an optically bleached and irradiated sample. For details of the simulation, see Pagonis et al. [148]

11.11 Simulation of a Pulse Annealing Experiment in Quartz

In an important pioneer experimental work for the development of the SAR protocol, Wintle and Murray [192] studied sensitivity changes that occurred during laboratory heating of a 30,000-year-old sedimentary quartz from Australia (laboratory code WIDG8). They used two aliquots, one of which was in its natural state and the other that had been optically bleached and had subsequently received in the laboratory a dose of 56 Gy. The experiments consisted of preheating and holding a single aliquot to a certain temperature for 10 s, then progressively heating to higher temperatures from 160 to 500 °C in increments of 10 °C, and measuring the OSL at 125 °C using a short 0.1 s stimulation. The 110 °C TL peak was used to record the sensitivity changes, by measuring the TL signal at each step after delivering a small test dose of 0.1 Gy.

Pagonis et al. [148] simulated the experimental protocol of Wintle and Murray [192], and the following code shows the steps in the simulation. The simulated results for the natural and the bleached quartz samples are shown in Fig. 11.11, and they verified the assumption that sensitivity changes are occurring during the heating and are not due to thermal transfer processes.

Code 11.8: Simulation of pulse annealing experiment with Bailey2001 model

```
#Simulation of pulse annealing experiment with
# Bailey2001 model
rm(list=ls())
library("deSolve")
source("Bailey01_Model.R")
setInis()
setPars ()
pars["N6"] <- 3e9
### Start natural sample simulations
irradiate(temp=20, tim=1000, doseRate=1)
heatAt(temp=20, tim=60)
heatTo(temp1=20, temp2=350, hRate=5)
heatTo(temp1=350, temp2=200, hRate=-5)
stimOSL(temp=200, tim=100, pValue=2.0, nChannel=10)
irradiate(temp=20, tim=51/1e-11, doseRate=1e-11)  #Burial 51 Gy
## End of natural history with natural irradiation
res<- heatAt(temp=20, tim=60)
storeNat<-inis
nTemps<-20
initTemp<-100
finalTemp<-350
actTemp<-seq(initTemp,finalTemp,by=(finalTemp-initTemp)/
(nTemps-1))
tlm<-vector(length=nTemps)
tlx<-tly<-matrix(nrow=1000,ncol=nTemps)
tl<-osl<-rep(0,nTemps)
par(mfrow=c(1,2))
for (i in seq(nTemps)) {
  heatTo(20,actTemp[i],hRate=1)
  heatAt(actTemp[i],tim=10)
  heatTo(actTemp[i],20,hRate=-1)
  heatTo(20,125,hRate=1)
  resOSL<-stimOSL(temp=125, tim=0.1, pValue=2.0, nChannel=10)
  osl[i]<-sum(resOSL[,"osly"])
  heatTo(125,20,hRate=-1)
  irradiate(temp=20,tim=0.1,doseRate=1)
  heatAt(20,tim=60)
  resTL<-stimTL(20,160,hRate=1,nChannel=110)
  tl[i]<-max(resTL[,"tly"])}
xlabel<-expression("Activation T [""^""o""*""C]")
plot(actTemp,osl/tl,xlab=xlabel,
ylim=c(0,200000),ylab="Sensitivity corrected OSL/TL")
legend("topright",bty="n",legend=c("(a)"," ","Pulse annealing",
"Natural sample"))
#
```

```
inis<-storeNat   # Restore concentrations of natural sample
heatTo(20,125,hRate=1)
stimOSL(temp=125, tim=200, pValue=2.0, nChannel=10)
heatTo(125,20,hRate=-1)
irradiate(temp=20,tim=56,doseRate=1)   #56 Gy in lab
for (i in seq(nTemps)) {
  heatTo(20,actTemp[i],hRate=1)
  heatAt(actTemp[i],tim=10)
  heatTo(actTemp[i],20,hRate=-1)
  heatTo(20,125,hRate=1)
  resOSL<-stimOSL(temp=125, tim=0.1, pValue=2.0, nChannel=10)
  osl[i]<-sum(resOSL[,"osly"])
  heatTo(125,20,hRate=-1)
  irradiate(temp=20,tim=0.1,doseRate=1)
  heatAt(20,tim=60)
  resTL<-stimTL(20,160,hRate=1,nChannel=110)
  tl[i]<-max(resTL[,"tly"])}
plot(actTemp,osl/tl,xlab=xlabel,ylim=c(0,320000),
ylab="Sensitivity corrected OSL/TL")
legend("topright",bty="n",legend=c("(b)"," ",
"Irradiated sample"))
```

11.12   Simulation of the SAR Protocol in Quartz

Table 11.3 is an outline of the stages involved in the simulation of the TT-OSL dating procedure by Wang et al. [188], which was simulated by Pagonis et al. [149]. Specifically steps 1–4 are a simulation of the "natural" quartz sample as proposed by Bailey [9], and the rest of the steps are a simulation of the sequence of steps in the experimental protocol of Wang et al. [188].

The following code simulates the SAR protocol using the *Pagonis2008* code in the KMS programs, as described in *Template1* and *Template4* of the paper by Peng and Pagonis [151].

Figure 11.12 shows the OSL signal from the natural sample as a + symbol, the zero dose signal as a circle o, and the sequence of regenerative doses as triangles. The repeat dose signal is shown as an × symbol.

Table 11.3 Steps in the simulation of the TT-OSL protocol of Wang et al. [188], based on the simulations of Pagonis et al. [149]

Step	Description
1	Geological dose—irradiation of 1000 Gy at 1 Gy/s
2	Geological time—heat to 350 °C
3	Illuminate for 100s at 200 °C
4	Burial dose—200 Gy at 220 °C at 0.01 Gy/s
5	Regenerative dose D_i at 20 °C and at 1 Gy/s
6	Preheat to 260 °C for 10 s
7	Blue stimulation at 125 °C for 270 s
8	Preheat to 260 °C for 10 s
9	Blue stimulation at 125 °C for 90 s (L_{TT-OSL})
10	Test Dose = 7.8 Gy
11	Preheat to 220 °C for 20 s
12	Blue stimulation at 125°C for 90 s (T_{TT-OSL})
13	Anneal to 300 °C for 10 s
14	Blue stimulation at 125 °C for 90 s
15	Preheating at 260 °C for 10 s
16	Blue stimulation at 125 °C for 90 s (L_{BT-OSL})
17	Test Dose = 7.8 Gy
18	Preheat to 220 °C for 20 s
19	Blue stimulation at 125 °C for 90 s (T_{BT-OSL})
20	Repeat 1–19 for different regenerative doses $D_i = 0$–4000 Gy in step 5

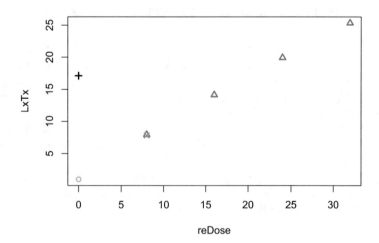

Fig. 11.12 Simulation of a sequence of thermal/optical events using the *Pagonis_Model* code in the KMS program (Peng and Pagonis [151])

Code 11.9: SAR protocol using the Pagonis model in KMS

```
### TEMPLATE1 in Peng and Pagonis (2016)
# SAR protocol using the Pagonis model in KMS
library(deSolve)
source("Pagonis_Model.R")
setPars()
setInis()
irradiate(temp=20,tim=50000/3.17e-11,doseRate=3.17e-11)
heatAt(temp=20,tim=60)
stimOSL(temp=20,tim=6000,pValue=4.73e16,nChannel=6000)
nCycle<-2
for (i in seq(nCycle)) {
irradiate(temp=20,tim=10/3.17e-11,doseRate=3.17e-11)
heatAt(temp=20,tim=60)
stimOSL(temp=20,tim=6000,pValue=4.73e16,nChannel=6000)
} #end for.
irradiate(temp=20,tim=20/3.17e-11,doseRate=3.17e-11)
heatAt(temp=20,tim=60)
### TEMPLATE4 in Peng and Pagonis (2016)
reDose<-c(1e-13,8,16,24,32,1e-13,8)
LxTx<-sLxTx<-vector(length=7)
for (i in seq(7)) {
irradiate(temp=20,tim=reDose[i]/0.1,doseRate=0.1)
heatAt(temp=20,tim=60)
heatTo(temp1=20,temp2=260,hRate=5)
heatAt(temp=260,tim=10)
heatTo(temp1=260,temp2=125,hRate=-5)
Lxdat<-stimOSL(temp=125,tim=100,pValue=4.73e16,nChannel=1000)
heatTo(temp1=125,temp2=20,hRate=-5)
irradiate(temp=20,tim=1/0.1,doseRate=0.1)
heatAt(temp=20,tim=60)
heatTo(temp1=20,temp2=220,hRate=5)
heatAt(temp=220,tim=10)
heatTo(temp1=220,temp2=125,hRate=-5)
Txdat<-stimOSL(temp=125,tim=100,pValue=4.73e16,nChannel=1000)
heatTo(temp1=125,temp2=20,hRate=-5)
LxTx[i]<-sum(Lxdat[1:5,"osly"])/sum(Txdat[1:5,"osly"])
} #end for.
plot(reDose,LxTx,pch=c(3,2,2,2,2,1,4),col=c(1,rep(2,4),3,4),
lwd=2)
```

11.13   Using the R package *RlumModel* to Simulate Quartz Luminescence Phenomena

The package *RLumModel* was developed by Friedrich et al. [48]. The various functions in this package are applicable to *any* of the models available in the package: *Bailey2001, Bailey2002, Bailey2004, Pagonis2007, Pagonis2008, Friedrich2017,* and *customized* [8–10, 49, 128, 149].

After the users choose the model for the simulation, they create sequences of events that can be thermal or optical treatments of the sample, or irradiation and relaxation stages. For all sequences, temperature differences between sequence steps are automatically simulated by using a cooling step in between. Also, after irradiating the sample, it is automatically kept at the irradiation temperature for further 5 s to allow the system to relax prior to the next step.

In the example shown in Fig. 11.13, we simulate a typical TL measurement in which the sample is given a dose of 10 Gy at a temperature of 20 °C and a dose rate of 1 Gy/s, followed by a TL measurement from 20 °C to 400 °C with a heating rate of 5 °C/s. The model is selected with the parameter *model = "Pagonis2008"*. Running this sequence with the function *model_LuminescenceSignals* produces a model output that is stored in the parameter *model.output*.

Code 11.10:   Sequence of thermal/optical events

```
# Sequence of thermal/optical events (RLumModel)
rm(list=ls())
suppressMessages(library(package = "RLumModel"))
sequence <- list(
  IRR = c(temp = 20, dose = 10, dose_rate = 1),
  TL = c(temp_begin = 20, temp_end = 400, heating_rate = 1))
model.output <- model_LuminescenceSignals(
model = "Pagonis2008",   sequence = sequence, main=" ",
verbose = FALSE)
```

The following code simulates the dose response of the TL signal in the laboratory for doses in the range 2–8 Gy. The doses are set up using the vector *Lab.dose*, and the *lapply* command is used to evaluate the TL glow curve at each dose, by using the function *model_LuminescenceSignals*. The results of the simulation are stored in the parameter *model.output*, and they are merged using the function *merge_RLum* in Fig. 11.14. The function *plot_RLum* is called to plot the merged TL glow curves.

Code 11.11: Dose response of TL (RlumModel package)

```
#Dose response of TL (RlumModel package)
rm(list=ls())
suppressMessages(library(package = "RLumModel"))
##set list with laboratory doses
Lab.dose <- seq(from = 2, to = 8, by = 2)
model.output <- lapply(Lab.dose, function(x){
  sequence <- list(
    IRR = c(20, x, 0.1),
    TL = c(20, 400, 1))
  TL_data <- model_LuminescenceSignals(
    sequence = sequence,
    model = "Bailey2001",
    plot = FALSE,
    verbose = FALSE)
  return(Luminescence::get_RLum(TL_data, recordType = "TL$",
drop = FALSE))
})
model.output.merged <- merge_RLum(model.output)
plot_RLum(
  object = model.output.merged,
  xlab = "Temperature [\u00B0C]",
  ylab = "TL signal [a.u.]",main=" ",lwd=3,lty=1:4,
  legend.text = paste(Lab.dose, " Gy"),
  combine = TRUE)
legend("top",bty="n",legend=c("TL dose response","RLumModel"))
```

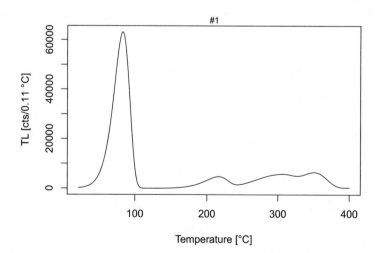

Fig. 11.13 Simulation of a sequence of thermal/optical events using the package *RLumModel*

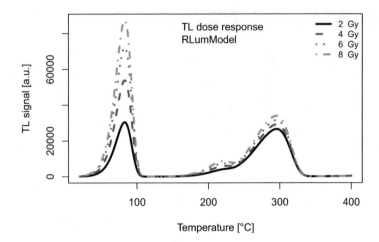

Fig. 11.14 Simulation of the dose response of the TL signal in the laboratory for doses in the range 2–8 Gy for a quartz sample, using the package *RLumModel*

In the following code, we simulate the effect of the burial temperature on the TL of a natural quartz sample, with no additional treatments in the laboratory. The structure of this code is very similar to the previous one, and the results are shown in Fig. 11.15. As the burial temperature is increased, the TL signal due to the shallower traps decreases, due to the lower thermal stability of these traps.

For a detailed simulation of the effect of burial temperatures on the estimated natural dose using the SAR protocol for quartz samples, see the study by Koul et al. [84].

For more examples of using the package *RLumModel*, the reader is directed to the paper by Friedrich et al. [48] and also the study of radiofluorescence (RF) signals in Friedrich et al. [49].

Code 11.12:   Effect of burial temperature on TL of quartz (RlumModel)

```
#Effect of burial temperature on TL of quartz (RlumModel)
rm(list=ls())
suppressMessages(library(package = "RLumModel"))
##set list with burial temperatures
burial.temp <- seq(from = 0, to = 30, by = 10)
model.output <- lapply(burial.temp, function(x){
  sequence <- list(
```

```
    IRR = c(x, 20, 1e-11),
    TL = c(20, 400, 1))
  TL_data <- model_LuminescenceSignals(
    sequence = sequence,
    model = "Bailey2001",
    plot = FALSE,
    verbose = FALSE)
  return(Luminescence::get_RLum(TL_data, recordType = "TL$",
drop = FALSE))
})
model.output.merged <- merge_RLum(model.output)
plot_RLum(
  object = model.output.merged,
  xlab = "Temperature [\u00B0C]",
  ylab = "TL signal [a.u.]",main=" ",lwd=3,lty=1:4,
  legend.text = paste(burial.temp, "\u00B0C"),ylim=c(0,70000),
  combine = TRUE)
legend("top",bty="n",legend=c("Burial Temperature","RLumModel"))
```

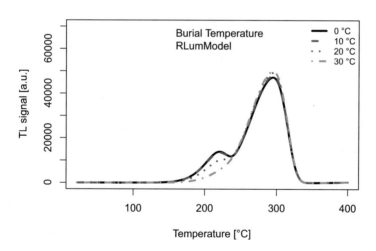

Fig. 11.15   Simulation of the effect of the burial temperature $T = 0–30\,°C$ on the TL of a natural quartz sample. For a detailed study, see Koul et al. [84]

Chapter 12
Comprehensive Models for Feldspars

12.1 Introduction

In this chapter we will look at comprehensive models that apply to materials exhibiting quantum tunneling phenomena. Although we will look at models that have been used specifically for feldspars, these models also apply to many materials used for luminescence dosimetry. Note that this chapter is essentially a continuation and generalization of Chap. 6, where we looked at four different models:

- ground state tunneling (GST) model,
- irradiation GST (IGST) model,
- excited state tunneling (EST) model, and
- thermally assisted tunneling (TA-EST) model.

These four models were shown schematically previously in Fig. 6.2.

We will summarize the mathematical description of these four models, point out their similarities and differences, and will present appropriate R functions to simulate a wide variety of processes in feldspars.

Section 12.2 contains simulations of ground state tunneling phenomena within the GST model, for both natural and laboratory irradiated samples. This is followed in Sect. 12.3 by simulations of simultaneous irradiation and ground state tunneling, using the IGST model.

Section 12.4 presents the mathematical description for quantum tunneling taking place from the excited state of the electron trap and is followed in Sects. 12.5 and 12.6, by simulations of CW-IRSL and TL signals from freshly irradiated samples.

Section 12.7 simulates a variety of multiple stage experiments, involving thermal and optical treatments of samples in the laboratory. Finally, Sect. 12.8 presents examples for the recent TA-EST model for low temperature thermochronology studies.

Table 12.1 The various FSF functions used in this chapter. The first column indicates the model for which each function can be used

Model	The FSF Functions
GST	*AFfortimeT(tim)*
	Sets parameter *distr* at the end of the anomalous fading period *tim* (in s)
IGST	*irradfortimeT(tirr)*
	Sets parameter *distr* at the end of the irradiation time *timCW* (in s)
EST	*CWfortimeT(timCW)*
	Sets parameter *distr* at the end of the IR stimulation time *timCW* (in s)
EST	*CWsignal(timCW)*
	Evaluates and returns the CW-IRSL signal
EST	*stimIRSL()*
	Calls *CWfortimeT* and *CWsignal*; sets *distr* and returns the CW-IRSL signal
EST	*heatTo(Tph)*
	Sets parameter *distr* at the end of preheating to temperature Tph (in °C)
EST	*heatAt(Tph,tph)*
	Sets parameter *distr* at the end of preheating for time *tph* (in s)
	at a temperature *Tph* (in °C)
EST	*TLsignal(temp)*
	Evaluates and returns the TL signal at temperatures *temp* (in °C)
EST	*stimTL()*
	Calls functions *heatTo* and *TLsignal* and sets *distr* and returns the TL signal
TA-EST	*irradandThermalfortimeT(tirr)*
	Sets parameter *distr* at the end of irradiation time *tirr* (in s), for a fixed
	sample temperature (*Tirr*)
TA-EST	*irradattemp(Tirr)*
	Sets parameter *distr* at the end of irradiation temperature *Tirr* (in °C), for
	a fixed irradiation time (*tirr*)

Table 12.1 shows the various feldspar simulation functions (FSFs) and summarizes their purpose. We will discuss each one of these functions in the subsequent sections.

12.2   Ground State Tunneling: The Anomalous Fading Effect

In the GŞT model of Fig. 6.2a, the tunneling mechanism takes place in a random distribution of holes and electrons in the crystal, and transitions take place directly from the ground state of the system. In this model, the distribution of electrons in the ground state $n(r', t)$ varies both with the distance parameter r' and with the elapsed time t. As we saw in Eq. (6.11) of Chap. 6, the total concentration of trapped electrons at time t is evaluated numerically by integrating $n(r', t)$ over all possible values of the variable r' (see Huntley [59])

$$n(t) = \int_0^\infty n\left(r',t\right) dr' = N \int_0^\infty g(r') \exp\left\{-s \exp\left[-\left(\rho'\right)^{-1/3} r'\right] t\right\} dr' \quad (12.1)$$

$$n(t) = N \int_0^\infty g(r') \exp\left\{-s_{eff}(r')\, t\right\} dr' \quad (12.2)$$

where $g(r') = 3\left(r'\right)^2 \exp\left[\left(r'\right)^3\right]$ is the nearest neighbor distribution of distances, and we have defined an effective frequency factor $s_{eff}(r')$ by

$$s_{eff}(r') = s \exp\left[-\left(\rho'\right)^{-1/3} r'\right] \quad (12.3)$$

Equation (12.2) can be interpreted as the sum of several decaying exponential functions $A \exp(-\lambda t)$, with each of these exponentials having a different amplitude given by $g(r')$, and with a different effective decay constant $\lambda = s_{eff}(r')$. In the code that follows, we will calculate the integral in Eq. (12.1) and in other similar equations, by using a *summation* over the different r' values, instead of a formal integration. For all practical purposes, the summation can be carried out up to a maximum distance parameter $r' = 2.2$, which is the upper limit of the extent of the function $g(r')$. For an implementation of this method using a Monte Carlo method, see the detailed study by Pagonis et al. [131]. In this method, one is evaluating the function $n(t)$ in Eq. (12.2) by adding several exponential functions; thus,

$$n(t) = N \sum_{r'=0}^{r'=2.2} 3\left(r'\right)^2 \exp\left[\left(r'\right)^3\right] \exp\left[-s_{eff}\left(r'\right) t\right] \Delta r' \quad (12.4)$$

The parameter $\Delta r'$ is an appropriate small interval, with a typical value of $\Delta r'=0.01$.

After defining the values of s and ρ', the fading times for the simulation are defined with the parameter $times AF = 0, 10^2, 10^4, 10^6$ in years after the start of the tunneling process. The value of $times AF = 0$ corresponds to the nearly symmetric unfaded distribution of distances r'. The vector $rprimes$ contains the values of r' from $r' = 0$ to a maximum of $r' = 2.2$, in steps of $dr' = 0.002$. This line of code also defines how many exponentials we are adding in order to calculate $n(t)$. For example, in the case of $r' = 0$ to $r' = 2.2$ in steps of $\Delta r' = 0.002$, we are adding a total of 1,100 exponential curves.

The first FSF function *AFfortimeT(tim)* in Table 11.1 calculates the distributions $n(r',t)$ indicated in Eq. (12.1). The line *sapply(times,AFfortimeT)* performs the summation over the distances r' indicated in Eq. (12.4). The advantage of the command *sapply* is that it automatically creates a matrix labeled *distribs*, which contains four columns corresponding to the four fading times. This simplifies and speeds up significantly the calculation of the total remaining charge $n(t)$ in

Eq. (12.4), by taking advantage of the vectorization of the R code. The command *matplot* is used to plot the four distributions of the distances r' at the fading times $times = 0, 10^2, 10^4, 10^6$ (years).

Figure 12.1 shows the distributions of distances r' obtained using the following code, and at a times $t = 0, 10^2, 10^4, 10^6$ years after the start of the tunneling process. The solid line indicates the initial unfaded peak-shaped symmetric distribution at time $t = 0$. The values of the parameters used in Fig. 12.1 are typical for ground state tunneling in feldspars, $\rho' = 1 \times 10^{-6}$ and $s = 3 \times 10^{15}$ (s^{-1}). The sharply rising dashed lines in Fig. 12.1 represent the "moving tunneling fronts" in the tunneling process, previously discussed in Chap. 6 (Pagonis and Kitis [134]).

Code 12.1: The nearest neighbor distribution at geological times

```
# The nearest neighbor distribution at geological times
rm(list=ls())
s<-3e15                    # frequency factor
rho<-1e-6                  # rho-prime values 0.005-0.02
rc<-0.0                    # for freshly irradiated samples, rc=0
timesAF<-3.154e7*c(0,1e2,1e4,1e6)         # times in seconds
rprimes<-seq(from=rc,to=2.2,by=0.002)     # rprime=0-2.2

##### function to find distribution of distances ###
AFfortimeT<-function(tim){3*(rprimes**2.0)*exp(-(rprimes**
   3.0))*
     exp(-exp(-(rho**(-1/3))*rprimes)*s*tim)}
######

distribs<-sapply(timesAF,AFfortimeT)
# Plots
cols=c(NA,NA,NA,"black","red","green","blue")
matplot(rprimes,distribs,xlab="Dimensionless distance r'",
ylab="Nearest neighbor Distribution g(r')",type="l",lwd=4)
legend("topright",bty="n", lty=c(NA,NA,NA,1,2,3,4), lwd=4,
col=cols,legend = c("Elapsed", "time"  ," ","t=0 s",
  expression("10"^"2"*" years"),
  expression("10"^"4"*" years"),expression("10"^"6"*" years")))
```

The phenomenon of anomalous fading (AF) was discussed in Chap. 6. We can simulate the AF process both in nature over a long geological period, as well as in the laboratory over a much shorter time period, by changing the argument *timAF* of the function, from a geological time period 0–10^4 years to a laboratory time of 0–10 days. The command *colSums* is used to calculate the sum over all distances r' indicated in Eq. (12.4), by adding the columns of the matrix *distribs*. The result of *colSums* is multiplied by the distance interval $\Delta r' = 0.002$, as defined by the parameters *rprimes* and *dr* in the code. The result of the code is shown in Fig. 12.2a, which shows that after 10^4 years, approximately 77% of the trapped

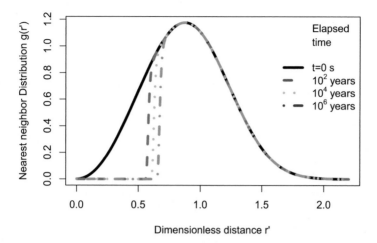

Fig. 12.1 Examples of the nearest neighbor distribution at different times $t = 0, 10^2, 10^4, 10^6$ years. The solid black line represents the *unfaded* nearly symmetric distribution at time $t = 0$. As time increases, the "tunneling front" is the almost vertical line that moves to the right, as more and more electrons are recombining at different distances r'. See also the discussion in Chap. 6

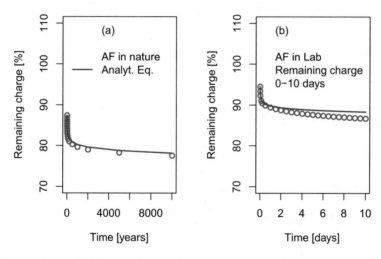

Fig. 12.2 (a) Simulation of long term anomalous fading in nature over a time period of 10^4 years, starting with an unfaded sample. The solid line indicates the approximate analytical Eq. (12.5). (b) Short term AF in the laboratory, over a period of 10 days after the end of irradiation. The parameters in the model are typical for feldspars

electrons remain in the sample. Figure 12.2b simulates the AF process after 10 days in the laboratory, with approximately 87% of the trapped electrons remaining in the sample. If desired, the user can add a few more lines of code, in order to obtain the g-factor characterizing the AF process (see, e.g. Pagonis and Kitis [134]).

Code 12.2: Anomalous fading at geological and laboratory times

```
# Anomalous fading  over geological and laboratory times
rm(list = ls(all=T))
rho<-1e-6              # Dimensionless acceptor density
dr<-.01               #Step in dimensionless distance r'
rprimes<-seq(0,2.2,dr)    #Values of r'=0-2.2 in steps of dr
s<-3e15
seff<-s*exp(-rprimes*(rho**(-1/3.0)))  # Effective A


#################### Functions  ################

#################### Anomalous fading Functions
#### AFfortimeT
AFfortimeT<-function(timAF){distr<-distr*exp(-(seff*timAF))}
#################### End of Functions

######################## Simulations ##############
par(mfrow=c(1,2))
#### Example : Anomalous fading
# Long term fading  0--10^4 years in nature
distr<<-3*rprimes^2*exp(-rprimes^3)
timesAF<-3.154e7*c(.1,.2,.5,1,2,5,10,20,50,100,200,500,1000,
2000,5000,1e4)
n<-dr*colSums(sapply(timesAF,AFfortimeT))
plot(timesAF/(3.154e7),100*n,typ="p",lwd=2,pch=1,col="red",
    ylim=c(70,110),xlab=expression("Time [years]"),
    ylab="Remaining charge [%]")
legend("topleft",bty="n",legend=c("(a)"," ","AF in nature",
"Analyt. Eq."),lwd=2,lty=c(NA,NA,NA,1), col=c(NA,NA,NA,"blue"))
lines(timesAF/3.154e7,y=100*exp(-rho*(log(1.8*s*timesAF)**3.0)),
    lwd=2,col="blue")
### Repeat for short term fading 0--10 days in lab
distr<<-3*rprimes^2*exp(-rprimes^3)  # unfaded distribution
timesAF<-3600*24*c(1e-4,1e-3,1e-2,.1,.2,.5,seq(1,10,.5))
n<-dr*colSums(sapply(timesAF,AFfortimeT))
plot(timesAF/(3600*24),100*n,typ="p",lwd=2,pch=1,col="red",
    ylim=c(70,110),xlab=expression("Time [days]"),
    ylab="Remaining charge [%]")
legend("topleft",bty="n",legend=c("(b)"," ","AF in Lab",
 "Remaining charge","0-10 days"))
lines(timesAF/(3600*24),
y=100*exp(-rho*(log(1.8*s*timesAF)**3.0)),lwd=2,col="blue")
```

We can compare the results of these simulations with the semi-analytical version of the model described in Huntley [59]. As we saw in Chap. 6, the following approximate analytical equation expresses the concentration $n(t)$ of charge carriers in the ground state during geological time scales:

$$n(t) = n_0 \exp\left(-\rho' \, \ln[z \, s \, t]^3\right) \tag{12.5}$$

A plot of this equation is shown by the solid lines in Fig. 12.2a, b. The analytical expression in Eq. (12.5) overestimates somewhat the remaining charge $n(t)$, in agreement with the detailed study by Pagonis and Kitis [134]).

Next we give examples of using the FSF functions for the IGST model.

12.3 Simultaneous Irradiation and Quantum Tunneling (IGST Model)

In this section we present a R function for the IGST model shown in Fig. 6.2b. Li and Li [98] applied the differential equation of Huntley and Lian [61] and carried out an extensive experimental and modeling study of both laboratory irradiated and naturally irradiated feldspars.

As we saw in Eq. (6.22) of Chap. 6, Li and Li [98] investigated the simultaneous effects of irradiation and tunneling. The total concentration of trapped electrons at time t is evaluated numerically by integrating $n(r', t)$ over all possible values of the variable r':

$$n(t) = \int_0^\infty Ng(r') \frac{\dot{D}}{D_0 s_{eff}(r') + \dot{D}} \left[1 - \exp\left\{-\left[\frac{\dot{D}}{D_0} + s_{eff}(r')\right]t\right\}\right] dr' \tag{12.6}$$

where we defined the effective frequency $s_{eff}(r')$ for the IGST process as

$$s_{eff}(r') = s \exp\left[-(\rho')^{-1/3} r'\right] \tag{12.7}$$

where s (s^{-1}) is the frequency characterizing the ground state tunneling process, $\dot{D}$ (Gy/s) is the dose rate, and D_0 (Gy) is the characteristic dose of the sample. This equation is valid for samples irradiated in nature with a very slow dose rate $\dot{D}$ (Gy/s) of the order of 10^{-11} Gy/s but also for samples irradiated with much higher dose rates of about 0.1 Gy/s used in the laboratory. The parameter $N(r') = N \, g(r')$ represents the total concentration of traps corresponding to a distance parameter r', and N is the total number of traps in the sample. The rest of the parameters in this equation have the same meaning as in the previous section.

The modeling results of Li and Li [98] were expressed in terms of integral equations, which require numerical integration over the distances r' in the model.

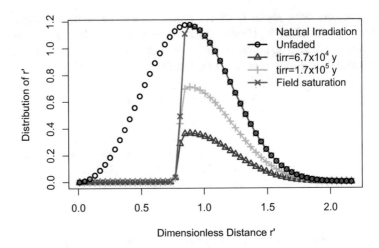

Fig. 12.3 Simulation of irradiation process in nature. As the irradiation time increases, the asymmetric distribution of distances r' approaches the field saturation distribution (×symbols). The symmetric curve indicates the initial distribution of distances, for the unfaded sample (o symbols). Compare with the results of Li and Li [98]

The R functions presented in this section circumvent the need for these numerical integrations, by replacing them with finite sums over the distances r'.

Equation (12.6) can be interpreted as the sum of several saturating exponential functions of the form $A\left(1 - \exp\left(-\lambda t\right)\right)$. Each of these exponentials has a different amplitude $A\left(r'\right) = g(r')\dot{D}/\left[D_0 s_{eff}\left(r'\right) + \dot{D}\right]$ and a different decay constant $\lambda(r') = \dot{D}/D_0 + s_{eff}\left(r'\right)$.

Typical values of the parameters used in the simulations by Li and Li [98] are $\rho' = 2 \times 10^{-6}$, $s = 3 \times 10^{15}\,\mathrm{s}^{-1}$, the natural dose rate $\dot{D} = 3\,\mathrm{Gy/ka}$, and the characteristic dose $D_0 = 538\,\mathrm{Gy}$ (Fig. 12.3).

The irradiation process is simulated by using the FSF function *irradfortimeT(tirr)* in Table 11.1, whose arguments are the irradiation time *tirr*, and the relevant input parameters in the model are ρ', s, $\dot{D}$, and D_0. The function *irradfortimeT(tirr)* evaluates the distribution of nearest neighbor distances at the end of the irradiation process. The results of the code are shown in Fig. 12.4, and they are identical to Fig. 5a in the paper by Li and Li [98]. As the irradiation time increases, Fig. 12.4 shows that the asymmetric distribution of distances r' approaches the distribution for a field saturated sample (× symbols). The symmetric curve indicates the initial distribution of distances for the unfaded sample at time $t = 0$ (o symbols).

Code 12.3: Feldspar irradiation in nature

```
## Distribution of distances for feldspar irradiation in nature
rm(list = ls(all=T))
rho<-2e-6                    # Dimensionless acceptor density
Do<-538                      #Do in Gy
yr<-365*24*3600             #year is seconds
Ddot<-3/(1e3*yr)           #Low natural dose rate = 3 Gy/kA
dr<-.04                     #Step in dimensionless distance r'
rprimes<-seq(0.01,2.2,dr)  #Values of r'=0-2.2 in steps of dr
s<-2e15                     # Frequency factors in s^-1
seff<-s*exp(-rprimes*(rho**(-1/3.0)))   # Effective s
tau<-1/seff
##################### Irradiation Function
#### irradfortimeT
irradfortimeT<-function(tirr){distr<<-unFaded*(Ddot*tau/
(Do+Ddot*tau))*  (1-exp(-(Do+Ddot*tau)/(Do*tau)*tirr))}
# function calculates new distribution at end of irradiation
#################### End of Functions
unFaded<-3*rprimes^2*exp(-rprimes^3)   # Unfaded sample
irrTimes<-c(6.67e4,1.67e5,1e6)*yr
distribs<-sapply(irrTimes,irradfortimeT)
########## plot distibutions
matplot(rprimes,distribs,xlab="Dimensionless Distance r'",
typ="o",lty="solid", ylab="Distribution of r'",pch=c(2,3,4),
lwd=2,col=c("blue","green","red"))
lines(rprimes,unFaded,col="black",pch=1,typ="p",lwd=2)
legend("topright",bty="n",legend=c(
  expression("Natural Irradiation",
"Unfaded","tirr=6.7x10"^4*" y", "tirr=1.7x10"^5*" y",
"Field saturation")),  pch=c(NA,1,2,3,4),
col=c(NA,"black","blue","green","red"),lwd=2)
```

Figures 12.4 and 12.5 show the result of simulating the irradiation process and the measurements of the dose response curves, in nature and in the laboratory, respectively. The command *colSums(distribs)* is used to produce the dose response plots. In the simulations of Fig. 12.5, the laboratory irradiations are of course carried out with a much higher dose rate of 0.1 Gy/s and for much shorter irradiation times $tirr = 1 - 10^6$ s. As the irradiation time increases in both Figs. 12.4 and 12.5, the asymmetric distribution of distances r' and the trap filling ratio $n(t)/N$ approach the saturation distribution for a field saturated sample.

Code 12.4: Feldspar irradiation in nature—dose response

```
## Feldspar irradiation in nature  - Dose response
rm(list = ls(all=T))
rho<-2e-6                      # Dimensionless acceptor density
s<-3e15                        # Frequency factors in s^-1
Do<-538                        #Do in Gy
yr<-365*24*3600                #year is seconds
Ddot<-2.85/(1e3*yr)            #Low natural dose rate = 2.85 Gy/Ka
dr<-.05                        #Step in dimensionless distance r'
rprimes<-seq(0.01,2.2,dr) #Values of r'=0-2.2 in steps of dr
seff<-s*exp(-rprimes*(rho**(-1/3.0)))  # Effective s
tau<-1/seff
##################### Irradiation Function
#### irradfortimeT
irradfortimeT<-function(tirr){distr<<-unFaded*(Ddot*tau/
  (Do+Ddot*tau))*  (1-exp(-(Do+Ddot*tau)/(Do*tau)*tirr))}
# function calculates new distribution at end of irradiation
##################### End of Function ##########
par(mfrow=c(1,2))
unFaded<-3*rprimes^2*exp(-rprimes^3)  # Unfaded sample
irrTimes<-10^seq(2,6,by=.2)*yr
distribs<-sapply(irrTimes,irradfortimeT)
########## plot distibutions
matplot(rprimes,distribs,typ="l",ylim=c(0,1.8),lty="solid",
xlab="Dimensionless Distance r'",  ylab="Distribution of r'",
lwd=2)
legend("topleft",bty="n",legend=c(expression("(a)",
"Irradiation in Nature","Distributions of r'",
"tirr=100-10"^"6"*"y")))
plot(irrTimes/yr,colSums(distribs)*dr,typ="p",lwd=2,
xlab="Time [y]",ylab="Trap filling ratio n(t)/N",ylim=c(0,1.1))
legend("topleft",bty="n",legend=c("(b)"," ",
"Trap filling n(t)/N",  "Analyt. Eq."),lwd=2,
lty=c(NA,NA,NA,1),pch=c(NA,NA,1),  col=c(NA,NA,NA,"red"))
lines(irrTimes/yr,(1-exp(-Ddot*irrTimes/Do))*exp(-rho*
(log(Do*s/Ddot)**3.0)),lwd=2,col="red")
```

We can compare the results in Figs.12.4b and 12.5b, with the analytical equation developed by Pagonis and Kitis [134]. As we saw in Chap. 6, the integral equations for the dose response curves in the model of Li and Li [98] can be replaced with the following approximate analytical equation:

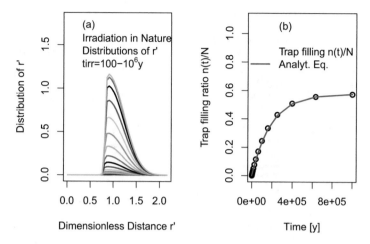

Fig. 12.4 Simulation of irradiation process in nature with a slow dose rate of 2.85 Gy/ka. (**a**) As the irradiation time increases, the asymmetric distribution of distances r' approaches the field saturation distribution. (**b**) The corresponding dose response

$$L_{FADED}(D) = N\,(1 - \exp[-D/D_0])\exp\left[-\rho' \ln\left(\frac{D_0 s}{\dot{D}}\right)^3\right] \qquad (12.8)$$

where the saturating exponential term $N\,(1 - \exp[-D/D_0])$ represents the luminescence signal that would have been obtained in the absence of any fading.

A plot of this equation is shown by the solid lines in Figs. 12.4b and 12.5b, showing excellent agreement between the analytical equation (12.8) and the R functions presented here.

12.4   Quantum Tunneling from the Excited State of the Electron Trap (EST Model)

In this section we use FSF functions for the EST model shown in Fig. 6.2c.

Pagonis et al. [131] discussed quantum tunneling taking place from the *excited* state of the electron trap, within the model of Jain et al. [62]. As we saw in Eq. (6.32) of Chap. 6, in the EST model the total concentration of trapped electrons at time t is found by integrating over all distances r':

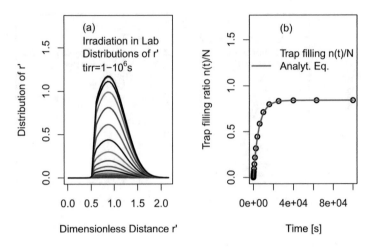

Fig. 12.5 Simulation of irradiation process in the *laboratory* with a dose rate of 0.1 Gy/s and for irradiation times $tirr = 1 - 10^6$ s. As the irradiation time increases, both the asymmetric distribution of distances r' in (**a**), and the trap filling ratio $n(t)/N$ in (**b**) approach saturation

$$n(t) = \int_0^\infty Ng\left(r'\right) \exp\left\{ -\frac{s_{tun}}{B} \exp\left[-\left(\rho'\right)^{-1/3} r'\right] \int_0^t A(t)\, dt \right\} dr' \qquad (12.9)$$

where $A(t)$ (s^{-1}) is the rate of optical or thermal excitation from the ground state to the excited state of the trapped electron, B (s^{-1}) is the rate of retrapping from the excited to the ground state, and s_{tun} (s^{-1}) represents the tunneling frequency from the excited state into the recombination center. As in the other models in this chapter, $n(r', t)$ is the concentration of trapped electrons at time t and at a distance r', and ρ' is the constant dimensionless density of acceptors in the material.

The excitation rate $A(t)$ is different for various types of experiments (TL, CW-IRSL, etc.) and is discussed in the next two sections. Once $n(r', t)$ is evaluated using the above equations, we can evaluate the luminescence intensity $I(t)$ using the equation:

$$I(t) = \int_0^\infty n\left(r', t\right) \frac{A(t)\, s_{tun}}{B} \exp\left[-\left(\rho'\right)^{-1/3} r'\right] dr' \qquad (12.10)$$

In all these different types of integral equations, the FSF functions replace the numerical integrations with appropriate sums.

12.5 Simulation of CW-IRSL Curves of Freshly Irradiated Samples

In a CW-IRSL experiment, the rate of excitation is given by $A_{CW} = \sigma I$, where σ (cm^2) represents the optical cross section for the process, and I (photons/(cm^{2}s) is the intensity of the excitation IR source. For these types of experiments, Eq. (12.9) becomes

$$n_{CW}\left(r',t\right) = N\, g\left(r'\right)\exp\left\{-\frac{s_{tun}\sigma I}{B}\exp\left[-\left(\rho'\right)^{-1/3}r'\right]t\right\} \qquad (12.11)$$

$$n_{CW}\left(r',t\right) = 3N\left(r'\right)^2\exp\left[-\left(r'\right)^3\right]\exp\left\{-A_{eff}(r')\,t\right\} \qquad (12.12)$$

where we define an effective infrared excitation rate $A_{eff}(r')$ (s^{-1}) for the CW-IRSL process, given by

$$A_{eff}(r') = \frac{s_{tun}\sigma I}{B}\exp\left[\left(\rho'\right)^{-1/3}r'\right] \qquad (12.13)$$

In the R codes we will replace the combination of frequency factors σI, s_{tun}, B with a frequency $s = s_{tun}\sigma I/B$. This does not affect the simulations, since these three factors appear as the combination $s_{tun}\sigma I/B$ in the model and not separately. In this case, s represents an effective total frequency factor for the CW-IRSL process.

The total concentration of trapped electrons at time t is found by integrating over all distances r':

$$n(t) = \int_0^\infty n\left(r',t\right)dr' = \int_0^\infty N g\left(r'\right)\exp\left\{-A_{eff}(r')\,t\right\}dr' \qquad (12.14)$$

This equation can again be evaluated as the sum of many exponentials, with amplitude $3\left(r'\right)^2\exp\left(r'\right)^3$ and decay constant $A_{eff}(r')$. One can then evaluate $n(t)$ during the CW-IRSL experiment by adding the exponentials in this form:

$$n(t) = N\sum_{r'=0}^{r'=2.2} 3\left(r'\right)^2\exp\left(r'\right)^3\exp\left[-A_{\text{eff}}(r')\,t\right]\Delta r' \qquad (12.15)$$

The following code shows an example of using the three FSF functions *CWfortimeT*, *CWsignal*, and *stimIRSL* from Table 11.1, to simulate a CW-IRSL experiment. An example of running the code is shown in Fig. 12.6. The code loads all the FSF functions from the appropriate R file using the command *source("PagonisFSF2020.R")*.

The first function *CWfortimeT(timCW)* evaluates the distribution of trapped charge $n(r',t)$ and stores it in the global parameter *distr* by using the scoping

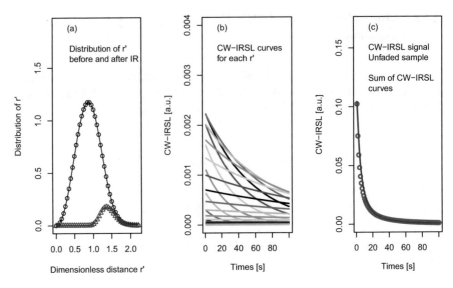

Fig. 12.6 Simulation of CW-IRSL experiment in the EST model, for freshly irradiated samples. (**a**) The distributions of trapped charge $n(r', t)$ at the beginning and at the end of the CW-IRSL experiment. (**b**) The CW-IRSL curves evaluated for each distance r'. (**c**) The sum of the curves shown in (**b**) yields the total CW-IRSL signal. The solid line in (**c**) is the analytical KP-CW equation developed by Kitis and Pagonis [73]. The parameters in the model are typical for feldspars

assignment «-. The second function *CWsignal(timCW)* evaluates the CW-IRSL signal based on Eq. (12.10). The third function *stimIRSL()* calls the other two functions and has a dual purpose: (a) it sets the distribution of trapped charge $n(r', t)$ using the first function *CWfortimeT* and (b) it returns the CW-IRSL intensity as the parameter *signal*.

The code evaluates and plots the unfaded distribution of distances, and the parameter *timesCW* defines the duration of the CW-IRSL simulation (1–100 s). The function *stimIRSL()*, which sets the new distribution of charges $n(r', t)$ at the end of the CW-IRSL experiment, also returns the CW-IRSL intensity as the parameter *signal*. The signal is stored in the parameter *IRsignal* for plotting purposes. The code also plots the *new* distribution of charges $n(r', t)$ at the end of the CW-IRSL experiment, plots all the individual exponential decay curves corresponding to different values of r', and finally plots the sum of the curves.

The solid line in Fig. 12.6c is the analytical KP-CW equation developed by Kitis and Pagonis [73]; this Eq. (6.43) was discussed previously in Sect. 6.9.

Code 12.5: Simulation of CW-IRSL signal from freshly irradiated feldspars

```
## CW-IRSL function for freshly irradiated feldspars (EST)
rm(list = ls(all=T))
source("Pagonis2020FSF.R")# Loads the functions
rho<-.013                  # Dimensionless acceptor density
dr<-.05                    # Step in dimensionless distance r'
rprimes<-seq(0,2.2,dr)     # Values of r'=0-2.2 in steps of dr'
A<-3                       # A=stun*sigma*I/B
Aeff<-A*exp(-rprimes*(rho**(-1/3.0)))   # Effective A
######################### Simulations ###############
par(mfrow=c(1,3))
#### Example : IRSL for unfaded sample
distr<<-3*rprimes^2*exp(-rprimes^3)   # unfaded distribution
distr1<-distr
timesCW<-seq(1,100)
IRsignal1<-stimIRSL()
distr2<-distr
########## End of Simulations, next Plot the results #######

plot(rprimes,distr1,ylim=c(0,1.9),typ="o",pch=1,col="black",
     ylab="Distribution of r'",xlab="Dimensionless distance r'")
lines(rprimes,distr2,ylim=c(0,1.9),typ="o",pch=2,col="blue",
     ylab="Distribution of r'",xlab="Dimensionless distance r'")
legend("topleft",bty="n",legend=c("(a)"," ",
" Distribution of r'", " before and after IR"))
matplot(timesCW,t(sapply(timesCW,CWsignal)),typ="l",lty="solid",
ylim=c(0,.004), lwd=2, xlab=expression("Times [s]"),
ylab="CW-IRSL [a.u.]")
legend("topleft",bty="n",legend=c("(b)","     ","CW-IRSL curves",
    "for each r'"))
plot(timesCW,IRsignal1,typ="p",lwd=2,pch=1,col="red",
ylim=c(0,.17),xlab=expression("Times [s]"),
ylab="CW-IRSL [a.u.]")
legend("topleft",bty="n",legend=c("(c)"," ","CW-IRSL signal",
   "Unfaded sample"," ","Sum of CW-IRSL","curves"))
lines(timesCW,3*rho*A*1.8*exp(-rho*(log(1+1.8*A*timesCW))**3.0)*
(log(1 +1.8* A*timesCW) ** 2.0)/(1 + 1.8*A* timesCW),lwd=2,
col="blue")
```

12.6 Simulation of TL Glow Curves in Freshly Irradiated Samples

In a TL experiment, the rate of excitation is $A_{TL}(t) = s_{th} \exp[-E/(k_B T)]$, where s_{th}, E are the thermal parameters of the trap and T is the temperature of the sample. For a TL experiment, we saw in Chap. 6 that the total concentration of trapped electrons is

$$n_{TL}\left(r',t\right) = N\, g\left(r'\right) \exp\left\{ -\frac{s_{tun} s_{th}}{B} \exp\left[-\left(\rho'\right)^{-1/3} r'\right] \int_0^t \exp\left[-E/(k_B T)\right] dt \right\}$$

(12.16)

$$n_{TL}\left(r',t\right) = N\, g\left(r'\right) \exp\left\{ -s_{\text{eff}}(r') \int_0^t \exp\left[-E/(k_B T)\right] dt \right\}$$

(12.17)

where we define an effective frequency factor $s_{\text{eff}}(r')$ (s^{-1}) for the TL process, given by

$$s_{\text{eff}}(r') = \frac{s_{tun} s_{th}}{B} \exp\left[-\left(\rho'\right)^{-1/3} r'\right]$$

(12.18)

Pagonis et al. [131] noted that the TL signal from feldspars can be calculated as the sum of several partial first order TL glow curves, with each of these partial TL glow curves corresponding to a different distance r'. Just as in the case of the CW-IRSL curves described in the previous section, the amplitude of these TL glow curves is proportional to the nearest neighbor distribution $3\left(r'\right)^2 \exp\left(r'\right)^3$, and the respective effective frequency constant $s_{eff}\left(r'\right)$ is given by Eq. (12.18).

Just as in the previous section, we will replace the combination of frequency factors s_{th}, s_{tun}, B in the R code with a frequency $s = s_{th} s_{tun}/B$. This will not affect the simulations, since these three frequencies appear as the combination $s_{th} s_{tun}/B$ in the model; in this case, s represents a frequency factor for the TL process in this model.

In the code that follows, we perform a summation over the different r' values, instead of a formal integration. By contrast to the CW-IRSL code in the previous section, instead of adding many exponentials, we are now adding many first order TL peaks. For an implementation of this method using a Monte Carlo method, see Figure 7 in Pagonis et al. [131]. In the R code, the first order glow peaks will be added; thus,

$$I(t) = N \sum_{r'=0}^{r'=2.2} 3\left(r'\right)^2 \exp\left[\left(r'\right)^3\right] s_{eff}\left(r'\right) e^{-E/kT} \exp\left[-s_{eff}\left(r'\right) \int_0^t e^{-E/kT} dt'\right]$$

(12.19)

The integral appearing in this equation will be approximated using the following well-known expression for the exponential integral when using a linear heating rate β (see, e.g. the book by Chen and Pagonis [38]):

$$\int_0^t e^{-E/kT} dt' = \frac{k_B T^2}{\beta E} \left[\exp\left(-\frac{E}{k_B T}\right) \left(1 - \frac{2k_B T}{E}\right) \right] \tag{12.20}$$

The following code provides an example of using the three FSF functions *heatTo*, *TLsignal*, and *stimIRSL* from Table 11.1, to simulate a TL experiment, and an example of running the code is shown in Fig. 12.7. The function *TLsignal(temp)* evaluates the TL signal based on Eq. (12.10). The function *stimTL()* calls the other two functions *heatTo(Tph)* and *TLsignal(temp)* and has a dual purpose: (a) it sets the distribution of trapped charge $n(r', t)$ using the first function *heatTo(Tph)* and (b) it also returns the TL intensity as the parameter *signal*. The two general functions, *heatAt* and *heatTo*, simply evaluate and store the new distribution of distances r' during the heating experimental stage. The new distribution is updated each time by setting the global parameter *distr*.

The model parameters in these simulations are ρ', E, s, β, D_0, $\dot{D}$, T_{PH}, and t_{PH}.

Figure 12.7 shows a simulation of heating a sample up to 380 °C, just below the high temperature end of the TL glow curve. Figure 12.7a shows the distributions of distances r' before and after heating the sample. The circles indicate the initial unfaded distribution before heating, and the triangles indicate the corresponding distribution after heating to 380 °C. As may be expected, after heating to 380 °C, only a few distant electrons remain trapped in the sample, corresponding to large values of $r' > 1.3$.

Figure 12.7b shows the partial TL glow curves, which are summed to produce the glow curve in Fig. 12.7c.

12.7 TL Signals from Thermally and Optically Treated Samples

The following R code provides several examples of applying the FSF functions for more complex protocols, which are frequently used in the laboratory. The parameters in the model are typical for feldspars. The curves 1–4 in Fig. 12.8 correspond to the following situations:

1. TL for freshly irradiated sample (unfaded distribution of distances r');
2. sample freshly irradiated, heated to a preheat temperature $Tph = 320$ °C, then TL measurement;
3. sample freshly irradiated, then preheated for time $tph = 30$ s at a preheat temperature $Tph = 320$ °C, then TL measurement;

Code 12.6: Simulation of TL signal from freshly irradiated feldspars

```
## Simple TL function for freshly irradiated feldspars
rm(list = ls(all=T))
source("Pagonis2020FSF.R") # Loads the FSF functions
rho<-.013                    # Dimensionless acceptor density
s<-3.5e12                    # Frequency factors in s^-1
E<-1.45                      # Energy in eV
Tph<-300                     #Preheat temperature (C)
tph<-10                      #Preheat time (s)
dr<-.1                       #Step in dimensionless distance r'
rprimes<-seq(0.01,2.2,dr)    #Values of r'=0-2.2 in steps of dr
kb<-8.617e-5                 #Boltzmann constant
beta<-1
Tpreheat<-273+320            # Preheat Temperature
seff<-s*exp(-rprimes*(rho**(-1/3.0)))  # Effective s
######################### Simulations ###############
par(mfrow=c(1,3))
# #### Example #1: TL for unfaded sample
distr<<-3*rprimes^2*exp(-rprimes^3)
distr1<-distr
temps<-273+seq(1:400)        # Temperatures for TL
manyTL<-t(sapply(temps,TLsignal))
TL1<-stimTL()                # Evaluate TL signal
distr2<-distr

########## End of Simulations, next Plot the results #######
plot(rprimes,distr1,ylim=c(0,1.9),typ="o",pch=1,col="black",
     ylab="Distribution of r'",xlab="Dimensionless distance r'")
lines(rprimes,distr2,typ="o",pch=2,col="red",
     ylab="Distribution of r'",xlab="Dimensionless distance r'")
legend("topleft",bty="n",legend=c(" (a)"," ",
"   Distribution of r'",  "  before and after", "TL"))
matplot(temps-273,manyTL,typ="l",lty="solid",
xlim=c(175,450),ylim=c(0,.03),lwd=2,
xlab=expression("Temperature ["^"o"*"C]"), ylab="TL [a.u.]")
legend("topleft",bty="n",legend=c(" (b)","     ","TL curves ",
  "for each r'"))
plot(temps-273,TL1,typ="l", lwd=2,pch=1,col="black",
xlim=c(175,450),ylim=c(0,.018),
xlab=expression("Temperature ["^"o"*"C]"),ylab="TL [a.u.]")
legend("topleft",bty="n",legend=c(" (c)"," ","Sum of all",
  "TL curves "))
```

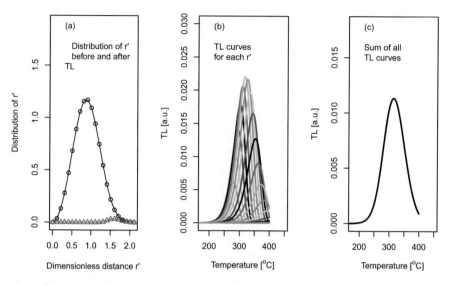

Fig. 12.7 Simulation of TL glow curve for freshly irradiated samples by heating up to 380 °C, just below the high temperature end of the TL glow curve. (**a**) The distributions of distances r' before and after heating the sample are shown as circles and triangles, respectively. (**b**) The partial first order TL glow curves. (**c**) The sum of the glow curves from (**b**)

4. sample freshly irradiated, then CW-IRSL excitation for 10 s, then TL measurement.

Figure 12.8a shows the distributions of distances before measurement of the TL glow curve in these four examples, while Fig. 12.8b shows the corresponding TL glow curves for each simulation.

Figure 12.8 demonstrates a general property of the TL glow curves in pretreated feldspar samples; the shape and location of the TL glow curves along the temperature axis reflect the underlying similar behavior of the distributions of distances r'. However, note that the horizontal axis in Fig. 12.8a (distance r') is physically very different from the horizontal axis in Fig. 12.8b (sample temperature). This similarity in the shapes of the TL glow curves and the distribution of distances r' is discussed in detail in Pagonis and Brown [123].

Code 12.7: TL from thermally/optically pretreated feldspar samples

```
## Examples of TL for thermally and optically treated samples
rm(list = ls(all=T))
source("Pagonis2020FSF.R")# Loads the FSF functions
rho<-.013                      # Dimensionless acceptor density
Po<-s<-3.5e12                  # Frequency factors in s^-1
E<-1.45                        # Energy in eV
Tph<-300                       #Preheat temperature (C)
tph<-10                        #Preheat time (s)
dr<-.05                        #Step in dimensionless distance r'
rprimes<-seq(0.01,2.2,dr)     #Values of r'=0-2.2 in steps of dr
kb<-8.617e-5                   #Boltzmann constant
beta<-1
seff<-s*exp(-rprimes*(rho**(-1/3.0)))  # Effective s
A<-5
Aeff<-A*exp(-rprimes*(rho**(-1/3.0)))  # Effective A
######################### Simulations ##############
par(mfrow=c(1,2))
    ## Example #1: TL of unfaded sample
distr<<-3*rprimes^2*exp(-rprimes^3)  # unfaded distribution
distr1<-distr                  # Store distr for plotting later
temps<-273+seq(1:500)          # Temperatures for TL
TL1<-stimTL()                  # Evaluate TL signal
    ## Example #2: Heat to temperature Tpreheat, then measure TL
distr<<-3*rprimes^2*exp(-rprimes^3)  # unfaded distribution
Tpreheat<-273+320              # Preheat Temperature
heatTo(Tpreheat)               # Heat to 320 degC
distr2<-distr                  # Store distr for plotting
TL2<-stimTL()                  # Store TL
## Example #3: Heat for 30 s at Tpreheat=320C, then measure TL
distr<<-3*rprimes^2*exp(-rprimes^3)
heatTo(Tpreheat)               # Heat to 320 degC
heatAt(Tpreheat,30)            # Heat for 30 s at 320C
distr3<-distr
TL3<-stimTL()
    ## Example #4: CW-IRSL excitation for 10 s, then measure TL
distr<<-3*rprimes^2*exp(-rprimes^3)  # unfaded distribution
timesCW<-seq(1,10)
distr4<-CWfortimeT(max(timesCW))
TL4<-stimTL()
########## End of Simulations, next Plot the results #######
plot(rprimes,distr1,ylim=c(0,1.9),typ="l",pch=1,col="black",
lwd=2,ylab="Distribution of r'",
xlab="Dimensionless distance r'")
legend("topleft",bty="n",legend=c("(a)","Various Examples",
 "of r' Distributions"))
text1 <- data.frame(x=c(.5,.8,1.,1.25),
 y=c(1,.7,.55,.3), labels=c("1","2","3","4"))
```

```
text(text1)
lines(rprimes,distr2,typ="l",col="red",lwd=2)
lines(rprimes,distr3,typ="l",col="blue",lwd=2)
lines(rprimes,distr4,typ="l",col="magenta",lwd=2)
plot(temps-273,TL1,xlim=c(200,450),ylim=c(0,.017),typ="l",
    col="black",lwd=2,
xlab=expression("Temperature ["^"l"*"C]"), ylab="TL [a.u.]")
lines(temps-273,TL2,typ="l",col="red",lwd=2)
lines(temps-273,TL3,typ="l",col="blue",lwd=2)
lines(temps-273,TL4,typ="l",col="magenta",lwd=2)
legend("topleft",bty="n",legend=c("(b)"," ",
 "Corresponding TL"))
text1 <- data.frame(x=c(270,310,330,350),
y=c(0.009,.006,.004,.002), labels=c("1","2","3","4"))
text(text1)
```

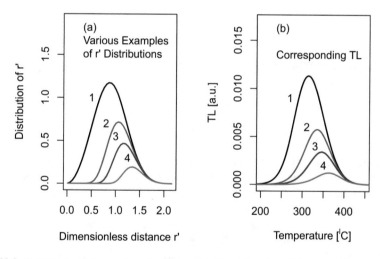

Fig. 12.8 Several examples of the simulation functions for thermally and optically treated samples. The parameters in the model are typical for feldspars. (**a**) The distribution of distances and (**b**) the corresponding TL glow curves, for the following processes: (1) TL for unfaded sample; (2) heat to a preheat temperature $T = 320\,°C$ and then measure TL; (3) heat for 30 s at $T = 320\,°C$ and then measure TL; and (4) CW-IRSL excitation for 10 s, and then measure TL

12.8   The Low Temperature Thermochronology Model by Brown et al.

In this section we present a R function for the TA-EST model shown in Fig. 6.2d. In Chap. 6 we saw that in the TA-EST model by Brown et al. [24], one investigates the simultaneous effects of irradiation and quantum tunneling on the TL glow curves. This model was developed for low temperature thermochronology; however, it is completely general and can be used for any thermally active dosimetric trap. In the TA-EST model, the total concentration of trapped electrons at time t is found from Eq. (6.53):

$$n(t) = \int_{0}^{\infty} Ng(r') \frac{\dot{D}}{D_0 P_{eff}(r') + \dot{D}} \left[1 - \exp\left\{-\left[\frac{\dot{D}}{D_0} + P_{eff}(r')\right] t\right\}\right] dr'$$

(12.21)

where $P(r')$ is the rate of excited state tunneling given by

$$P(r') = P_0 \exp\left[-(\rho')^{-1/3} r'\right]$$

(12.22)

and we define an effective tunneling probability $P_{eff}(r')$, which also depends on the sample temperature T:

$$P_{eff}(r') = \frac{P(r') s}{P(r') + s} \exp[-E/(k_B T)]$$

(12.23)

The parameters in this model are the tunneling frequency P_0 (s^{-1}), T represents the temperature of the sample, k_B is the Boltzmann constant, s (s^{-1}) is the trap frequency factor, and E (eV) is the thermal activation energy of the trap from the ground state to an (unspecified) higher energy state.

We simulate multiple irradiations in nature, by using the function *irradandThermalfortimeT*, to describe the trap filling process for various irradiation times t_{irr}, and for a fixed steady-state temperature $T_{IRR} = -4\,°C$. The result is shown in Fig. 12.9. As the irradiation time increases, the distribution of trapped electrons $n(r', t)$ at the end of the irradiation shifts toward higher distances and increases overall, eventually reaching the distribution of the field saturated sample. At large doses, more traps get filled at larger distances r', and the overall distribution shifts to higher values of r' until equilibrium is reached.

The input parameters in this TA-EST model are ρ', P_0, E, s, D_0, $\dot{D}$, T_{irr}, and t_{irr}.

The following code simulates the dose response during natural irradiation. Once more, the dose response is obtained by adding the columns of the distributions corresponding to each value of r', with the command *rowSums*. From a physical point of view, it is very interesting to compare the shape of the distributions in

Fig. 12.9 with those in Fig. 12.4, which were obtained using a much lower value of the acceptor density parameter $\rho' = 10^{-6}$ and for a thermally stable trap. In Fig. 12.9 there is no obvious critical radius r'_c similar to the one shown in Fig. 12.4.

Code 12.8: Dose response of feldspar in the TA-EST model of Brown et al.

```
#Dose response of feldspars, irradiation in nature (TA-EST)
rm(list = ls(all=T))
rho<-3e-3                    # Dimensionless acceptor density
Po<-s<-2e15                  # Frequency factors in s^-1
E<-1.3                       # Energy in eV
Do<-1600                     #Do in Gy
yr<-365*24*3600              #year is seconds
Ddot<-2.85/(1e3*yr)          #Low natural dose rate = 1 Gy/kA
dr<-.05                      #Step in dimensionless distance r'
rprimes<-seq(0.01,2.2,dr)    #Values of r'=0-2.2 in steps of dr
kb<-8.617e-5                 #Boltzmann constant
seff<-s*exp(-rprimes*(rho**(-1/3.0)))   # Effective s
Tirr<-273-4
Peff<-(1/(1/s+1/seff))*exp(-E/(kb*Tirr))

########## Irradiation functions ##########
#### irradfortimeT
irradandThermalfortimeT<-function(tirr){distr<<-distrUnfaded*
  (Ddot*rprimes/(Do*Peff+Ddot))*(1-exp(-(Ddot/Do+Peff)*tirr))}
##################### End of Functions ##########
par(mfrow=c(1,2))
distrUnfaded<-3*rprimes^2*exp(-rprimes^3)  # Unfaded sample
irrTimes<-10^seq(2,6,by=.2)*yr
distribs<-sapply(irrTimes,irradandThermalfortimeT)

########## plot distibutions
matplot(rprimes,distribs,typ="l",ylim=c(0,.45),lty="solid",
xlab="Dimensionless Distance r'",
ylab="Distribution of r'",lwd=2)
legend("topleft",bty="n",legend=c(expression("(a)",
"Irradiation in nature",
"Distributions of r'","tirr=10"^3*"-10"^6*"y")))
plot(irrTimes/yr,colSums(distribs)*dr,typ="o",lwd=2,
 xlab="Time [y]",ylab="Trap filling ratio n(t)/N",
 ylim=c(0,.27))
legend("topleft",bty="n",legend=c(" ","(b)",
 "Trap filling n(t)/N"))
```

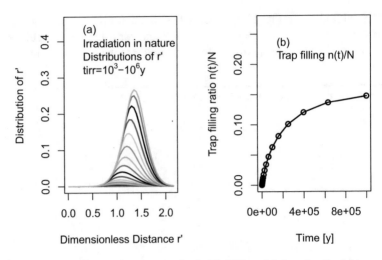

Fig. 12.9 Simulation of irradiations in nature in the TA-EST model, for a fixed burial temperature −4 °C. (**a**) The distributions of the distance parameter r' at various irradiation times. Compare the shape of these distributions with Fig.12.4. (**b**) The corresponding dose response, shown as the trap filling ratio $n(t)/N$

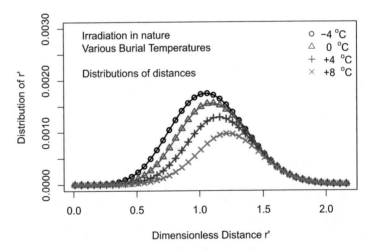

Fig. 12.10 Multiple irradiations in nature using the TA-EST model, for variable burial temperatures $T_{irr} = -4, 0, 4, 8$ °C, and for a fixed irradiation time $t_{irr} = 10^3$ y

We next simulate multiple irradiations in nature, with variable steady-state temperatures $T_{irr} = -4, 0, 4\,8$°C, and for a fixed irradiation time $t_{irr} = 10^3$ s. The result is shown in Fig. 12.10. As the burial temperature increases, the distribution of trapped electrons $n\left(r', t\right)$ at the end of the irradiation shifts toward higher distances and decreases overall. This is because more thermally assisted tunneling takes place

at elevated burial temperatures; the closest neighbors recombine first in time, and
the overall distribution shifts to higher values of r'.

Code 12.9: Irradiations at various steady-state temperatures (TA-EST model)

```
## Multiple feldspar irradiations at steady-state temperatures

rm(list = ls(all=T))
rho<-1e-2                    # Dimensionless acceptor density
Po<-s<-2e15                  # Frequency factors in s^-1
E<-1.3                       # Energy in eV
Do<-1600                     #Do in Gy
yr<-365*24*3600              #year is seconds
Ddot<-2.85/(1e3*yr)          #Low natural dose rate = 1 Gy/kA
dr<-.05                      #Step in dimensionless distance r'
rprimes<-seq(0.01,2.2,dr)    #Values of r'=0-2.2 in steps of dr
kb<-8.617e-5                 #Boltzmann constant
seff<-s*exp(-rprimes*(rho**(-1/3.0)))   # Effective s
# Peff<-(1/(1/s+1/seff))*exp(-E/(kb*Tirr))    Effective P

########## Irradiation functions ##########
#### irradatsometemp
irradatsometemp<-function(Tirr){
  Peff<-(1/(1/s+1/seff))*exp(-E/(kb*Tirr))
  distr<<-distrUnfaded*
  (Ddot*rprimes/(Do*Peff+Ddot))*(1-exp(-(Ddot/Do+Peff)*tirr))}
##################### End of Functions ##########
tirr<-1e3*yr
Tirrs<-273+c(-4,0,4,8)       # Burial temperatures
distrUnfaded<-3*rprimes^2*exp(-rprimes^3)   # Unfaded sample
distribs<-sapply(Tirrs,irradatsometemp)

########## plot distributions
cols=c("black","red","blue","magenta")
pchs=c(1,2,3,4)
matplot(rprimes,distribs,typ="o",pch=pchs,col=cols,
ylim=c(0,.003),lty="solid",
xlab="Dimensionless Distance r'", ylab="Distribution of r'",
lwd=2)
legend("topleft",bty="n",legend=c("Irradiation in nature",
"Various Burial Temperatures"," ","Distributions of distances"))
legend("topright",bty="n",legend=c(expression("-4 "^o*"C",
 "  0  "^o*"C"," +4  "^o*"C"," +8  "^o*"C")),pch=pchs,col=cols)
```

References

1. G Adamiec, R M Bailey, X L Wang, and A G Wintle. The mechanism of thermally transferred optically stimulated luminescence in quartz. *Journal of Physics D: Applied Physics*, 41(13):135503, 2008.
2. G Adamiec, A Bluszcz, R Bailey, and M Garcia-Talavera. Finding model parameters: Genetic algorithms and the numerical modelling of quartz luminescence. *Radiation Measurements*, 41(7–8):897–902, 2006.
3. G Adamiec, M Garcia-Talavera, R M Bailey, and P I de La Torre. Application of a genetic algorithm to finding parameter values for numerical simulation of quartz luminescence. *Geochronometria*, 23:9–14, 2004.
4. M S Akselrod, N A Larsen, V Whitley, and S W S McKeever. Thermal quenching of F-center luminescence in Al_2O_3:C. *Journal of Applied Physics*, 84(6):3364–3373, 1998.
5. L J S Allen. *An introduction to stochastic processes with applications to biology*. Chapman and Hall/CRC, 2010.
6. V Anechitei-Deacu, A Timar-Gabor, K J Thomsen, J-P Buylaert, M Jain, M Bailey, and A S Murray. Single and multi-grain OSL investigations in the high dose range using coarse quartz. *Radiation Measurements*, 120:124–130, 2018.
7. M Autzen, A S Murray, G. Guérin, L. Baly, C. Ankjærgaard, M. Bailey, M. Jain, and J.-P. Buylaert. Luminescence dosimetry: Does charge imbalance matter? *Radiation Measurements*, 120:26–32, 2018.
8. R Bailey. Simulations of variability in the luminescence characteristics of natural quartz and its implications for estimates of absorbed dose. *Radiation Protection Dosimetry*, 100:33–38, 2002.
9. R Bailey. Paper I - Simulation of dose absorption in quartz over geological timescales and its implications for the precision and accuracy of optical dating. *Radiation Measurements*, 38:299–310, 2004.
10. R M Bailey. Towards a general kinetic model for optically and thermally stimulated luminescence of quartz. *Radiation Measurements*, 33(1):17–45, 2001.
11. R M Bailey, B W Smith, and E J Rhodes. Partial bleaching and the decay form characteristics of quartz OSL. *Radiation Measurements*, 27(2):123–136, 1997.
12. I K Bailiff. The pre-dose technique. *Radiation Measurements*, 23(2):471–479, 1994.
13. H G Balian and N W Eddy. Figure-of-merit (FOM), an improved criterion over the normalized chi-squared test for assessing goodness-of-fit of gamma-ray spectral peaks. *Nuclear Instruments and Methods*, 145:389–395, 1977.
14. G W Berger. Regression and error analysis for a saturating-exponential-plus-linear model. *Ancient TL*, 8(3):23–25, 1990.

© The Author(s), under exclusive license to Springer Nature Switzerland AG 2021
V. Pagonis, *Luminescence*, Use R!, https://doi.org/10.1007/978-3-030-67311-6

15. G W Berger and R Chen. Error analysis and modelling of double saturating exponential dose response curves from SAR OSL dating. *Ancient TL*, 29(1):9–14, 2011.

16. R H Biswas, F Herman, G E King, and J Braun. Thermoluminescence of feldspar as a multi-thermochronometer to constrain the temporal variation of rock exhumation in the recent past. *Earth and Planetary Science Letters*, 495:56–68, 2018.

17. A Bluszcz. Exponential function fitting to TL growth data and similar applications. *Geochronometria*, 13:135–141, 1996.

18. A Bluszcz and G Adamiec. Application of differential evolution to fitting OSL decay curves. *Radiation Measurements*, 41(7–8):886–891, 2006.

19. L Bøetter-Jensen, S W S McKeever, and A G Wintle. *Optically Stimulated Luminescence Dosimetry*. Elsevier Science, 2003.

20. A J J Bos. Thermoluminescence as a research tool to investigate luminescence mechanisms. *Materials (Basel, Switzerland)*, 10, November 2017.

21. A J J Bos, T M Piters, J M Gómez-Ros, and A Delgado. An intercomparison of glow curve analysis computer programs: I. synthetic glow curves. *Radiation protection dosimetry*, 47:473–477, 1993.

22. J J Bosken and C Schmidt. Direct and indirect luminescence dating of tephra: A review. *Journal of Quaternary Science*, 35(1–2):39–53, 2020.

23. S G E Bowman and R Chen. Superlinear filling of traps in crystals due to competition during irradiation. *Journal of Luminescence*, 18–19:345–348, 1979.

24. N D Brown, E J Rhodes, and T M Harrison. Using thermoluminescence signals from feldspars for low-temperature thermochronology. *Quat. Geochronol.*, 42:31–41, 2017.

25. R K Bull. Kinetics of the localised transition model for thermoluminescence. *Journal of Physics D: Applied Physics*, 22(9):1375–1379, sep 1989.

26. E Bulur. An alternative technique for optically stimulated luminescence (OSL) experiment. *Radiation Measurements*, 26(5):701–709, 1996.

27. E Bulur. A simple transformation for converting CW-OSL curves to LM-OSL curves. *Radiation Measurements*, 32(2):141–145, 2000.

28. E Bulur and H Y Göksu. Infrared (IR) stimulated luminescence from feldspars with linearly increasing excitation light intensity. *Radiation Measurements*, 30:505–512, 1999.

29. E Bulur and A Yeltik. Optically stimulated luminescence from BeO ceramics: An LM-OSL study. *Radiation Measurements*, 45:29–34, 2010.

30. I F Chang and P Thioulouse. Treatment of thermostimulated luminescence, phosphorescence, and photostimulated luminescence with a tunneling theory. *Journal of Applied Physics*, 53(8):5873–5875, 1982.

31. R Chen. Glow curves with general order kinetics. *Journal of The Electrochemical Society*, 116:1254, 1969.

32. R Chen. On the calculation of activation energies and frequency factors from glow curves. *Journal of Applied Physics*, 40:570–585, 1969.

33. R Chen and Y Kirsh. *Analysis of Thermally Stimulated Processes*. Oxford: Pergamon Press., 1981.

34. R Chen, N Kristianpoller, Z Davidson, and R Visocekas. Mixed first and second order kinetics in thermally stimulated processes. *Journal of Luminescence*, 23:293–303, 1981.

35. R Chen, J L Lawless, and V Pagonis. Competition between long time excitation and fading of thermoluminescence (TL) and optically stimulated luminescence (OSL). *Radiation Measurements*, 136:106422, 2020.

36. R Chen and S W S McKeever. Characterization of nonlinearities in the dose dependence of thermoluminescence. *Radiation Measurements*, 23(4):667–673, 1994.

37. R Chen and S W S McKeever. *Theory of thermoluminescence and related phenomena*. World Scientific, Singapore, 1997.

38. R Chen and V Pagonis. *Thermally and Optically Stimulated Luminescence: A Simulation Approach*. John Wiley & Sons, Chichester, 2011.

39. M L Chithambo. The analysis of time-resolved optically stimulated luminescence: I. theoretical considerations. *Journal of Physics D: Applied Physics*, 40(7):1874, 2007.

40. M L Chithambo. The analysis of time-resolved optically stimulated luminescence: II. computer simulations and experimental results. *Journal of Physics D: Applied Physics*, 40(7):1880, 2007.

41. M L Chithambo, C Ankjærgaard, and V Pagonis. Time-resolved luminescence from quartz: An overview of contemporary developments and applications. *Physica B: Condensed Matter*, 481:8–18, 2016.

42. M L Chithambo and R B Galloway. A pulsed light-emitting-diode system for stimulation of luminescence. *Measurement Science and Technology*, 11(4):418–424, mar 2000.

43. R J Clark and I K Bailiff. Fast time-resolved luminescence emission spectroscopy in some feldspars. *Radiation Measurements*, 29(5):553–560, 1998.

44. R M Corless, G H Gonnet, D G E Hare, D J Jerey, and D E Knuth. On the Lambert W function. *Advances in Computational Mathematics*, 5:329–359, 1996.

45. R M Corless, D J Jerey, and D E Knuth. A sequence series for the Lambert W function. *In Proceedings of the International Symposium on Symbolic and Algebraic Computation , ISSAC*, pages 133–140, 1997.

46. M Duval. Dose response curve of the ESR signal of the aluminum center in quartz grains extracted from sediment. *Ancient TL*, 30(2):1–9, 2012.

47. J M Edmund. *Effects of temperature and ionization density in medical luminescence dosimetry using Al2O3:C (PhD Thesis, Riso, Denmark)*. PhD thesis, Risø National Laboratory, 2007. Riso-PhD-38(EN).

48. J Friedrich, S Kreutzer, and C Schmidt. Solving ordinary differential equations to understand luminescence: 'RLumModel', an advanced research tool for simulating luminescence in quartz using R . *Quaternary Geochronology*, 35:88–100, 2016.

49. J Friedrich, V Pagonis, R Chen, S Kreutzer, and C Schmidt. Quartz radiofluorescence: a modelling approach. *Journal of Luminescence*, 186:318–325, 2017.

50. M Fuchs, S Kreutzer, D Rousseau, P Antoine, C Hatté, F Lagroix, O Moine, C Gauthier, J Svoboda, and L Lisá. The loess sequence of Dolní Věstonice, Czech Republic: A new OSL-based chronology of the last climatic cycle. *Boreas*, 42(3):664–677, 2013.

51. G F J Garlick and A F Gibson. The electron trap mechanism of luminescence in sulphide and silicate phosphors. *Proceedings of the Physical Society*, 60:574–590, 1948.

52. J M Gómez-Ros and G Kitis. Computerized glow-curve deconvolution using mixed and general order kinetics. *Radiation Protection Dosimetry*, 101:47–52, 2002.

53. H Gould, J Tobochnik, D C Meredith, S E Koonin, S R McKay, and W Christian. *An Introduction to Computer Simulation Methods: Applications to Physical Systems, 2nd Edition*, volume 10. Addison-Wesley;, 1996.

54. G Grolemund. *Hands-on programming with R: write your own functions and simulations.* "O'Reilly Media, Inc.", 2014.

55. A Halperin and A A Braner. Evaluation of thermal activation energies from glow curves. *Physical Review*, 117:408–415, 1960.

56. A Halperin and R Chen. Thermoluminescence of semiconducting diamonds. *Phys. Rev.*, 148:839–845, Aug 1966.

57. Y Horowitz, E Fuks, H Datz, L Oster, J Livingstone, and A Rosenfeld. Mysteries of LiF TLD response following high ionization density irradiation: Glow curve shapes, dose response, the unified interaction model and modified track structure theory. *Radiation Measurements*, 46(12):1342–1348, 2011.

58. Y S Horowitz. The theoretical and microdosimetric basis of thermoluminescence and applications to dosimetry. *Physics in medicine and biology*, 26:765–824, Sep 1981.

59. D J Huntley. An explanation of the power-law decay of luminescence. *Journal of Physics: Condensed Matter*, 18(4):1359, 2006.

60. D J Huntley and M Lamothe. Ubiquity of anomalous fading in K-feldspars and the measurement and correction for it in optical dating. *Canadian Journal of Earth Sciences*, 38(7):1093–1106, 2001.

61. D J Huntley and Olav B Lian. Some observations on tunnelling of trapped electrons in feldspars and their implications for optical dating. *Quaternary Science Reviews*, 25(19–20):2503–2512, 2006.

62. M Jain, B Guralnik, and M T Andersen. Stimulated luminescence emission from localized recombination in randomly distributed defects. *Journal of Physics: Condensed Matter*, 24(38):385402, 2012.

63. M Jain, R Sohbati, B Guralnik, A S Murray, M Kook, T Lapp, A K Prasad, K J Thomsen, and J P Buylaert. Kinetics of infrared stimulated luminescence from feldspars. *Radiation Measurements*, 81:242–250, 2015.

64. R H Kars and J Wallinga. IRSL dating of K-feldspars: Modelling natural dose response curves to deal with anomalous fading and trap competition. *Radiation Measurements*, 44(5):594–599, 2009.

65. R Katz. Track structure theory in radiobiology and in radiation detection. *Nuclear Track Detection*, 2(1):1–28, 1978.

66. G E King, B Guralnik, P G Valla, and F Herman. Trapped-charge thermochronometry and thermometry: A status review. *Chemical Geology*, 44:3–17, 2016.

67. G E King, F Herman, R Lambert, P G Valla, and B Guralnik. Multi OSL thermochronometry of feldspar. *Quaternary Geochronology*, 33:76–87, 2016.

68. G Kitis, C Furetta, and V Pagonis. Mixed-order kinetics model for optically stimulated luminescence. *Modern Physics Letters B*, 23(27):3191–3207, 2009.

69. G Kitis and J M Gómez-Ros. Glow curve deconvolution functions for mixed order kinetics and a continuous trap distribution. *Nucl. Instrum. Methods A 440*, 440:224–231, 1999.

70. G Kitis, J M Gómez-Ros, and J W N Tuyn. Thermoluminescence glow curve deconvolution functions for first, second and general order kinetics. *J. Phys. D: Appl. Phys.*, 31:2666–2646, 1998.

71. G Kitis, N Kiyak, G S Polymeris, and N C Tsirliganis. The correlation of fast OSL component with the TL peak at in quartz of various origins. *Journal of Luminescence*, 130:298–303, 2010.

72. G Kitis and V Pagonis. Computerized curve deconvolution analysis for LM-OSL. *Radiation Measurements*, 43:737–741, 2008.

73. G Kitis and V Pagonis. Analytical solutions for stimulated luminescence emission from tunneling recombination in random distributions of defects. *Journal of Luminescence*, 137:109–115, 2013.

74. G Kitis and V Pagonis. Properties of thermoluminescence glow curves from tunneling recombination processes in random distributions of defects. *Journal of Luminescence*, 153:118–124, 2014.

75. G Kitis and V Pagonis. Localized transition models in luminescence: A reappraisal. *Nuclear Instruments and Methods in Physics Research Section B: Beam Interactions with Materials and Atoms*, 432:13–19, 2018.

76. G Kitis, V Pagonis, H Carty, and E Tatsis. Detailed kinetic study of the thermoluminescence glow curve of synthetic quartz. *Radiation protection dosimetry*, 100(1–4):225–228, 2002.

77. G Kitis, V Pagonis, and R Chen. Comparison of experimental and modelled quartz thermal-activation curves obtained using multiple-and single-aliquot procedures. *Radiation Measurements*, 41:910–916, 2006.

78. G Kitis, G S Polymeris, and V Pagonis. Stimulated luminescence emission: From phenomenological models to master analytical equations. *Applied Radiation and Isotopes*, 153:108797, 2019.

79. G Kitis, G S Polymeris, V Pagonis, and N C Tsirliganis. Thermoluminescence response and apparent anomalous fading factor of Durango fluorapatite as a function of the heating rate. *Physica Status Solidi (A) Applications and Materials Science*, 203(15):3816–3823, 2006.

80. G Kitis, G S Polymeris, E Sahiner, N Meric, and V Pagonis. Influence of the infrared stimulation on the optically stimulated luminescence in four K-feldspar samples. *Journal of Luminescence*, 176:32–39, 2016.

81. G Kitis, G S Polymeris, I K Sfampa, M Prokic, N Meriç, and V Pagonis. Prompt isothermal decay of thermoluminescence in Mg_4BO_7: Dy, Na and Li_4BO_7:Cu,In dosimeters. *Radiation Measurements*, 84:15–25, 2016.

82. G Kitis and N D Vlachos. General semi-analytical expressions for TL, OSL and other luminescence stimulation modes derived from the OTOR model using the Lambert W-function. *Radiation Measurements*, 48:47–54, 2013.

83. M J Klein. Principle of detailed balance. *Phys. Rev.*, 97:1446–1447, Mar 1955.
84. D K Koul, V Pagonis, and P Patil. Reliability of single aliquot regenerative protocol (SAR) for dose estimation in quartz at different burial temperatures: A simulation study. *Radiation Measurements*, 91:28–35, 2016.
85. S Kreutzer, Christoph Burow, Michael Dietze, Margret C Fuchs, Manfred Fischer, and Christoph Schmidt. Software in the context of luminescence dating: status, concepts and suggestions exemplified by the R package Luminescence. *Ancient TL*, 35(2), 2017.
86. S Kreutzer, M Dietze, C Burow, M C Fuchs, C Schmidt, M Fischer, and J Friedrich. Luminescence: Comprehensive Luminescence Dating Data Analysis. *CRAN*, version 0.7.5, 2017. Developer version on GitHub: https://github.com/R-Lum/Luminescence.
87. S Kreutzer, J Friedrich, V Pagonis, and C Schmidt. Rlumcarlo: Tedious features-fine examples. *Simulation*, 1:1–1E5, 2021.
88. S Kreutzer, C Schmidt, M C Fuchs, M Dietze, M Fischer, and M Fuchs. Introducing an R package for luminescence dating analysis. *Ancient TL*, 30(1):1–8, 2012.
89. R N Kulkarni. The development of the Monte Carlo method for the calculation of the thermoluminescence intensity and the thermally stimulated conductivity. *Radiation protection dosimetry*, 51(2):95–105, 1994.
90. David P L and K Binder. *A guide to Monte Carlo simulations in statistical physics*. Cambridge University Press, 2014.
91. M Lamothe, M Auclair, C Hamzaoui, and S Huot. Towards a prediction of long-term anomalous fading of feldspar IRSL. *Radiation Measurements*, 37(4):493–498, 2003.
92. A Larsen, S Greilich, M Jain, and A S Murray. Developing a numerical simulation for fading in feldspar. *Radiation Measurements*, 44(5):467–471, 2009.
93. J L Lawless, R Chen, and V Pagonis. On the theoretical basis for the duplicitous thermoluminescence peak. *Journal of Physics D: Applied Physics*, 42(15), 2009.
94. J L Lawless, R Chen, and V Pagonis. Sublinear dose dependence of thermoluminescence and optically stimulated luminescence prior to the approach to saturation level. *Radiation Measurements*, 44(5):606–610, 2009.
95. J L Lawless, R Chen, and V Pagonis. Inherent statistics of glow curves from small samples and single grains. *Journal of Luminescence*, 226:117389, 2020.
96. P W Levy. Overview Of Nuclear Radiation Damage Processes: Phenomenological Features Of Radiation Damage In Crystals And Glasses. In Paul W. Levy, editor, *Radiation Effects on Optical Materials*, volume 0541, pages 2–24. International Society for Optics and Photonics, SPIE, 1985.
97. B Li, Z Jacobs, and R G Roberts. Investigation of the applicability of standardised growth curves for OSL dating of quartz from Haua Fteah cave, Libya. *Quaternary Geochronology*, 35:1–15, 2016.
98. B Li and S H Li. Investigations of the dose-dependent anomalous fading rate of feldspar from sediments. *Journal of Physics D: Applied Physics*, 41(22):225502, oct 2008.
99. S E Lowick, F Preusser, and A G Wintle. Investigating quartz optically stimulated luminescence dose response curves at high doses. *Radiation Measurements*, 45(9):975–984, 2010.
100. A Mandowski. The theory of thermoluminescence with an arbitrary spatial distribution of traps. *Radiation Protection Dosimetry*, 100:115–118, 2002.
101. A Mandowski. Semi-localized transitions model for thermoluminescence. *Journal of Physics D: Applied Physics*, 38:17, 2005.
102. A Mandowski. Calculation and properties of trap structural functions for various spatially correlated systems. *Radiation protection dosimetry*, 119:85–88, 2006.
103. A Mandowski. How to detect trap cluster systems? *Radiation Measurements*, 43(2):167–170, 2008.
104. A Mandowski and A J J Bos. Explanation of anomalous heating rate dependence of thermoluminescence in YPO4:Ce^{3+}, Sm^{3+} based on the semilocalized transition (SLT) model. *Radiation Measurements*, 46:1376–1379, 2011.
105. A Mandowski and J Świątek. Monte Carlo simulation of thermally stimulated relaxation kinetics of carrier trapping in microcrystalline and two-dimensional solids. *Philosophical Magazine B*, 65:729–732, 1992.

106. A Mandowski and J Światek. Monte Carlo simulation of TSC and TL in spatially correlated systems. In *Electrets, 1994.(ISE 8), 8th International Symposium on*, pages 461–466. IEEE, 1994.

107. A Mandowski and J Światek. On the influence of spatial correlation on the kinetic order of TL. *Radiation protection dosimetry*, 65:25–28, 1996.

108. A Mandowski and J Światek. The kinetics of trapping and recombination in low dimensional structures. *Synthetic Metals*, 109:203–206, 2000.

109. A Mandowski and Józef Światek. Thermoluminescence and trap assemblies: results of Monte Carlo calculations. *Radiation Measurements*, 29:415–419, 1998.

110. C E May and J A Partridge. Thermoluminescent kinetics of alpha-irradiated alkali halides. *The Journal of Chemical Physics*, 40(5):1401–1409, 1964.

111. S W S McKeever. Thermoluminescence of solids. Cambridge University Press, 1985.

112. S W S McKeever and R Chen. Luminescence models. *Radiation Measurements*, 27(5):625–661, 1997.

113. E F Mische and S W S McKeever. Mechanisms of Supralinearity in Lithium Fluoride Thermoluminescence Dosemeters. *Radiation Protection Dosimetry*, 29(3):159–175, 11 1989.

114. P Morthekai, J Thomas, M S Pandian, V Balaram, and A K Singhvi. Variable range hopping mechanism in band-tail states of feldspars: A time-resolved irsl study. *Radiation Measurements*, 47(9):857–863, 2012.

115. R Nanjundaswamy, K Lepper, and SWS McKeever. Thermal quenching of thermoluminescence in natural quartz. *Radiation protection dosimetry*, 100(1–4):305–308, 2002.

116. S V Nikiforov, V S Kortov, and M G Kazantseva. Simulation of the superlinearity of dose characteristics of thermoluminescence of anion-defective aluminum oxide. *Physics of the Solid State*, 56(3):554–560, March 2014.

117. S V Nikiforov, I I Milman, and V S Kortov. Thermal and optical ionization of F-centers in the luminescence mechanism of anion-defective corundum crystals. *Radiation Measurements*, 33(5):547–551, 2001. Proceedings of the International Symposium on Luminescent Detectors and Transformers of Ionizing Radiation.

118. A S Novozhilov, G P Karev, and E V Koonin. Biological applications of the theory of birth-and-death processes. *Briefings in Bioinformatics*, 7(1):70–85, 03 2006.

119. V Pagonis, C Ankjærgaard, M Jain, and R Chen. Thermal dependence of time-resolved blue light stimulated luminescence in Al$_2$O$_3$:C. *Journal of Luminescence*, 136:270–277, 2013.

120. V Pagonis, C Ankjærgaard, M Jain, and M L Chithambo. Quantitative analysis of time-resolved infrared stimulated luminescence in feldspars. *Physica B: Condensed Matter*, 497:78–85, 2016.

121. V Pagonis, C Ankjærgaard, A S Murray, M Jain, R Chen, J Lawless, and S Greilich. Modelling the thermal quenching mechanism in quartz based on time-resolved optically stimulated luminescence. *Journal of Luminescence*, 130(5):902–909, 2010.

122. V Pagonis, L Blohm, M Brengle, G Mayonado, and P Woglam. Anomalous heating rate effect in thermoluminescence intensity using a simplified semi-localized transition (SLT) model. *Radiation Measurements*, 51–52:40–47, 2013.

123. V Pagonis and N Brown. On the unchanging shape of thermoluminescence peaks in preheated feldspars: Implications for temperature sensing and thermochronometry. *Radiation Measurements*, 2019.

124. V Pagonis, N Brown, G S Polymeris, and G Kitis. Comprehensive analysis of thermoluminescence signals in Mg$_4$BO$_7$: Dy, Na dosimeter. *Journal of Luminescence*, 213:334–342, 2019.

125. V Pagonis and R Chen. Monte Carlo simulations of TL and OSL in nanodosimetric materials and feldspars. *Radiation Measurements*, 81:262–269, 2015.

126. V Pagonis, R Chen, and G Kitis. On the intrinsic accuracy and precision of luminescence dating techniques for fired ceramics. *Journal of Archaeological Science*, 38(7):1591–1602, 2011.

127. V Pagonis, R Chen, C Kulp, and G Kitis. An overview of recent developments in luminescence models with a focus on localized transitions. *Radiation Measurements*, 106:3–12, 2017.

128. V Pagonis, R Chen, and J L Lawless. A quantitative kinetic model for Al2O3: C: TL response to ionizing radiation. *Radiation Measurements*, 42(2):198–204, 2007.

129. V Pagonis, R Chen, Maddrey J W, and Sapp B. Simulations of time-resolved photoluminescence experiments in Al_2O_3:C. *Journal of Luminescence*, 131(5):1086–1094, 2011.

130. V Pagonis, Gochnour E, Hennessey M, and Knower C. Monte Carlo simulations of luminescence processes under quasi-equilibrium (QE) conditions. *Radiation Measurements*, 67:67–76, 2014.

131. V Pagonis, J Friedrich, M Discher, A Müller-Kirschbaum, V Schlosser, S Kreutzer, R Chen, and C Schmidt. Excited state luminescence signals from a random distribution of defects: A new Monte Carlo simulation approach for feldspar. *Journal of Luminescence*, 207:266–272, 2019.

132. V Pagonis, M Jain, K J Thomsen, and A S Murray. On the shape of continuous wave infrared stimulated luminescence signals from feldspars: A case study. *Journal of Luminescence*, 153:96–103, 2014.

133. V Pagonis and G Kitis. Prevalence of first-order kinetics in thermoluminescence materials: An explanation based on multiple competition processes. *Physica Status Solidi B*, 249:1590–1601, 2012.

134. V Pagonis and G Kitis. Mathematical aspects of ground state tunneling models in luminescence materials. *Journal of Luminescence*, 168:137–144, 2015.

135. V Pagonis, G Kitis, and R Chen. A new analytical equation for the dose response of dosimetric materials, based on the Lambert W function. *Journal of Luminescence*, 225:117333, 2020.

136. V Pagonis, G Kitis, and R Chen. Superlinearity revisited: A new analytical equation for the dose response of defects in solids, using the Lambert W function. *Journal of Luminescence*, 227:117553, 2020.

137. V Pagonis, G Kitis, and C Furetta. *Numerical and practical exercises in thermoluminescence.* Springer Science & Business Media, 2006.

138. V Pagonis, G Kitis, and G S Polymeris. On the half-life of luminescence signals in dosimetric applications: A unified presentation. *Physica B: Condensed Matter*, 539:35–43, 2018.

139. V Pagonis, S Kreutzer, A R Duncan, E Rajovic, C Laag, and C Schmidt. On the stochastic uncertainties of thermally and optically stimulated luminescence signals: A Monte Carlo approach. *Journal of Luminescence*, 219:116945, 2020.

140. V Pagonis and C Kulp. Monte Carlo simulations of tunneling phenomena and nearest neighbor hopping mechanism in feldspars. *Journal of Luminescence*, 181:114–120, 2017.

141. V Pagonis, C Kulp, C Chaney, and M Tachiya. Quantum tunneling recombination in a system of randomly distributed trapped electrons and positive ions. *Journal of physics. Condensed matter : an Institute of Physics journal*, 29:365701, September 2017.

142. V Pagonis, J L Lawless, R Chen, and ML Chithambo. Analytical expressions for time-resolved optically stimulated luminescence experiments in quartz. *Journal of Luminescence*, 131(9):1827–1835, 2011.

143. V Pagonis, S M Mian, M L Chithambo, E Christensen, and C Barnold. Experimental and modelling study of pulsed optically stimulated luminescence in quartz, marble and beta irradiated salt. *Journal of Physics D: Applied Physics*, 42(5):055407, 2009.

144. V Pagonis, P Morthekai, Singhvi A K, Thomas J, Balaram V, Kitis G, and Chen R. Time-resolved infrared stimulated luminescence signals in feldspars: Analysis based on exponential and stretched exponential functions. *Journal of Luminescence*, 132(9):2330–2340, 2012.

145. V Pagonis, H Phan, D Ruth, and G Kitis. Further investigations of tunneling recombination processes in random distributions of defects. *Radiation Measurements*, 58:66–74, 2013.

146. V Pagonis, G S Polymeris, and G Kitis. On the effect of optical and isothermal treatments on luminescence signals from feldspars. *Radiation Measurements*, 82:93–101, 2015.

147. V Pagonis and P Truong. Thermoluminescence due to tunneling in nanodosimetric materials: A Monte Carlo study. *Physica B: Condensed Matter*, 531:171–179, 2018.

148. V Pagonis, A G Wintle, and R Chen. Simulations of the effect of pulse annealing on optically-stimulated luminescence of quartz. *Radiation Measurements*, 42(10):1587–1599, 2007.

149. V Pagonis, A G Wintle, R Chen, and X L Wang. A theoretical model for a new dating protocol for quartz based on thermally transferred OSL (TT-OSL). *Radiation Measurements*, 43:704–708, 2008.

150. J Peng, Z Dong, and F Han. tgcd: An R package for analyzing thermoluminescence glow curves. *SoftwareX*, 5:112–120, 2016.

151. J Peng and V Pagonis. Simulating comprehensive kinetic models for quartz luminescence using the R program KMS. *Radiation Measurements*, 86:63–70, 2016.

152. S A Petrov and I K Bailiff. Thermal quenching and the initial rise technique of trap depth evaluation. *Journal of Luminescence*, 65(6):289–291, 1996.

153. S A Petrov and I K Bailiff. Determination of trap depths associated with tl peaks in synthetic quartz (350–550 k). *Radiation Measurements*, 27(2):185–191, 1997.

154. G S Polymeris. OSL at elevated temperatures: Towards the simultaneous thermal and optical stimulation. *Radiation Physics and Chemistry*, 106:184–192, 2015.

155. G S Polymeris, V Pagonis, and G Kitis. Thermoluminescence glow curves in preheated feldspar samples: An interpretation based on random defect distributions. *Radiation Measurements*, 97:20–27, 2017.

156. G S Polymeris, V Pagonis, and G Kitis. Investigation of thermoluminescence processes during linear and isothermal heating of dosimetric materials. *Journal of Luminescence*, 222:117142, 2020.

157. G S Polymeris, N Tsirliganis, Z Loukou, and G Kitis. A comparative study of the anomalous fading effects of TL and OSL signals of durango apatite. *Physica Status Solidi (a)*, 203(3):578–590, 2006.

158. N R J Poolton, K B Ozanyan, J Wallinga, A S Murray, and L Bøtter-Jensen. Electrons in feldspar II: a consideration of the influence of conduction band-tail states on luminescence processes. *Physics and Chemistry of Minerals*, 29(3):217–225, 2002.

159. N R J Poolton, J Wallinga, A S Murray, E Bulur, and L Bøtter-Jensen. Electrons in feldspar I: on the wavefunction of electrons trapped at simple lattice defects. *Physics and Chemistry of Minerals*, 29(3):210–216, April 2002.

160. F Preusser, M Chithambo, T Götte, M Martini, K Ramseyer, E J Sendezera, G Susino, and A G Wintle. Quartz as a natural luminescence dosimeter. *Earth-Science Reviews*, 97(1):184–214, 2009.

161. J T Randall and M H F Wilkins. Phosphorescence and electron traps. I. The study of trap distributions. *Proceedings of the Royal Society of London A: Mathematical, Physical and Engineering Sciences*, 184(999):366–389, 1945.

162. M S Rasheedy. On the general-order kinetics of the thermoluminescence glow peak. *J. Phys.: Condens. Matter*, 5:633–636, 1993.

163. E Şahiner, G Kitis, V Pagonis, N Meriç, and G S Polymeris. Tunnelling recombination in conventional, post-infrared and post-infrared multi-elevated temperature IRSL signals in microcline K-feldspar. *Journal of Luminescence*, 188:514–523, 2017.

164. N Salah. Nanocrystalline materials for the dosimetry of heavy charged particles: A review. *Radiation Physics and Chemistry*, 80(1):1–10, 2011.

165. D C W Sanderson and R J Clark. Pulsed photostimulated luminescence of alkali feldspars. *Radiation Measurements*, 23(2):633–639, 1994.

166. I K Sfampa, G S Polymeris, N Tsirliganis, V Pagonis, and G Kitis. Prompt isothermal decay of thermoluminescence in an apatite exhibiting strong anomalous fading. *Nuclear Instruments and Methods in Physics Research Section B: Beam Interactions with Materials and Atoms*, 320:57–63, 2014.

167. J S Singarayer and R M Bailey. Further investigations of the quartz optically stimulated luminescence components using linear modulation. *Radiation Measurements*, 37(4):451–458, 2003.

168. J S Singarayer and R M Bailey. Component-resolved bleaching spectra of quartz optically stimulated luminescence: preliminary results and implications for dating. *Radiation Measurements*, 38:111–118, 2004.

169. L L Singh and R K Gartia. Theoretical derivation of a simplified form of the OTOR/GOT differential equation. *Radiation Measurements*, 59:160–164, 2013.
170. K Soetaert, T Petzoldt, and R Setzer. Solving differential equations in R: Package deSolve. *Journal of Statistical Software, Articles*, 33(9):1–25, 2010.
171. Karline Soetaert, Jeff Cash, and Francesca Mazzia. *Solving differential equations in R.* Springer Science & Business Media, 2012.
172. H Sun and Y Sakka. Luminescent metal nanoclusters: controlled synthesis and functional applications. *Science and Technology of Advanced Materials*, 15(1):014205, 2014.
173. C M Sunta, W E F Ayta, J F D Chubaci, and S Watanabe. A critical look at the kinetic models of thermoluminescence: I. first-order kinetics. *Journal of Physics D: Applied Physics*, 34(17):2690–2698, aug 2001.
174. C M Sunta, W E F Ayta, J F D Chubaci, and S Watanabe. A critical look at the kinetic models of thermoluminescence II. non-first order kinetics. *Journal of Physics D: Applied Physics*, 38(1):95–102, dec 2004.
175. M Tachiya and A Mozumder. Decay of trapped electrons by tunnelling to scavenger molecules in low-temperature glasses. *Chemical Physics Letters*, 28(1):87–89, 1974.
176. M Tachiya and A Mozumder. Kinetics of geminate-ion recombination by electron tunnelling. *Chemical Physics Letters*, 34(1):77–79, 1975.
177. P Thioulouse, E A Giess, and I F Chang. Investigation of thermally stimulated luminescence and its description by a tunneling model. *Journal of Applied Physics*, 53(12):9015–9020, 1982.
178. J S Thomsen. Logical relations among the principles of statistical mechanics and thermodynamics. *Phys. Rev.*, 91:1263–1266, Sep 1953.
179. A Timar-Gabor, D Constantin, J P Buylaert, M Jain, A S Murray, and A G Wintle. Fundamental investigations of natural and laboratory generated SAR dose response curves for quartz OSL in the high dose range. *Radiation Measurements*, 81:150–156, 2015.
180. A Timar-Gabor, A Vasiliniuc, D A G Vandenberghe, C Cosma, and A G Wintle. Investigations into the reliability of SAR-OSL equivalent doses obtained for quartz samples displaying dose response curves with more than one component. *Radiation Measurements*, 47(9):740–745, 2012.
181. F Trompier, C Bassinet, S Della Monaca, A Romanyukha, R Reyes, and I Clairand. Overview of physical and biophysical techniques for accident dosimetry. *Radiation protection dosimetry*, 144:571–574, March 2011.
182. S. Tsukamoto, P.M. Denby, A.S. Murray, and L. Bøetter-Jensen. Time-resolved luminescence from feldspars: New insight into fading. *Radiation Measurements*, 41(7):790–795, 2006.
183. D A G Vandenberghe, M Jain, and A S Murray. Equivalent dose determination using a quartz isothermal TL signal. *Radiation Measurements*, 44(5):439–444, 2009.
184. R Visocekas. Tunneling radiative recombination in K-feldspar sanidine. *Nuclear Tracks and Radiation Measurements*, 21:175–178, 1993.
185. R Visocekas, V Tale, A Zink, and I Tale. Trap spectroscopy and tunnelling luminescence in feldspars. *Radiation Measurements*, 29:427–434, 1998.
186. M P R Wáligorski and R Katz. Supralinearity of peak 5 and peak 6 in TLD-700. *Nuclear Instruments and Methods*, 172(3):463–470, 1980.
187. M P R Wáligorski, P Olko, P Bilski, M Budzanowski, and T Niewiadomski. Dosimetric Characteristics of LiF:Mg,Cu,P Phosphors - A Track Structure Interpretation. *Radiation Protection Dosimetry*, 47(1–4):53–58, 05 1993.
188. X L Wang, A G Wintle, and Y C Lu. Thermally transferred luminescence in fine-grained quartz from Chinese loess: Basic observations. *Radiation Measurements*, 41(6):649–658, 2006.
189. A Wieser, Y Göksu, D F Regulla, and A Waibel. Unexpected superlinear dose dependence of the El' centre in fused silica. *International Journal of Radiation Applications and Instrumentation. Part D. Nuclear Tracks and Radiation Measurements*, 18(1):175–178, 1991.
190. A G Wintle. Thermal Quenching of Thermoluminescence in Quartz. *Geophysical Journal International*, 41(1):107–113, 1975.

191. A G Wintle and A S Murray. The relationship between quartz thermoluminescence, photo-transferred thermoluminescence, and optically stimulated luminescence. *Radiation Measurements*, 27(4):611–624, 1997.

192. A G Wintle and A S Murray. Towards the development of a preheat procedure for OSL dating of quartz. *Radiation Measurements*, 29(1):81–94, 1998.

193. X H Yang and S W S McKeever. The pre-dose effect in crystalline quartz. *Journal of Physics D: Applied Physics*, 23(2):237, 1990.

194. D Yossian and Y S Horowitz. Mixed-order and general-order kinetics applied to synthetic glow peaks and to peak 5 in LiF:Mg, Ti (TLD-100). *Radiation Measurements*, 27(3):465–471, 1997.

195. E G Yukihara and S W S McKeever. Optically stimulated luminescence. Book, 2011.

196. E G Yukihara, V H Whitley, J C Polf, D M Klein, S W S McKeever, A E Akselrod, and M S Akselrod. The effects of deep trap population on the thermoluminescence of Al_2O_3:C. *Radiation Measurements*, 37(6):627–638, 2003.

197. J Zimmerman. The radiation-induced increase of the $100\,^\circ C$ thermoluminescence sensitivity of fired quartz. *Journal of Physics C: Solid State Physics*, 4(18):3265–3276, 1971.

Index

Printed in the United States
by Baker & Taylor Publisher Services